Dialectics, Dialogue and Argumentation
An Examination of Douglas Walton's Theories
of Reasoning and Argument

Tributes Series Editor
Dov Gabbay   dov.gabbay@kcl.ac.uk

Dialectics, Dialogue and Argumentation
An Examination of Douglas Walton's Theories of Reasoning and Argument

edited by

Chris Reed

and

Christopher W. Tindale

ISBN 978-1-84890-005-9

College Publications
Scientific Director: Dov Gabbay
Managing Director: Jane Spurr
Department of Computer Science
King's College London, Strand, London WC2R 2LS, UK

http://www.collegepublications.co.uk

Original cover design by orchid creative   www.orchidcreative.co.uk
Printed by Lightning Source, Milton Keynes, UK

CONTENTS

PREFACE

This volume started out as a way of marking Douglas Walton's retirement, which was to have been in 2008. With his move to the University of Windsor in that year, however, retirement was no longer a mandatory requirement: this was, of course, great for Doug and superb news for argumentation in both its philosophical and computational guises. But the change did raise a question about the role of this book.

On reflection though, even if he had not moved, it is impossible to imagine Doug retiring and simply deciding to stop work. His commitment, his dedication and his sheer hard work would not have allowed him to do that. So in many ways this book was never going to mark a cessation in Doug's research. Instead, what we are trying to achieve with this festschrift is a celebration of a truly remarkable career.

The papers collected here attest to both the breadth and the depth of Doug's influence. In the study of argumentation, he has had indelible impact on theories of dialogue, fallacies and presumptive inference. In jurisprudence, he has developed accounts of evidence and of burden of proof. In computer science and artificial intelligence, he has not only striven to introduce modern theories of argument into computational research, but he has forged collaborations that have helped to define a new emerging field of research.[1]

It is relatively easy for eminent academics to encourage new talent in non-specific and general terms. It is quite another to dedicate time to nurture and hone young scholars, working with them and publishing with them, inculcating skills and attitudes of the highest academic order. It is a testament to Doug's tireless efforts that many of the authors here owe an enormous debt of gratitude to him for helping them establish and develop their academic voices and professional careers.

Finally, this book will be presented to Doug at the 7th Conference on Argumentation of the International Society for the Study of Argumentation, ISSA 2010, in Amsterdam. The event is scheduled for July 2, 2010, just a couple of weeks after Doug's 68th birthday. So our festschrift could be seen as a birthday gift. And like any birthday present, it is ultimately a way of expressing our respect and fond regards for a good friend – and a remarkable scholar.

Chris Reed and Chris Tindale
Dundee and Windsor
Spring, 2010

---

[1] Computational models of argument, as an area of research, now has dedicated to it a biennial conference series, COMMA, www.comma-conf.org and a journal, *Argument & Computation*, www.tandf.co.uk/journals/tarc, published by Taylor & Francis / Routledge.

CHAPTER 1

INTRODUCTION

Chris Tindale and Chris Reed

1.1 Introduction

To track the career of argumentation scholar Douglas N. Walton is at the same time to track the path of development and innovation in the field itself. And on that path, Walton has been a trailblazer rather than a follower or fellow traveler. Through the early work on fallacies with John Woods, best captured in their textbook (Woods and Walton 1982) and collected papers (Woods and Walton 1989), to the engagement with dialectics and the production of a series of studies on specific fallacies, through to the later work on argumentation schemes, particularly in law and artificial intelligence, Walton's scholarship has set the standard at which others would later aim. The questions he has asked have been the questions taken up and explored by those who followed. The problems he has uncovered have formed in large part the research agenda of the field in general. And the solutions he has proposed have served as a valued corpus of ideas for others to adopt, critically investigate, and push beyond. He has produced a range of groundbreaking monographs as well as several important textbooks (1989, 2006c), thus deeply influencing both the theory and the pedagogy of the field.

Of the many issues and concepts he has explored, we might closely attach to his name three ideas that have most felt the influence of his originality: a detailed and comprehensive account of 'fallacy'; a dialectical approach to argumentation; and an abiding interest in argumentation schemes.

1.2 Fallacy Theory

A large portion of Walton's output is taken up with his treatments of fallacy, as a general concept and through individual studies. Twelve books are devoted in their entirety to this area and several others, as well as many published papers, address the fallacies in some way. The roots of this engagement – and undoubtedly many other aspects of his thought – can be traced to the early partnership with John Woods that produced the papers (Woods and Walton 1989) in which they began to rethink the tradition of the fallacies and tried to give some substance and coherence to a field that clearly lacked both.

For Woods and Walton, an understanding of the fallacies involved moving beyond a fixation on formal deductive logic to a different sense of formal concerned with matters of form itself (Woods and Walton 1989, p. 249). The theory of the fallacies that then emerged was a "more or less formal theory" (p. 250) depending on the complexity of the structures involved. This observation sheds considerable light on the theoretical treatments gathered in the collected essays. Here, the authors dismiss any thorough formalism (Woods and Walton 1989, p.

xvii) while striving to fix the underlying structure that captures the nature of each fallacy. They also give considerable attention to the dialectical approach to fallacies through formal dialogue games or games of dialectic (pp. 8-9; 61; 143ff; 253ff). [1]

It is not difficult to see here the interests that will develop and reappear later in Walton's extensive treatment of argumentation schemes and his new dialectic, including the dialectical treatments of the fallacies. We see such interests emerging, for example, in *Arguments from Ignorance* (1996c), where considerable energy is spent on uncovering the structure of the *Argumentum ad Ignorantiam* while at the same time gradually shifting the focus on to the way in which it is embedded in sequences of dialogue. Walton writes: "Starting out this book, I based my approach on the methods used traditionally in logic, whereby evaluation of the argument from ignorance concentrated on the reasoning in the argument and the propositional form" (1996c, p. 168). He comes to see, however, that the focus must be on a longer sequences of dialogue exchanges, observing that the "use of the dialogue concept to analyze a sequence of multisentence units in an interactive argumentation between two speakers is a radical departure from these previously dominant semantic traditions in logic" (p. 170). This understanding characterizes the detailed treatments of individual fallacies: the appeal to pity (1997b), the appeal to expert knowledge, (1997a), the *ad hominem* (1998a), the appeal to popularity (1999a), and the *ad baculum* and related tactics (2000c). It is also evident in the central work that informs all these studies: the pragmatic theory of fallacy (1995).

Walton's general approach to fallacy arises from his reaction to the treatment found in Hamblin (1970). While other informal logicians may have read Hamblin as denouncing the tradition of fallacies conveyed in his so-called Standard Treatment, Walton (1991b) insists otherwise. Hamblin did find the Standard Treatment superficial and lacking rigorous analyses of the fallacies, but the traditional divisions were worth preserving (1991b, p. 335). Accordingly, Walton looked to redefine the traditional categories of the informal fallacies within a dialectical account that was initially influenced by the Dutch school of pragma-dialectics, with which he was closely associated during this period. However, in returning to the Aristotelian source of the Standard Treatment he unearthed a pragmatic concept of fallacy that differed from the Dutch dialectical model in several important ways.

Fallacies in general come to be viewed as unfair argumentation tactics deployed in the context of a dialogue, thus any semantic view is rejected and replaced by a full-fledged pragmatic theory of fallacy (1995). In *A Pragmatic Theory of Fallacy*, the concept of fallacy has two aspects to it. In the first instance, the type of argument that is involved is identified through its underlying structure or form. Then, a further aspect involves the misuse of this argument form in a dialogue. Walton explains it this way: "the fallacies are first and foremost identified as being certain distinctive types of arguments, as indicated by being instances of their characteristic argumentation schemes ... Then the fallacy is analyzed as a certain type of misuse of the argumentation scheme" (1995, p. 17). Not evident here is any explicit reference to a dialogue, but the second aspect of the concept assumes this feature. The misuse of arguments was apparent in Walton's early work (1992c), where the correct use refers to the appropriate type of dialogue involved. A fallacy, then, for Walton, involves there having been an "underlying dialectical shift from one type of dialogue to another" (1995, p. 250). Beyond its other distinguishing features, it is this aspect which marks the originality of Walton's account and most clearly separates it from those of competitor treatments

---

[1] The Woods-Walton approach to fallacies is discussed in (van Eemeren et al. 1996, pp. 236-38).

like that of the pragma-dialecticians. Given that the Dutch model directs its focus on the critical discussion and Walton's model – as we will see below – includes many other types of dialogue, then the difference is actually written into the very concept of fallacy as Walton explains it.

1.3   Dialectics and Pragmatics

The authors of *Fundamentals of Argumentation Theory* (van Eemeren et al. 1996, p. 74) identified a pragmatic shift in Walton's work with the appearance of *Arguer's Position* (1985a). But by the publication of *Informal Fallacies* in 1987, van Eemeren and Grootendorst judged there to be still too much logic in his pragmatism (van Eemeren and Grootendorst 1989, p. 103). He had become a dialectician, they noted, but his dialectics remained too formal and the concessions made to pragmatism seemed limited to the everyday origins of his source examples. "Walton is still a logician who has now become a dialectician. In order to become a pragmatist as well, he still has some way to go. What is needed for the development of the study of argumentation and fallacies is, in our view, a radical pragma-dialectic approach. Walton is on his way. He only needs a last small push" (p. 105).

Evidence of such a shift did appear in key features of his work throughout the 1990s. We see, for example, a conception of argument that viewed it as "a social and verbal means of trying to resolve, or at least contend with, a conflict or difference that has arisen between two (or more) parties. An argument necessarily involves a claim that is advanced by at least one of the parties" (Walton 1990b, p. 411). This dialectical conception was noted by Hitchcock (2007a, p. 117) as fitting the disputational sense of argument better than the reason-giving sense insofar as it excludes written texts that have not been challenged or rejected as well as excluding claims already accepted. In point of fact, by defining argument in terms of the resolution of a conflict Walton was working quite clearly on pragma-dialectic territory. Previously (1989), he had introduced rules and stages of discussion that corresponded with those in the established pragma-dialectical account, judging the persuasion dialogue to be "the single most significant type of dialogue" (1989, p. 9).[2]

As he developed his conception of fallacy as an illicit dialectical shift, the pragma-dialectical understanding of fallacy as the violation of dialectical rules became insufficient for the range of ideas that were emerging.[3] On Walton's reading, not all violations of rules are fallacies (1991a, p. 217-8), and the persuasion dialogue, while important, is only one among several types of dialogues in which fallacies can occur. In fact, the presence of a fallacy will depend in part on the type of dialogue involved.

Walton's new dialectic fully captures his dialogue approach to informal reasoning. Building on the conception of argument that was noted above, involving interactions between participants, a taxonomy of dialogue-types was developed. The number of dialogue types involved ranged from five to ten over the course of the account's development, with the core group consisting of personal quar-

---

[2]A fuller treatment of Walton's engagement with pragma-dialectics and his movement beyond it can be found in (Tindale 1997).

[3]Such a rejection, of course, does not go uncontested. In their (2007a), Frans van Eemeren and Peter Houtlosser contrast the illicit dialectical shift concept to the view of fallacies in pragma-dialectics. As explored there, the root of the disagreement lies in part with the types of dialogue identified in Walton's new dialectic.

rel, forensic debate, persuasion dialogue, inquiry, negotiation (1989, p. 3-6).[4] On the terms set out in the new dialectic, "arguments have a dialectical rather than a monolectical structure, involve defeasible rather than deductive reasoning, and are used for human rather than abstract purposes" (Godden 2007, p. 1).[5] It is within this new dialectic that the full force of Walton's treatments of individual fallacies comes to bear the fruits they do.

1.4   Schemes and Critical Questions

A final piece in the innovative architecture of Walton's theorizing that we will consider here is his occupation with argumentation schemes.[6] Across the later stages of his work, Walton has been concerned to identify schemes of argument representing common patterns of defeasible reasoning.[7] As noted earlier, part of Walton's understanding of fallacies lies in the misuse of argumentation schemes (1995, p. 17). In that work, he gave twenty-five examples of such schemes, including structures like the argument from analogy, argument from vagueness, and the argument from cause to effect. By (2008), working with Reed and Macagno, Walton had identified 65 such schemes, including categories of schemes like Causal Argumentation Schemes and Knowledge-related Schemes. Walton's goal in this part of his overall program has been to "construct a systematic methodology for classifying schemes and analyzing the formal structure of each scheme in a precise system" (Walton 2005f, p. 2). Each scheme is accompanied by another Walton-innovation — a set of critical questions for its assessment. The use of the questions is a key tool in deciding whether a particular instance of a scheme is a good argument. For example, the argumentation scheme for the argument from commitment can be given as the following: "a is committed to proposition A (generally, or in virtue of what she said in the past). Therefore, in this case, a should support A" (Walton 1996b, p. 56). To assist in the assessment of particular cases of the argument from commitment, critical questions ask (1) whether a is really committed to A, and what evidence supports this; (2) whether if the evidence for commitment is indirect or weak there might be contrary evidence or room for rebuttal to show this case as an exception; and (3) whether A is the same proposition in the conclusion as in the premise (or what relationship do the two propositions have if this is not the case). While questions remain about the nature and source of such critical questions, they have become an important tool in the argumentation theorist's repertoire and promise to be of significant pedagogical value.[8]

1.5   Interdisciplinary Ambassador and Collaborator

Walton is also a past master at working across interdisciplinary boundaries, to the benefit of both his own work, and the fields with which he is collaborating. Perhaps his early research in medical ethics (such as Walton (1983)) gave

---

[4]Godden (2007, p. 1) adds deliberation dialogue to the core group. In (1990b) Walton adds three more: the planning committee dialogue, the pedagogical dialogue, and the dialogue of expert consultation; and in (1991a, p. 43) he introduces the deliberation dialogue along with the interview. In (1996c, p. 190) all ten are presented in a table identifying goals and benefits of each type.

[5]A critical appraisal of Walton's approach to dialectic and the dialogue types involved is advanced in (Krabbe and van Laar 2007).

[6] In all his work on schemes, Walton acknowledges a debt to Hastings (1963). But he has made considerable advances on that early material.

[7]Walton's treatment of defeasible reasoning is discussed in (Blair 2007).

[8]For a critical appraisal of the critical questions approach see (Kock 2007).

him a taste for such interdisciplinary research, and with his work on argumentation in jurisprudence he found an exceedingly fertile area (Walton 2002a, 2007a, 2008c), with a particular focus on the ways in which evidence is adduced and judged. Since 2000, he has also been collaborating with computer scientists in exploring applications of his theories of argumentation in artificial intelligence. One of the starting points for this collaboration was his development of argumentation schemes, which appeals to many computationally minded researchers because the approach strikes a happy balance between theoretical soundness (and, indeed, familiarity, since there are strong resonances between argumentation schemes and nonmonotonic and defeasible AI logics (Gordon et al. 2007)). The computational ramifications of argumentation schemes continue to be the focus for a great deal of energy (see the papers by Prakken, Atkinson & Bench-Capon and Reed in this volume, for example).

The pedagogic role of schemes also led to computational uptake, initially in the analysis software, Araucaria (Reed and Rowe 2004), and subsequently in other software systems[9], which were then explored back in philosophical contexts (Rowe et al. 2006)).

One particularly important area in AI over the past decade has been multi-agent systems, a type of distributed computer software which has enormous benefits of robustness and flexibility. Much of the power of these systems arises from the protocols used to structure the communication between the software components or agents and unfortunately these protocols can be fiendishly difficult to specify. It has turned out that argumentation theoretic models of dialogue are particularly well suited to the task, and Walton's work with Krabbe (1995) has been particularly influential in driving forward theoretical developments in this area. Walton's argumentation schemes have also had concrete benefits in the knowledge representation techniques used by agents in these systems (Reed and Walton 2005).

As the computational links continued to develop, it was in some senses only natural for Walton to bring together the two interdisciplinary links he had made: there has been a strong AI & Law community since at least the mid 1980's, and connecting with that community turned out to be very productive. Walton (2005a) brought together many of his techniques for this community, including mechanisms for capturing evidential generalisations (Bex et al. 2003) for handling corroborative evidence (Walton and Reed 2008), and for tracking the burden of proof (Walton 2007d).

1.6   Synopsis of the book

The following papers conduct explicit or implicit explorations of numerous facets of Walton's work and comprise contributions from colleagues and collaborators across the world and across areas of the field in which his ideas have had influence. In some cases they develop Walton's work, in others they contest aspects of it, and a few fill in gaps in the multiple stories that thread through his considerable output.

In her paper, 'Reflections on Fact, Value, and Argument', Trudy Govier takes Walton's work as an occasion to explore the fact/value distinction in argument. She recommends against a deep division between the two, arguing for a position that would seem to be consistent with Walton's pragmatic approach to argumentation. Erik Krabbe and Jan Albert van Laar take up Walton's recent discussion of an examination dialogue in their contribution, 'Examining the Examination Dialogue'. They examine the dialogue type to determine its characteristics and

---

[9]See, for example, `http://compendium.open.ac.uk/compendium-arg-schemes.html`

how it fits into the wider typology of dialogues developed previously by Walton and Krabbe (Walton and Krabbe 1995). The role of plausible and presumptive arguments is explored by Fabrizio Macagno in his paper, 'Dialectical and Heuristic Arguments: Presumptions and Burden of Proof', which places stress on the dialectical effects of a conclusion rather than its truth. By focusing on cases in law, he shows how this involves crucial shifts in the burden of proof. More interest in dialogical argument is the occasion for Fabio Paglieri's contribution: 'Committed to Argue: On the Cognitive Roots of Dialogical Commitment'. He compares Walton and Krabbe's notion of dialogical commitment with that of social commitment used in the field of Distributed Artificial Intelligence. On the basis of this comparison, it is argued that dialectical commitment must take into account the mental states of the arguer (while not being reduced to them). In 'Walton's Theory of Argument and its Impact on Computation Models', Chris Reed looks at two aspects of Walton's work that he feels have most impacted computational models: argumentation schemes and the structures of dialogue. In both cases, the nature and importance of the influences are stressed and their value emphasized. Finally, in Part One, Eddo Rigotti and Rudi Palmieri explore financial argumentation, another area in which, they observe, Walton's work is relevant. In their paper, 'Evaluating Argumentation in an Economic/Financial Context: An Example Related to the Current Global Crisis,' the authors focus on a key case study that they believe bears on the economic crisis facing the world at the time of writing. Their reconstruction and evaluation of this case as a piece of complex argumentation raises some interesting questions.

Part Two shifts attention more directly to issues related to argument schemes and the fallacies. In 'Argumentation Schemes: From Informal Logic to Computational Models', Trevor Bench-Capon and Katie Atkinson illustrate how Walton's schemes can be developed into the more precise formulations required for computational contexts. A different approach to Walton's work is taken by Frans van Eemeren, Peter Houtlosser, Constanza Ihnen and Marcin Lewinski in 'Contextual Considerations in the Evaluation of Argumentation'. They provide a deep investigation of Walton's approach to contextual factors in the assessment of arguments, and follow this by comparing it with the pragma-dialectical approach. Jean Goodwin's interest in 'Trust and Experts' is on what Walton has had to say about expert opinion in argument. She develops his work by grounding his critical questions in a more general theory of appeals to expertise. An important recent project for Walton has been the development of the Carneades system. Thomas Gordon gives an overview of what is involved in 'An Overview of the Carneades Argumentation Support System'. He explains how the tools involved can be used to support various argumentation tasks. In 'The Generation of Argumentation Schemes', David Hitchcock poses the question of where argumentation schemes come from. He gives a number of possibilities for their generation and in the end favours generalizing schemes by combining a framework of types of statements and of reasonable inference with an empirical base of actual arguments. The last paper in this section is Henry Prakken's contribution, 'On the Nature of Argument Schemes'. He uses insights from AI to understand the nature of argument schemes as a means to evaluate arguments, and he compares his findings with Walton's account of argument schemes.

Part Three focuses more closely on Walton's contributions to the theory of reasoning. J. Anthony Blair's paper, 'Reflecting on the Connections among Reasoning, Argument and Logic,' examines core logical concepts like 'reasoning' and then explores Walton's position on reasoning in light of the results. In 'Corroboration Evidence', David Godden studies the strengthening function of corroborative evidence in order to develop a model that does not follow one of Walton's related conclusions and accept the fallaciousness of double counting. In his paper, 'Walton's Equivocal Views about Informal Logic,' Ralph Johnson takes on the task of trying to sort out the meaning of 'informal logic' through an exploration of the various statements that Walton has made on the subject. Christopher Tindale follows one of Walton's forays into the history of argumentation in 'Walton and the Tradition of Plausibility Arguments'. He traces the pre-Aristotelian roots of plausibility arguments to the Sophists' use of likelihood, thus filling a gap in the history provided by Walton and noting connections between ancient and contemporary uses of this reasoning. The final paper in this section and the collection as a whole is 'Defeasible Reasoning,' in which John Woods starts from a consideration of Walton's work on defeasible reasoning and then proceeds to an examination of the matter that gives rise to a far more detailed account, including a series of rules for governing it.

These papers give but a hint of the depth and breadth of Walton's interests and the topics he has influenced. But they do reveal something of what is involved in taking that work seriously and responding to it. They also give testimony to a body of work that, while it can be applauded and honoured in this way, is essentially unfinished. The ground has been prepared for further research that will be both useful and fertile still, and there seems little doubt that Douglas Walton will himself continue to be a leader in that work.

In a very real sense, Walton's career has involved the repeated opening of numerous doors, some of which he was the first to recognize. Many scholars and students have now passed through those doors. What we have found on the other side may not have been what Douglas Walton expected, as each as adapted the terrain to her or his own interests and capabilities. But each of us owes him a debt for opening those doors, for showing us the way in a field that was new and unfamiliar and largely unexplored. And for doing so with an energy and passion that attest to the unique pleasures that a life dedicated to scholarship can provide.

Part I

Types of Dialogue and their Social Context

CHAPTER 2

REFLECTIONS ON FACT, VALUE AND ARGUMENT

Trudy Govier

In his work Douglas Walton has not sought to draw a firm distinction between fact and value. And indeed, to do so would not easily fit his generally pragmatic and flexible approach to issues of philosophy and argument. Walton has distinguished different sorts of contexts in which arguments are used. He lists quarrel, critical discussion, negotiation, inquiry, deliberation, information-seeking, and consulting an expert (Walton 1992b, 1996a, 1995)[1]. If we consider these contexts, it would appear that Walton shown little interest in drawing a firm distinction between normative and non-normative matters so far as the nature and importance of argument is concerned. One can, after all, have a *critical discussion* about almost any topic: the question can concern a matter of fact, a matter of law, a matter of scientific theory, personal hygiene, prudence, ethics or social policy. The idea of *inquiry* suggests inquiry into a matter of fact or theory as to what is the case; on the other hand, the notion of *negotiation* suggests debate as to what should be done about some contested matter; this may be policy, ethics, or prudence. *Deliberation* suggests decision-making or choice: one could be deliberating about what to do or which option to select. On the other hand, one might also deliberate about the implications that some new piece of evidence has on a standing scientific theory. *Consulting an expert* suggests a quest for knowledge, implying a cognitive context, but then too, one might consult an expert for advice on some matter of finance or personal relationships, asking him, in effect, 'what should I do?'

In a recent article treating persuasive definitions and considering the work of C.L. Stevenson, Douglas Walton alludes to Stevenson's emphasis on the distinction between differences in *belief* and differences in *attitude*, and expresses concern that Stevenson has lapsed into a false dichotomy regarding cognition and emotion. This dichotomy, Walton says, is one that amounts to "bifurcation" that "polarizes the discussion," implying that the dichotomy is exaggerated and will not stand up to scrutiny (Walton 2005c).

On the other hand, Christian Kock has recently relied on this very dichotomy, in arguing that a firm grasp of the fact/value distinction is fundamentally important for the understanding of arguments. Kock argues, in particular, that the distinction between factual judgments and normative judgment is crucial for understanding the many arguments admitting of pro and con reasoning. He urges that normative claims must be recognized and handled as such. These sorts of claims, he urges, do not admit of clear truth conditions, and those who have insisted on representing them as having some identifiable propositional content, (insisting on what Kock calls "assertive reconstruction") are making a mistake

---

[1] I have discussed some of these distinctions in (Govier 2005a)

(Kock 2009).[2] In its marked divergence from Walton's approach, Kock's account poses a challenge to much established practice in argumentation theory and informal logic. Many issues are raised by this challenge, which I will seek to explain, consider, and finally reject here.

2.1 Fact and Value

In a recent discussion of the fact/value dichotomy, Hilary Putnam argues that clarification is needed on both sides of the split (Putnam 2002). Putnam maintains that we tend to have an impoverished notion of facts, stemming from logical positivism. In particular, he states that we fail to understand how facts and values are inter-related. For one thing, it is necessary to rely on values to establish facts. To even begin investigating, we need to consider which things are important, which are plausible, and what will count as evidence and as good evidence. To acquire knowledge and regard ourselves as knowing something, we need to make judgments about the merits, consistency and implications of the evidence we have and make decisions as to whether that evidence entitles us to claim knowledge.

A fundamental point about abductive arguments can be cited in support of Putnam's position on the inter-relation of facts and values. Abductive arguments have the following basic structure:

1. Data D exist

2. Hypothesis A is the best explanation of D
   Therefore, probably

3. Hypothesis A is true

In scientific theory selection, notions such as 'best explanation' are crucial; the notion of a 'best explanation' clearly requires a judgment of value. The second premise in this schema will obviously need support; various factors such as simplicity, compatability with established theory, explanatory power, and fruitfulness will be appealed to in order to show that Hypothesis A is better than its competitors. This means, in effect, that abductive arguments, or inferences to the best explanation, require a sub-argument structure. The sub-argument will be conductive in nature: separately relevant factors will be cited in an effort to show that this hypothesis is the best one available. To say that a hypothesis offers the *best* explanation, or that some hypothesis offers a *better* explanation than some competitor, is clearly to evaluate it.[3] Abductive arguments in favour of scientific hypotheses or theories require conductive sub-arguments to support an evaluative premise.

Putnam reminds us that not all values are ethical values. As the case of abductive argumentation illustrates, there are epistemic values crucial for scientific reasoning. (In the judgment of explanatory hypotheses, simplicity is an epistemic value, not an ethical one.) According to Putnam, facts and values are entangled with each other and cannot be separated.[4] Putnam argues that the notion of a dichotomy between fact and value is ideological. It has sometimes served to excuse people from examining their values, implying that serious reflection on

---

[2]I am grateful to Professor Kock for supplying me with a copy of his paper, which is available electronically with the proceedings of that conference.

[3]This example is my own. It is defended and explained in Chapter Ten of (Govier 2005b).

[4]*Ibid*

them is not possible because in the end we can only make subjective choices.[5] Few who do not take reflection about values seriously are prepared to give up on facts; they adopt a non-cognitive or subjectivist stance on value judgments while maintaining that objectivity and knowledge are possible in contexts of factual judgment. In this sort of view, a legacy of positivism, a dichotomy of fact and value is required, in order to allow cognition of facts while maintaining a non-cognitive account of value.

2.2 Direction of Fit

The fact/value dichotomy is often defended by referring to direction of fit. The suggestion underlying the direction-of-fit metaphor is that there are two and only two exclusive possibilities so far as the relation between claims and the world is concerned. Either our words should fit the world, *or* the world should fit our words. The either/or is supposed to amount to an exclusive disjunction, or true dichotomy. The idea here is that in factual claims, about matters of belief, the words should fit the world, whereas in value claims, expressive of our desires, we intend that the world should fit our words. To say 'Winnipeg is in Manitoba' is to make a factual claim, one which will count as true if the world fits our words, in the sense that the city referred to is spatially located within the geographic area referred to. If those relationships described by our words are characteristic of the non-verbal world, then our words appropriately fit the world as it is, so that our claim, made when we perform the speech act of asserting, is true. The city of Winnipeg really does lie within the province of Manitoba; there is a word to world 'fit' in the case. On this view, factual claims are to be contrasted with value judgments, with regard to direction of fit. Consider, as representative, a value judgment such as 'Winnipeg should be the capital of Canada.' On the direction-of-fit view, someone who would make such a claim would be expressing her desire that the world be in one way rather than another – that this particular city be designated as Canada's capital. (Various distinct reasons could be cited to support the view in a conductive or 'good reasons' argument in which a number of distinct properties of Winnipeg were cited; there are also, obviously, powerful reasons against it.) In her statement that Winnipeg should be Canada's capital, a speaker is expressing a desire and indicating that the world should conform to this desire. That is to say, the world is to fit her words. And this is the other direction of fit. There are, on this view, two and only two directions of fit — and on this basis we may explain a dichotomy of fact and value.

The fact/value dichotomy, then, is to be buttressed by another dichotomy. *Either* our words are supposed to fit the world *or* the world is supposed to fit our words. In explaining his position on fact and value, Kock relies heavily on this notion of direction of fit, opposing any suggestion that claims committing one to values should be assertively re-cast. He says:

> Assertives have a word-to-world direction-of-fit. Directives and commissives have a world-to-word direction of fit. The difference, to use a formulation by Bernard Williams (1966), is between discourse which "has to fit the world" and discourse which "the world has to fit." Humberstone (1992, 60) has proposed the terms "thetic" for the word-world direction of fit and "telic" for the world-to-word directions of fit; commissives and directives are telic, whereas assertives are thetic; that is why the illocutionary point of a directive or a com-

---

[5]These comments do not, in fact, apply to Christian Kock.

missive is not a commitment to any "identifiable propositional content." [6]

If statements of value are commissives or directives — representable though on Kock's view improperly represented as assertives — then to interpret conductive arguments and other arguments as providing *reasons* for *claimed assertions* is a deep mistake. It is not only pragma-dialecticians who advocate what Kock calls assertive reconstruction.

The example given above, in which the conclusion of a conductive argument is expressed as 'Winnipeg should be the capital of Canada' would probably be understood by Kock as incorporating an assertive reconstruction. The representation given here is not one in which there are reasons for 'would that Winnipeg were the capital' or 'make Winnipeg the capital' or 'I desire that Winnipeg be the capital.' Rather, as it appears here, the example is one of somebody *claiming* that Winnipeg *should* be the capital. Kock commits himself to some version of non-cognitivism when he says that normative claims have "no clear truth conditions." He does not wish to endorse wholesale relativism or subjectivism. Rather, he wishes to mark contexts in which arguments (and especially conductive arguments in which separately relevant premises are put forward to support normative judgments) serve to articulate and mark disagreements, and are used in contexts of choice where there is no provable right or wrong.[7] Philosopher kings and queens do not exist, Kock reminds us, and they cannot demonstrate truths about matters of value; logicians formal or informal, and argumentation theorists, pragma-dialectical or not, cannot offer algorithms proving value judgments that (on this view) do not admit of truth or falsity in the first place.

Kock does not advocate across-the-board non-cognitivism and subjectivity; he is treating matters of value in ethics and politics and is thinking particularly on those matters where we find deep disagreement. But problems lurk here: given the ubiquity of conductive argumentation, the existence of pros and cons and the need for judgment in many contexts, and the frequent entanglement of fact and value. If disagreement, the presence of value judgments, and the lack of demonstrable proof show subjectivity of value, non-cognitivism and subjectivity will not be contained with the realms of ethics and politics. As Putnam argued, and as the case of abductive arguments serves to illustrate, value judgments about which there is serious disagreement may also be found in logic, epistemology, and science itself.

Clearly many philosophical questions are raised here. Central is the issue of whether a fact/value dichotomy can be shown to exist, based on the argument from direction-of-fit. We are looking at a situation in which one supposed dichotomy (fact/value) is to be buttressed by another (direction of fit). Now the contrast between two directions of fit may at first appear convincing as a dichotomy. A true dichotomy is supports an exhaustive and exclusive disjunction. Truly dichotomous categories are such that every item to be classified by these categories will fit into either one or the other (the categories will *exhaust* the possibilities) and will not fit into both (the categories will be *exclusive*). In other words, there will be at least two and at most two classificatory categories: a bifurcated classificatory system is achieved. If the dichotomy of direction of fit is a true one, then for every speech act, it must either be the case that one's words should fit

---

[6]Christian Kock, *Op. cit.*

[7]Kock, correspondence. See also his review of a recent book by Harald Wohlrapp. (*Der Begriff des Arguments: Uber die Beziehungen zwischen Wissen, Forschen, Glauben, Subjektivitat und Vernunft*, published in Wurzburg by Konigshausen und Neumann 2008). I am grateful to Christian Kock for providing me with a copy of this review.

the world or that the world should fit the words. And it should never be the case that both demands are made.

Do considerations about directions of fit support a true dichotomy? To explore the matter, we may first ask whether the disjunction 'this speech act aims at word-to-world fit or at world-to-word fit' is exhaustive. We can readily see that it is not: there are verbal utterances that count as speech acts in which it is neither the case that our words are intended to fit the world nor the case that we wish the world to fit our words. Exclamations such as 'ouch' and 'ugh' provide examples. Second, we may ask whether this disjunction is *exclusive*. It is not: some claims combine the two directions: they are *both* world-guided and action-guiding.[8] For example, a claim such as 'Joe is generous' is descriptive *and at the same time* has evaluative implications. The positive overtones of the word "generous" convey the evaluative message that there is something positive in Joe's character, which in the same word is described as one that is giving to an extent greater than normal. In recent discussions, terms like "generous" have been called thick moral terms and the related concepts, 'thick concepts.' Alan Gibbard defines a thick concept in this way, "A term stands for a thick concept if it praises or condemns an action as *having a certain property*."(Gibbard 1992, My emphasis) These terms, he says, have both descriptive and evaluative meaning. As examples of thick terms, Gibbard cites 'cruel,' 'decent', 'hasty,' 'lewd,' 'petty', 'sleazy,' and 'uptight.' Many other examples can be given: 'friendly,' 'sullen', 'racist,' 'compassion,' 'sympathy,' 'gratitude,' 'bullying,' 'kind,' 'victim,' 'capacity,' to cite just a few. One cannot actually explain what a word like "cruel" means, and how it can describe an action without also explaining its evaluative and prescriptive elements.[9] To use another example, there is a matter of fact as to whether someone is a victim, but when we say he is a victim, we also imply certain evaluative facts as to his innocence in the context of the wrong committed.

2.3   Thick Terms and Their Implications

This phenomenon of combined description and valuing, is of course a familiar one that has been much considered by philosophers. Their explorations extend back to C.L. Stevenson and, before him, to Hume and even Aristotle.[10] What is said about ethical values could also be said about other values — logical, epistemic, and aesthetic values, for instance. To say that a theory is coherent is *both* to make a claim about its characteristics (the claims made are consistent with each other) *and* to value it in a particular way. Now within meta-ethics, defenders of the thick – sometimes humorously called 'thickies' – maintain that when we use thick terms we both say what the world is like and express an evaluation. To use without qualification a term with both descriptive and evaluative implications is to embrace the thick conception or grasping it in an engaged way.[11] According to 'thickies', there is no dichotomy of fact and value.

It is fairly common to make claims of fact and of value in the same words. We can describe the world in value-laden terms and we can use words to indicate our approval of certain matters of fact.[12] With due respect to J.L. Austin, it would

---

[8]These terms are used by Christine Korsgaard in (Korsgaard 2003). See, for instance (Gunnarson 2008, Goldie 2007)

[9]Other defenders of the thick include Philippa Foot, Iris Murdoch, John MacDowell, David Wiggins, and Mary Midgley.

[10]For pertinent articles by Douglas Walton, see (Walton 2005a, Macagno and Walton 2008)

[11]Goldie, *Op. cit., 136.*

[12]This matter was emphasized by Julius Kovesi in (Kovesi 1967) and, echoing him, by Mary Midgley in (Midgley 1981)

appear that we can do these things with words and do that at the same time, with the same words. Describing and committing are not mutually exclusive, then. Neither are describing and prescribing. A thick concept is *both* responsive to the way the world is *and* gives the agent using it a defeasible reason for action. If an item X falls under a positive thick concept, then X is good, *ceteris paribus* or, as some prefer to put it, good *pro tanto*. This does not entail that X is good in toto — good all things considered (Tappolet 2004). Many words enable us to employ language to do several things at once, even when the evaluative overtones of language are not obvious. Consider, for instance, the word "toy." If we say to a child 'that vase is not a toy,' we are telling him that the vase is not to be played with, in other words, he should not play with the vase; that is a prescriptive remark. We see here how a description, in context, can imply a prescription.[13]

The existence of thick terms shows that asserting facts as to how the world is and committing ourselves to values are not mutually exclusive activities. The dichotomy supposedly provided by the metaphor of direction of fit breaks down and if it breaks down, it cannot buttress a dichotomy of fact and value. These phenomena related to think terms would appear to undermine the view that there is a clear dichotomy of fact and value, and to support the considered pragmatism of Douglas Walton, rather than the views of Christian Kock.

But that conclusion is not drawn by everyone. Critics of 'the thick' have contested the notion that one can generate reliable evaluative conclusions by the application of thick terms. They have, among other things, pointed to the *malleability* of language. We can use thick terms so as to dissociate ourselves from their normal evaluative implications. To say that an action is generous is to describe it as giving in a helpful way and to commend it as worthy other things being equal. We generally imply a positive evaluation of Joe's character if we say he is generous – but we can deny such an implication if we choose. An action might appear generous and nevertheless not to be recommended for some other reason: one might, for instance, judge that the recipient of someone's generosity is becoming dependent on it and could thus be harmed by it, questioning on these grounds whether generosity in such a case is really a good thing. In these cases, the positive evaluative implications normally associated with the thick terms "generous" can be withdrawn; a common strategy would be to add the word "too." We may say, "Joe is too generous" or "Joe is unwisely generous", and these claims would make sense. It may appear, then, that we can use a thick term without committing ourselves to the associated value. Modifiers are one way; we can also deny the normal evaluative implications by employing a questioning or sarcastic tone or a device such as scare quotes.

Suppose that someone disapproves of a man's actions and calls them *cruel*, indicating his disapproval. Another person who approves them can seek another descriptive term with different evaluative implications; he might for instance call the actions *strict*. The disagreement about the moral quality of this man's actions will not be resolved by the fact that "cruel" is a thick term. The word "obese" is another thick term, combining description with negative evaluation. If we use the term "obese" to describe someone, we are describing her so as to convey a harsh negative evaluation. That harsh evaluation can be avoided by selecting other words. This person might have been described as 'pleasantly plump,' 'husky,' 'stocky,' or 'chunky.' The descriptive notions of being wider and heavier than normal are carried over in these less insulting uses of language, while the harshly negative emotional connotations of the word "obese" are absent or

---

[13]The toy example comes from Colin Hirano. The point that many terms combine both factual and evaluational aspects has been argued by many including Julius Kovesi, Mary Midgley, and Bernard Williams.

at least diminished.[14] The linguistic fact that the term "obese" conveys both a description and a negative evaluation does not dictate an irresistable set of social norms. The negative emotive and evaluative implications of the word "obese" do not and could not compel us to denigrate and stigmatize every person whose weight and apparent size are (more neutrally) 'large.' Speaking and using the language of description which serves and conveys our interests, we often use language so as to communicate value judgments; however we have the linguistic means to distance ourselves from those judgments. Linguistic usage is resistable — at least to some extent, and some of the time.

Critics of the view that there are thick terms have also noted that terms having evaluative implications may go *out of usage* when moral practices change. Consider "chaste," "fornication," and "living in sin" as examples; in many contemporary sub-cultures such terms could now be used only ironically. Another example is "cute" as applied to women. That might have seemed like praise on the basis of appealing attributes, some decades back, but now is understood as a put-down. (Think about the implications of calling a female Supreme Court Justice "cute.") Critics have also urged that due to their evaluative implications, thick concepts are likely to be contestable concepts. Disputes about whether someone's actions are, in context, generous may lead to further disputes about whether what we have regarded as commendable giving (praiseworthy and praised under the name "generosity") genuinely does merit our praise. That matter will be and remain debatable whether we engage in assertive reconstruction and use thick terms or not. The existence of thick terms is not going to make moral disputes go away. Readers will note that even to speak of whether some actions *merit our praise* is to use two thick terms, "merit" and "praise." It is clear, then, that disagreement about moral matters, and other matters of value, will not be eliminated by the fact that thick terms exist. The fact that such terms do exist shows that in many speech acts we both describe and endorse, but it does not show that, or how, we can demonstrate value judgments to be true.[15]

2.4   Disentangling and its (supposed) benefits

One can insist that when thick terms are used, *two quite distinct things* are going on at the same time: *describing and appraising*. One can insist that although these 'distinct' things are often going on together, they really *are* distinct, and they need to be *disentangled*. It is merely that we project our attitudes onto the world when we describe the world, and because we do this, our linguistic practices allow us to do two distinct (and separable) things at the same time. This approach is taken by Simon Blackburn and others, who claim that we can and should preserve the fact/value dichotomy by *disentangling* descriptive and evaluative elements that present themselves in entangled form in thick terms (Blackburn 1992).[16] The idea here is that even though we often use thick terms so as to indicate *both* our beliefs (world-guided) *and* our desires (action-guiding), the entanglement of normative and descriptive elements appears only at one level. These elements can be separated, on analysis, and when the entangled descriptive and evaluative elements are (as it were) 'factored out,' the fact/value dichotomy will re-assert itself.

The idea would be roughly something like this: it is one thing to describe Joe as giving to others and another thing to appraise Joe's giving as worthy; when

---

[14]Current concerns about weight and health are such that these other terms seem to be going out of use.

[15]A major concern of Kock's is to deny such presumption.

[16]An interesting and useful article here is Blackburn's "Disentangling Disentangling", which may be found at http://www.phil.cam.ac.uk/~swb24/papers/disentangling.pdf.

we call Joe generous we are doing both things at once, but these two things are distinct and separable. And, it may be submitted, they should be separated. In this debate, we have seen that, to the successful questioning of a dichotomy at one level, people may respond by appealing to a dichotomy at another level, clinging stubbornly to what they originally understood to the basic 'opposition' between fact and value.

There are thick terms used in the interpretation and evaluation of argumentative discourse. Consider, for example, 'credible' and 'persuasive.' One claim, X, may be said to make another claim, Y, credible in the sense that those accepting X will find Y easier to believe. This is a psychological, descriptive sense of credibility. People may also be said to be credible, meaning that they are believable in the sense of being likely to believe. One might say, for instance, 'With her moderate speech and serious demeanor, she will make a credible witness.' But there is also, importantly, a normative sense of credibility. In personal contexts, a witness who was in this sense credible would be one who ought to be believed, who merits belief. Thinking again of claims as distinct from the people make those claims, to judge that X makes Y credible in a normative sense is to say that X provides good reason or evidence to support Y, making Y more worthy of belief. Credibility, then, may imply *ease* of belief or *worthiness* of belief. The first notion is broadly empirical; the second is normative, and terms such as "credibility" and "credible" are often used so as to bind these the empirical and the normative together. Advocates of disentangling insist these elements be separated in the interests of accuracy and critical clarity.

A well-told anecdote might make a general claim credible in the sense of rendering it *believable* (its acceptance likely); to attribute that sort of credibility of an anecdote is often plausible. If, however, one says that an anecdote is credible in the sense of rendering a general claim *worthy* of belief, that is quite another matter.[17] Similar points can be made about the term "persuasive" which is open to both psychological and normative interpretations.

2.5   Arguing with and about Words

Whether in moral, politics, or argument analysis, thick terms should not be used to evade discussion and debate. That possibility has been a matter of concern for some critics of 'lovers of the thick.' For discussion and criticism of moral and other values, we need to be alert to the implications of thick terms and there will be many contexts in which we seek to spell out evaluative implications and consider whether the facts in the case render our evaluations appropriate. The attitudes we presume on the basis of descriptions may seriously merit examination: when they do they should receive it.

Some critics of 'thickies' contend that thick moral terms amount to nothing more and nothing less than loaded language, that is to say language incorporating emotional responses and enabling speakers to avoid justifying those responses.[18] We can note here that the term "loaded language" as commonly used in textbook accounts is itself thick (or, one might say, loaded), serving to convey a negative evaluation. If such language is used to avoid offering reasons for judgments, there is a defect in argument in the sense that in a context where justification is needed, it is not offered. Emotive language is supposed to carry a

---

[17]This thickness of the notion of credibility is apparent in (Oldenburg and Leff 2009). Oldenburg and Leff, commenting on an anecdote told by George W. Bush in the 2004 U.S. presidential campaign, stated that it 'lends credibility' to conclusions being drawn from the case. I am grateful to the authors for providing me with a copy of this paper prior to the electronic publication of the proceedings.

[18]For my own discussion of loaded language see *A Practical Study of Argument*, Seventh Edition, Chapter Three.

contestable claim and may lead us to ignore the fact that no justificatory reasons are offered. The strategy of labeling by negatively loaded language, and deriving a value judgment from the application of a term, without argument, has been called 'Argument by Epithet.' The strategy will be familiar to many readers.

Here is an example. Person A has a quarrel with the Dean. His department chair, Person B, thinks that A is partially right in the matter, but believes that A is exaggerating his troubles and acted unwisely in various ways; believing this, B fails to give his wholehearted support to A. A is offended and hurt, and wishing to convey his beliefs and attitudes to departmental colleagues. He attacks B roundly and, among other things, tells his colleagues that B is engaging in 'Quisling behavior.' There is no argument in this description — just a label attached to that behavior. But for those who know of Quisling's collaborationist stance, facilitating Nazi control of Norway, the expression "Quisling behavior" conveys a powerful negative message about B. We can attribute to A the following argument:

1. B has engaged in Quisling behavior.

2. Quisling behavior is seriously wrong. (implicit, justified by negative overtones of term)
   Therefore

3. B has engaged in seriously wrong behavior.

We have here a textbook case of manipulative discourse. We might say A has no argument where he should have one, or we might attribute to A the above argument, which commits the fallacy of Argument by Epithet. Here it appears as a version of begging the question, because the conclusion is presumed in the label used in the first premise. Instead of showing that B has done anything wrong, A simply applies an epithet or *loaded term* to communicate his view. A word intended both as descriptive and as evaluative is supposed to do the work instead. One might call "Quisling" a thick term at this point, or loaded term. The case of Argument from Epithet illustrates the great potential for manipulative use of thick terms to convey claims in the absence of supportive argumentation.[19] We must recall, however, that although thick terms *can be misused*, that is not to say that to use a thick term is always to misuse a term.

In persuasive definitions, the referent of a term is made to shift while its attitudinal connotations are pulled along, implying a negative or positive evaluation. As with thick terms, we have a combination of evaluational, factual, and emotive aspects, and language will permit us to do several things at once with our words. In persuasive definitions we re-define so shift cognitive content but maintain the emotional content. A Quisling, on an implicit re-definition, is no longer one collaborating with an enemy force; rather he is one who fails to side with a colleague in a dispute with the dean. But the negative overtone is retained; if we accept the implied persuasive definition, we are going along with the idea that it is seriously, shockingly, wrong not to side with one's colleague against the dean in a case like this. Now Walton wants to say that sometimes, such shifts can be justified. He and his co-author Fabrizio Macagno acknowledge and explain that persuasive definitions are often used in misleading and fallacious ways. They cite many interesting cases including the following.

'Taxation is a form of theft.' (This statement can be understood as asserting an implicit re-categorization of taxation as the unauthorized claiming by one

---

person or institution or an entitlement to the property of another person or institution.)

'A satisfied society is one in which people are not protesting.' (This statement can be understood as tacitly presuming a definition of what it is for a society to be satisfied; satisfaction is to be the absence of protest.)

'To love only one woman would be unfair to the others.' (This statement can be understood as classifying monogamy and loyalty – normally assumed to be good – as unfairness – normally assumed to be bad.)

'Selfishness is acting according to one's own desires.' (This statement classifies selfishness – normally presumed bad – acting according to one's own desires, thus giving selfishness a more 'pro' description and making it seem more virtuous.)

'True temperance is a bottle of claret with each meal and three double whiskies after dinner.' (True temperance – good – presumes quite a bit of drinking – something which otherwise would have seemed bad.) [20]

But while acknowledging that persuasive definitions are often used in an effort to convey a point without argumentative support, Walton and his co-author Fabrizio Macagno also contend that they can play a legitimate role in argument in some cases. The matter, they say, needs to be addressed in a case by case way. Few details are given, but it is in this context that Walton remarks that to fully separate matters of belief from matters of attitude — as would be implied if we accepted a clear dichotomy of fact and value – would be to commit oneself to an over-simplification and a mistake.

2.6   Implications

We must acknowledge that the nature of our language and the practices of discourse permit us both to describe and to value, using the same words and sentences. We must further acknowledge that these aspects of discourse make it possible for us both to verbally insert into the public domain both our belief about a certain matter and to convey our emotional attitude to it. We can insist on disentangling, and in some contexts, that will be of critical importance. But in the end, insistence on disentangling will amount to an insistence that descriptive language be shaved in a most unrealistic way. We human beings describe things with a view to our purposes and problems; given that fact, much of our descriptive language is shot through with value.[21] Thick terms are pervasive in language and even terms such as "toy," which display no obvious thickness can be used in context to convey prescriptions and descriptions at the same time. Calling something a toy can have prescriptive implications; similarly, saying that something is a chair or a knife ultimately incorporates values. There is some point in having knives and chairs; there is some human purpose that these things serve and must be able to serve to merit their description. Given the pervasiveness of the thick, disentangling cannot be complete.

Clearly it would be hasty to presume that thick terms provide an objective basis for moral judgments, demonstrating that disagreements about value can (eventually) be resolved by argument and disproving the sorts of non-cognitivism and subjectivism endorsed by Christian Kock. There are nevertheless important

---

[20]Macagno and Walton, *Op.cit.*

[21]See Julius Kovesi (1967) *Moral Notions*, for a sustained presentation of this view.

implications of challenging a fact/value dichotomy based on a presumed dichotomy of direction of fit. Non-cognitivism about value will, however, threaten cognitivism about facts, if values and facts are entangled. The ways in which we can do things with words, show that our words allow us to entangle fact and value. The sort of dichotomy between fact and value that is required by Kock's account is not a feature of our language.

To resist the fact/value dichotomy, as I have recommended here, is not to refute non-cognitivism but rather to raise issues about its scope. If there is no genuine dichotomy of fact and value, then non-cognitivism about values will have implications of non-cognitivism for fact. Conversely, one could argue that cognitivism for facts will have implications of cognitivism for values. Another conclusion is available: the debate about cognitivism and non-cognitivism will lose much of its point. [22]

Arguments about the distinction and relation between fact and value are likely to continue in philosophy and elsewhere. Those convinced of a deep division will likely insist that there are two and only two distinct directions of fit and address the phenomenon of ubiquitous thick terms by insisting on disentangling. If one is convinced that somehow there must be dichotomy underneath an entangled reality insist on 'exposing' it – but its status as something discovered rather than constructed will remain in question.

---

[22] In a useful article on Moral Cognitivism versus Non-cognitivism in the *Stanford Encyclopedia of Philosophy*, Mark van Rooven states that the cognitivism/non-cognitivism dichotomy in philosophy is becoming rather blurred, as non-cognitivists adapt their accounts to accommodate the roles of evaluative statements in reasoning and argument. Van Rooven comments that if too many domains of discourse are going to be analyzed from a non-cognitivism point of view, the contrast between cognitive and non-cognitive domains will lose its significance and be hard to maintain.

CHAPTER 3

EXAMINING THE EXAMINATION DIALOGUE

Jan Albert van Laar and Erik Krabbe

3.1 Introduction

An examination dialogue is a dialogue between two participants, in which one of them is being examined. As such it is a recognizable type of dialogue, though its manifestations in various contexts and institutional settings are very diverse. Examination dialogues are of interest for theories of argumentation and reasoning, not only because they form themselves a setting for argumentation and reasoning, but also because of their connections with other types of dialogues which do so. The subject has been recently studied by Douglas Walton Walton (2006b), and has also got attention from the side of artificial intelligence (Bench-Capon et al. 2008, Dunne et al. 2005).

In this paper, we want to explore in some more detail what the common characteristics of different kinds of examination dialogues are, and how the examination dialogue can best be conceived and fitted within a general classification of types of dialogues (main type, subtype, or mixed type). We are not trying to construct normative models for examination dialogues, but restrict ourselves to an investigation of examination dialogue as an activity type. As we shall explain in Section 2, we see no contradiction between the theory of dialogue types (Walton and Krabbe 1995) and the pragma-dialectical theory of argumentative activity types (van Eemeren and Houtlosser 2005), and will use elements of both.

In Section 3 we approach the question of whether the examination dialogue should be considered as a main type of dialogue or as a subtype or mixed type, presenting some pros and cons for these positions. Section 4 focuses on several subtypes (and subsubtypes) of examination dialogues: expert examination (including ancient peirastic discussions), witness examination, academic examination, and various kinds of interview. In Section 5 we try to describe the general structure of examination dialogues in terms of five parameters. As prominent features we see the use of dialogues of various types as tests and the inherent double intention when dialogues are used as tests, which may become explicit in so-called 'intermediate checks'. Section 6 presents some conclusions.

3.2 Types of Dialogue and Activity Types

In their study into argumentative commitments, Walton and Krabbe distinguished, without attempting to establish a full-fledged taxonomy, between six main types of dialogue: persuasion dialogue, negotiation, deliberation, inquiry, eristics and information seeking dialogue. In addition to this, they identified mixed types of dialogue, such as that of a debate or of a committee meeting, where main types are blended, for example due to the embedding of dialogues of one type in a dialogue of another type (Walton and Krabbe 1995, p. 82). Even though Walton and Krabbe distinguished between descriptive and normative approaches (Walton and Krabbe 1995, pp. 82, 174-177), the distinction remained insufficiently clear and was not consistently applied.

Van Eemeren and Houtlosser were right to emphasize the importance of the distinction between normative models and characterizations of argumentative activity types. A normative model of critical discussion 'is a design of what argumentative discourse would be like if it were optimally and solely aimed at methodically resolving a difference of opinion on the merits' (Walton and Krabbe 1995, p. 75). By contrast, argumentative activities form 'cultural artefacts that can be identified on the basis of careful empirical observation of argumentative practice' (Walton and Krabbe 1995, p. 76). We take this distinction to be applicable to all types of dialogue.

For each type of dialogue, seen as an activity type (whether argumentative or not) a normative model can be developed that provides the rules that, in the view of the theorist, are to be followed in order for the participants of the dialogue to realize, in the best possible way, what according to the theorist should be the main goal of this type of dialogue.[1] This also applies to examination dialogue. One can propose norms for parties who want to turn their examination dialogue into a success. In this paper, however, we will focus on examination dialogue as an activity type. Even though a full descriptive characterization of this activity type will take into account the rules and conventions that participants typically impose upon one another, these rules and conventions need not coincide with the prescriptive norms that, in our eyes, are to be followed by the participants in order to realize this dialogue type's main goal. Along what lines can we develop such a descriptive model of examination dialogue?

Walton and Krabbe classify six main types by making distinctions with respect to the initial situation, which includes the so-called concessions (Walton and Krabbe 1995, pp. 133, 155), the participants' individual aims, and the shared main goal of the dialogue. A persuasion dialogue is characterized as starting from conflicting points of view such that the participants aim at persuading each other and where the dialogue, as a whole, heads for a resolution of the conflict by verbal means (Walton and Krabbe 1995, p. 66). An inquiry is characterized as starting from a question or problem such that both parties cooperate to establish or 'prove' propositions that provide a stable and generally agreed upon answer or solution (Walton and Krabbe 1995, p. 72). As Walton and Krabbe state, such general characterizations can be specified into elaborate normative models by laying down rules for each particular type of conversation (Walton and Krabbe 1995, pp. 66-67). They offer two normative models for persuasion dialogue by way of precisely formulated rules. Locution rules determine the available speech acts. Commitment rules determine the effects of particular locutions on the commitment stores of the participants. Structural rules determine the available options for making a move in a given situation. Win-and-loss rules determine in what case a participant has won or lost (Walton and Krabbe 1995, Chapter 4). We

---

[1] The main goal of a type of dialogue, assigned by a theorist, is to be distinguished from the aims participants have in dialogues of that type. For instance in persuasion dialogue, the main goal is resolution of a difference of opinion, whereas each participant's aim is to persuade the other.

We do not think that the activity type in which parties happen to be engaged, determines in any straightforward manner the goal and the norms by which the contributions by the parties are to be evaluated. The evaluator's choice of the goal and the norms is to be determined by his interests. If he wishes to evaluate the discourse from the perspective of inquiry, he can do so by applying the norms that make up a normative model for inquiry. If he is interested in reasoning from the perspective of dispute resolution, he should apply norms that make up a normative model for persuasion dialogue, e.g. the model known as 'critical discussion'. This can be done whatever argumentative activity type is involved.

would like to add that descriptions of rules and conventions are also needed in order to arrive at a more elaborate descriptive model of any dialogue type.

The parameters Walton and Krabbe use to characterize dialogue types resemble, but do not coincide with, those that are used by van Eemeren and Houtlosser in order to arrive at their characterizations of the argumentative activity types of adjudication, mediation and negotiation: *initial situation, starting points, argumentative means* and *outcome* (2005). These four parameters, which correspond to the four stages of the normative model of critical discussion, are selected in order to arrive at a description of activity types that brings the activity's argumentative features to light. They characterize, for example, adjudication in the following fourfold manner. In the *initial situation* of a case of adjudication there is a dispute between two parties in the presence of a third party who has the jurisdiction to settle the issue, in the end. Among the *starting points* of this kind of dialogue, we can expect to find explicitly codified procedural rules, but also the admitted evidence. The *argumentative means* are arguments that are based on an interpretation of concessions in terms of available facts and evidence. The outcome of a case of adjudication is the settlement of the dispute by the third party.

In section 5 we shall characterize examination dialogue as an activity type, making use of both systems, by specifying: the initial situation, which here does not include the concessions or shared starting points; the starting points, containing both concessions and procedural arrangements; the participants' individual aims; the method of proceeding, which, in the case of examination dialogue can but does not need to include argumentative means; and the desired outcome.

3.3   Is Examination Dialogue One of the Main Types of Dialogue?

There are some grounds for taking examination dialogue as a main type of dialogue, missing from Walton and Krabbe's typology. Dunne et al. (2005) are taking this approach. There are also reasons to understand examination dialogue as constituting a mixture of the other types or as a subtype of one of them. Walton develops the conception of the examination dialogue as a mixed type (Walton 2006b).

As regards the initial situation, examination dialogue centers round a special kind of issue: the issue is always concerned with one of the participants, and the question is whether, or to what extent, he or she satisfies a particular set of conditions. For example: Is the professor a real expert in the field? Does the candidate satisfy the job requirements? Does the student grasp the topic of tuition? Is the witness sufficiently reliable? Perhaps then, examination dialogue is a subtype of persuasion dialogue or inquiry. Which one would it be? Examination dialogues can start from a matter of contention, but also from an open issue. The applicant can express his opinion that he is fit for the job. He can also choose an explorative stance on the issue. Sometimes students are best seen as trying to convince their teachers of their skills, and sometimes as merely taking part in an attempt to establish whether or not their progress is up to standard. Presumably then, some examination dialogues are persuasive, in case parties adopt different positions with respect to the issue, while others are inquisitive in nature, in case none of the parties adopts a standpoint with respect to the issue.

But then, in both kinds of case the issue concerns one of the participants. This can be viewed as an inessential feature of what still is to be seen as basically persuasion dialogue or inquiry, but it might also amount to a distinctive initial situation that defines a seventh main type.

As regards the starting points, in many cases, the rules, conventions and pro-cedures that govern the dialogue are determined to a high degree by the exam-iner or the institution that employs the examiner. This asymmetry can be seen as speaking in favor of examination dialogue as a distinct main type. But then, some examinations are more informal and allow the examinee much more room for in-fluencing the procedural arrangements. What is more, asymmetry as a feature of dialogues is not incompatible with belonging to one of the main types of inquiry or persuasion dialogue. Institutions, such as courts or legislative committees, do often one-sidedly determine the procedural setting for trials or hearings, in-cluding those parts we would not hesitate to classify as belonging to these main types. As regards the individual aims of the parties, things are equally indeci-sive. On the one hand, both parties can be seen as dealing with the examinee's status as part of an inquiry or as part of a persuasion dialogue. If the issue forms an open problem, they are cooperating in establishing the premises with which to solve it. If it forms a difference of opinion, they are attempting to reach an agreement about it, each party trying to achieve an agreement in its own favor. This would speak for considering examination dialogue as a subtype of either inquiry or persuasion dialogue. On the other hand, at some points, the aims of examiner and examinee are more idiosyncratic, the examiner aiming for a correct judgment about the examinee, whereas the examinee is merely trying to pass the tests. This particular asymmetry is untypical for either inquiry or persuasion dialogue.

As regards the method of proceeding, examination dialogue employs, as a necessary part, one or more testing methods, and many of these methods are themselves to be understood in terms of the various types of dialogue. As we shall discuss below, the testing method can be a kind of persuasion dialogue, or a consistency probe, or a simulated negotiation, and so forth. During a dialogical test the examiner and examinee often must adopt, if only temporarily, individual aims that fit the roles that need to be played in order to proceed with the test. This speaks against examination dialogue as either an autonomous dialogue type or a subtype, and in favor of examination as being inherently mixed. But then, the execution of the testing method is instrumental to making headway in solving the issue with which the examination dialogue had started, and so, possibly the mixed nature of this kind of dialogue should not lead us away from conceiving it either as autonomous or as inquiry or persuasion dialogue.

The desired outcome of examination dialogue amounts to a solution of the ini-tial issue: does the examinee fit the requirements? This outcome is based on the results of the tests, and these results are, typically, determined unilaterally by the examiner's interpretation. In this respect examination dialogue is very different from inquiry or persuasion dialogue, where no single party has a one-sided com-petence to interpret or control, let alone fix the outcome. Possibly then, examina-tion is a separate main type of dialogue or a particular mixture of dialogue types. But notwithstanding the examiner's competence to determine the test results, the desired outcome of the examination dialogue as a whole can still be seen as the shared solution of a problem regarding the examinee's status, and thereby as an outcome that fits an inquiry or, if the parties were originally opposed, as the res-olution of a difference of opinion that fits a persuasion dialogue. If we adopt this stance, the unilateral competence of the examiner remains restricted to the results of the tests, and the final outcome is a more cooperative result based upon these tests. The tests, then, provide some of the agreed premises needed for determin-ing a generally agreed upon outcome of the examination dialogue considered as a inquiry or a persuasion dialogue.

To sum up, it seems not really necessary to deal with examination dialogue as a separate main type of dialogue, even though a case could be made for such a choice. Prompted by considerations of conceptual parsimony, we shall, in section 5, elaborate only the point of view that examination dialogue is to be understood, not as a seventh main type, but rather as a type having instances both among persuasion dialogues and inquiries, and as often being mixed by having functional embeddings of various dialogical types.

3.4 Subtypes of Examination Dialogue

In this section we shall discuss a number of common subtypes of what we see as the general activity type of examination dialogue. We do so before presenting a characterization of the shared structure of examination dialogues; rather the sketches that follow must be seen as steppingstones towards such a characterization (Section 5).

The number of subtypes we shall discuss is four: expert examination, witness examination, academic examination, and examination by interview. We do not claim that this classification is either exhaustive or exclusive. In considering these activity types we shall characterize each type by its initial situation, its starting points, its participants' aims, its method of proceeding (which may include argumentation), and its desired outcomes — the shared main goal being to get one such outcome (van Eemeren and Houtlosser 2005, Walton and Krabbe 1995).

Expert examination. By 'expert examination' we refer to such dialogues as the examination parts (interludes or intervals) of expert consultation dialogues, extensively discussed by Walton (2006b). Though these intervals occur notably as part of an expert consultation dialogue, they can also occur in the context of another kind of dialogue (Walton 2006b, pp. 757-758), or even in isolation. As Walton points out, the roots of such intervals go back to ancient times. Basically, what is at issue is that one may doubt whether someone claiming to be an expert in some field is really an expert in that field, or merely pretending. This initial situation is familiar from many Socratic dialogues. Take, for instance the rhapsodist Ion's claim in his answer to Socrates:

> SOCRATES: ..., but not before you answer me this: on which of Homer's subjects do you speak well? I don't suppose you speak well on all of them.
> ION: I do, Socrates, believe me, on every single one.
> SOCR: Surely not on those subjects you happen to know nothing about, even if Homer does speak of them.
> ION: And these subjects Homer speaks of, but I don't know about – what are they? (Plato 1997, p. 944, Ion 536e)

Since Homer talks about chariot-driving, medicine, angling, divination, navigation, military strategy, and many other things, Ion has implicitly claimed to be an expert in all these fields. Examined by Socrates, and after having been forced to admit the superiority of professionals in many fields, Ion's claim becomes ludicrous when applied to military strategy:

> SOCRATES: Are you also a general Ion? Are you the best in Greece?
> ION: Certainly, Socrates. That, too, I learned from Homer's poetry. (Plato 1997, p. 948, Ion 541b)

The next question is why Ion, being the best general in Greece, has never commanded an army. Thus, Ion's claims are step by step demolished by showing for each subject the implausibility of his having a better view of it than that of the professionals in the field. In this dialogue, no questions are asked about these subjects or fields themselves, but in other Socratic expert examination dialogues there are questions pertaining to the field at issue (usually moral philosophy).[2]

Walton discusses Socratic peirastic in the Laches (Walton 2006b, pp. 759-60) and also Aristotle's contribution: 'Aristotle took a more systematic approach by trying to classify examination as a type of dialogue' (Walton 2006b, p. 762). Indeed, Aristotle may have been the first to distinguish expert examination as a specific kind of dialogue. In Sophistical Refutations Aristotle distinguishes arguments (logoi) of a kind he calls 'examination-arguments' (perastikoi). Elsewhere we argued that this kind of arguments may be supposed to be characteristic for a type of dialogue: peirastic discussion (Krabbe and van Laar 2007, p. 37). Peirastic discussion, whether Socratic or Aristotelian, is just an ancient version of expert examination. We think this type of dialogue is still of tremendous importance: for as Aristotle points out in Sophistical Refutations, this is not a kind of discussion between academics, but an instrument for laymen:

> ...for neither is the art of examination of the same nature as geometry but it is an art which a man could possess even without any scientific knowledge. For even a man without knowledge can examine another who is without knowledge, if the latter makes concessions based not on what he knows nor on the special principles of the subject but on the consequential facts, which are such that, though to know them does not prevent him from being ignorant of the art in question, yet not to know them necessarily involves ignorance of it. (Aristotle 1965, 172a21-27)

Thus a good peirastic instrument will be a great help to solve the problem of evaluation of expertise: how can the layman (say, a voter or a judge) decide whether it would be right to follow some (allegedly) expert opinion, for instance on economical matters or on issues of forensic medicine? Whole or even partial solutions to this problem are of vital significance for a democratic society. According to Aristotle, what the layman can use to examine an alleged expert are the consequential facts (ta hepomena), mentioned above, which we may figure to be facts or phenomena explained by science but ascertainable by all. Besides, the layman can make use of the common principles (ta koina), which, in contrast to the consequences (Hasper 2009), are general principles that hold for all science and dialectic. The quoted passage continues:

> Clearly, therefore, the art of examination is not knowledge of any definite subject, and it therefore follows that it deals with every subject: for all the arts employ also certain common principles. Accordingly, everyone, including the unscientific, makes some kind of use of dialectic and the art of examination; for all, up to a certain point, attempt to test those who profess knowledge. Now this is where the common principles come in; for they know these of themselves just as well as the scientists, even though their expression of them seems to be very inaccurate. (*idem*, 172a27-34)

---

[2]Gentzler (1995, pp. 231-33) distinguishes three 'sorts of peirastic cross-examinations' found in these dialogues. Following Aristotle, she uses the term 'peirastic' to refer 'to cross-examinations used for the purposes of testing a claim to knowledge' and she introduces the term 'Socratic peirastic' (Walton and Krabbe 1995, p. 227 and Note 4). The three kinds of Socratic peirastic are exemplified by the Hippias Minor the Laches and the Ion. Of these the peirastic in the Ion seems least akin to the Aristotelian version, whereas the other kinds, in which the questioning relates to the field at issue, appear more so. Though the Socratic and the Aristotelian peirastic are obviously related, they cannot be straightforwardly identified.

The common principles, we might say, provide common starting points for the peirastic discussion. The consequential facts or consequences, on the other hand, belong rather to the method of proceeding. It is by showing ignorance of one of these that the would-be expert gives away his ignorance of the field. The participants' aims are opposed: the examiner (the questioner) trying to expose the suspected ignorance of the examinee (the answerer) and the latter trying to avoid being exposed as a pretender (whether he really is one or not). The outcome can be either that the examinee has been shown to be ignorant of the field or that he has proven himself to be able to parry some critical questioning. The first outcome is clearer than the second, because the examinee's correct answers could be based on acquaintance with the consequences rather than with the field, as Aristotle indicates. But even so the second outcome could count as a sign that the would-be expert is actually an expert.

Aristotelian peirastic discussion is, we may say, a method of testing an alleged expert in some field by having a conversation about substantial issues within that field. As such, it can usually be reconstructed as a persuasion dialogue with the alleged expert as the protagonist of the standpoint that he is indeed an expert and the other as the antagonist who doubts that standpoint. That is, the expert examination dialogue can usually be characterized as a persuasion dialogue starting from a difference of opinion about the status of one of the participants. If it cannot be so characterized, because none of the participants has committed himself to a standpoint about his own status or that of the other, the question of expertise will be an open problem, and the type of dialogue that of inquiry. Clearly, expert examination, whether it appears as a persuasion dialogue or as an inquiry, can also be conducted in other ways, perhaps supplementary to Aristotelian peirastic discussion: the antagonist could ask for credentials of expertise, such as degrees or professional qualifications, or testimony of peers (what happens in the Ion is close to this way of proceeding). In fact, such questions and many of the other kinds of critical questions that are raised to challenge arguments from authority can be applied in a direct examination of an alleged expert (Walton 1997a, p. 223). Here we shall not discuss these any further.

Witness examination. By 'witness examination' we refer to the examinations and cross-examinations of witnesses in court. Clearly, this may overlap with expert examinations, since many witnesses are called to testify in their capacity as an expert in some field (Walton 1997a, Ch. 6). In a sense, every witness is a kind of expert in so far as he or she is in a position to know some facts. Thus we may expect the structure of witness examination to resemble that of expert examination. Yet the examination of witnesses displays some features of its own. The main difference is that an examination of a witness is usually concerned with establishing or rejecting evidence. That is, it is primarily an information-seeking dialogue of the kind that may be called interrogation (Walton & Krabbe, 1995: 76).[3] But we shall focus on those parts of witness examinations, mainly cross-examinations, that are directly concerned with the quality of the witness. The initial problem, or difference of opinion, in these parts concerns not only the knowledgeability of the examinee, but also his honesty and good character. The starting points include the rules of the courtroom, and are therefore much more institutionalized than in the general case of expert examination. In examinations by the lawyer who called the witness to testify, it may be that examiner and examinee cooper-

---

[3]Interrogation dialogue is characterized by Walton as a kind of information-seeking dialogue, embedded in deliberation dialogue, which contains critical examination of answers given, and always involves deception and coercion, being controlled by the questioner (Walton 2003, 1797-8). Walton (2006b, p. 761) sees interrogation dialogue also as a 'degeneration of examination dialogue'. Above, we are, however, not using 'interrogation' in a pejorative sense, but merely to stress the information-seeking aspects of witness examinations.

ate to establish the good qualities of the witness, but in cross-examinations the aims of the participants are again opposed: the examiner tries to detract from the witness's supposed knowledge or position to know as well as from his character, the witness tries to uphold his position in both respects. The method of proceeding includes such devices as unearthing hidden contradictions, weaknesses of memory, or doubtful antecedents. The outcome may be a complete collapse of the testimony or a strengthening of the presumption that the witness is reliable, or something in between.[4]

Academic examination. By 'academic examination' we refer to exams in which a student's knowledge and abilities are assessed. Walton and Krabbe (1995: 76) classified these among the inquiry dialogues, and not among the information-seeking dialogues for the reason that 'one cannot suppose the examinee to have a good estimate [...] himself' of his 'level of abilities and knowledge'. In some cases, however, one could nevertheless reconstruct academic examinations, not as information-seeking dialogues or inquiries, but as persuasion dialogues, where the examinee (the student) acts as the protagonist of the standpoint that she deserves at least a pass. If there is no such standpoint to begin with, the examination starts from an open problem as its initial situation. The starting points include the presumption that the examiner (the professor) has the knowledge and the abilities to make the assessment, but there is no presumption that the student has the knowledge and the abilities to get a pass. The participants' aims are ideally not opposed: both would prefer that a deserved pass can be extended to the student. The student, though, may have as a subsidiary aim to get an undeserved pass, which aim is normally not shared by the professor. The method of examination may include open and closed questions, as well as other assignments. The dialogue character of the examination is explicit only in a viva voce situation, but some implicit dialogue can be reconstructed from written answers and essays. The outcome is a pass or a fail, usually detailed by some mark according to some locally or nationally accepted system.

Examination by interview. Here we refer mainly to job interviews, but also to similar situations that include examination, such as certain political interviews (Bench-Capon et al. 2008), interrogation of suspects, and even dating. (Walton and Krabbe 1995, p. 76) classify interviews together with interrogations among the information-seeking dialogues. This, however, does not hold for such interviews that try for more than just getting information from the interviewee. The problem could be: Is this person fit for this job? Is this politician really committed to peace? Is the suspect guilty? Is this man or woman a good partner for later life? The interviewee may not know this, or not know it for sure, though he or she could have a standpoint. The goal of the interview is to find out, and perhaps to resolve a difference of opinion, if a standpoint has been taken. The starting points include a shared willingness to have the interview, and the aims of the participants need not be opposed. But they might be, for instance if a politician wants to pretend to be wholeheartedly committed to peace, whereas the interviewer tries to unveil the politician's true ambitions. The method is generally that of question and answer, but various types of dialogue could be used to test an applicant's abilities and even nonverbal tests could be included. The outcome can be of a pass or fail type, but it can also be more of a surprise for everyone: the applicant might be found not to be the best one for the job in ques-

---

[4]How about examination or interrogation of suspects? Interrogating a suspect can sometimes best be seen as an information-seeking dialogue, although one where the suspect can be expected to be reluctant to provide the requested information if that would be incriminating. However, if the questioner conducts a consistency probe, where the suspect's failure to answer consistently is seen as indicative for the position that the suspect is guilty, the dialogue can also be seen as a peculiar kind of examination by interview (see below).

tion, but for another job for which he did not apply, the politician may start to rethinking his position from scratch, and the dating partner could become one's most appreciated critic for life.

3.5 The Structure of Examination Dialogue

Keeping in mind the considerations of Section 3 and the subtypes described in Section 4, we shall now characterize the examination dialogue, as a general activity type, by specifying its initial situation, its starting points, the participants' aims, its method of proceeding, and the desired outcome. An example will illustrate these features.

Initial situation
An examination dialogue originates in an occasion that fixes its initial situation. In the initial stage of the examination dialogue, the participants may make this initial situation wholly or partially explicit. The occasion can arise in a dialogue, of whatever type, or in a course of non-dialogical events. In a situation where a professor tries to provide information, a student may start to wonder whether the professor is genuinely knowledgeable. In a situation where a new employee must be hired, the committee may wonder whether the candidate satisfies the job requirements.

The issue in an examination dialogue pertains to one of the participants of the dialogue. An investigation into whether a third, absent, party suits a particular standard is a form of inquiry, or possibly a persuasion dialogue, but not an examination dialogue, as we understand this notion. More particularly, the issue is whether or to what extent this participant satisfies a standard or a set of conditions.

The parties in an examination dialogue can adopt different positions regarding the issue, either of taking a standpoint or of expressing doubt, in which case the dialogue starts from a difference of opinion, or the issue can remain open such that no participant expresses a standpoint about it. The applicant can, but need not, express her opinion that she is a suitable candidate. A student can express an open attitude with respect to the issue whether he deserves a pass. The expert witness can provide the requested information about her antecedents, without being willing to defend an opinion about whether or not she ought to be regarded as an appropriate expert witness, in the legal sense. If the initial situation is an open problem, the examination dialogue falls under the inquiry. If it starts from a difference of opinion, it falls under the persuasion dialogue, as in the following example:

Example:	*A job interview*
Vacancy:	salesman
Interviewer:	Welcome Mr. X, did you read our advertisement?
Mr. X:	Sure, I feel I am the man you are looking for.
Comment:	*This dialogue starts from a difference of opinion.*

Starting points
An examination dialogue has a kind of opening stage. In order for the exami-

nation to get on its way, there must be a, possibly implicit, agreement by both
parties to go ahead with an examination procedure in order to arrive at the intended outcome. Further, a decision must be made with respect to the testing
methods that are to solve the issue. In most cases this decision is made by the
examiner, or by the institution that he or she represents, but sometimes the examinee gets a voice as well. In addition, if persuasion dialogues are part of the
method of testing, the initial concessions that figure in these dialogues must be
determined. In general, this holds for points of agreement that are needed for the
execution of the tests, whatever the method of testing. Also, it must be clear that
it is up to the examiner, or the committee or institution of which he is a member,
to adopt standards by which to judge the examinee, and to decide upon the procedure that is to be followed when executing the tests. Lastly, in addition to the
results of the tests, other points of agreement may be needed in order to arrive at
a shared solution or resolution of the initial issue.

Example	*(continued)*
Interviewer:	This interview is precisely meant to become sure about that. Are you prepared to answer some questions?
Mr. X:	Sure man.
Interviewer:	Are you aware that for this job you need some negotiation skills?
Mr. X:	I ain't gonna disappoint you. I worked in the motor trade.
Comment:	*A test about negotiation skills has been covertly announced.*

Participants' aims
If the examination dialogue falls under the main type of inquiry, both participants aim at finding and establishing the propositions with which to decide the
issue in a way that satisfies them both. If it falls under that of persuasion dialogue, the participants adopt the role of protagonist and antagonist, each trying
to resolve the issue in his or her own favor.

In addition, there is a division of the roles of examiner and examinee. The
examinee is the interlocutor whose status happens to be the topic at issue in the
dialogue. These roles come to the fore in the so-called executive stage, where
the examiner puts the examinee through one or more tests. During these tests,
the examiner tries to determine whether the examinee passes the tests, while the
examinee tries to pass them.

Example	*(continued)*
Comment:	*In the example, Mr. X takes on the role of protago-nist and the interviewer the role of antagonist. It's also clear that in the executive stage Mr. X will act as the examinee and the interviewer as the ex-aminer.*

Method of proceeding
An examination dialogue has an executive stage in which the parties execute the

tests selected in the opening stage. Idiosyncratic for the examination dialogue is that the dialogical testing methods can, in principle, be taken from all types of dialogue, non-mixed or mixed.

Consequently, examination dialogues diverge, depending upon which testing method one selects. Still, whatever the adopted method, it will supposedly serve the purpose of finding out whether the examinee satisfies certain requirements. For instance, one simple method of testing would be to imitate an information-seeking dialogue and just ask simple factual questions to see whether the examinee can answer them correctly. The same method can also be used to see whether the examinee can maintain consistency in his answers (this could be called a 'consistency probe'). In Aristotelian peirastic the issue of consistency is of paramount importance: there the way to unmask a would-be expert is to refute his initial thesis on the basis of his answers to questions, i.e. to deduce the opposite of his thesis from the answers given, which amounts to showing his position to be inconsistent. In fact, the issue of consistency is important in all the subtypes of examination dialogue, from expert examination to dating, for whatever the quality one is testing for in these dialogues, manifest inconsistencies in the answers of the examinee will be a sure sign of a lack of knowledge and of logical acumen.

Another method would be to set up a persuasion dialogue in which the alleged expert (or the applicant) could demonstrate his arguing skills showing them to be up to standard for an expert (or for someone getting the job). One might even ask the expert to argue in favor of his expertise, or the applicant to argue in favor of the point of view that he is fit for the job. In those cases, the arguments put forward by the examinee may serve two distinct purposes, one being the purpose of displaying argumentative skill in order to pass the test. But if the occasion happens to amount to a difference of opinion about the status of the examinee, the arguments may also serve the purpose of directly persuading the examiner of the acceptability of the examinee's standpoint about his own status.

Alternatively, an applicant can be requested to argue in favor of a completely different proposition, for example about a product manufactured by the company where he applies for a job. This resembles the kind of test that students undergo, when asked to write argumentative essays. In such cases, the examinee is not so much trying to persuade the examiner of the acceptability of a standpoint, but rather showing that he possesses the skills needed to qualify for the job or for a pass.

The testing method need not be argumentative in nature, even if the occasion is constituted by a difference of opinion. Above we already mentioned informative queries and consistency probes. The negotiation skills of an applicant can be examined by entering a negotiation. This can be a simulation, as would be the case when the examinee is taking the role of salesman and the examiner that of a client. Instead, the examiner can choose, possibly secretly, to start a real negotiation, for example about the wages the applicant would earn if she were given the job, with the (hidden) agenda of testing her negotiation skills. Or a would-be expert can be requested to solve an open problem in order to demonstrate her expertise. Or the candidate could be subjected to an IQ test. The testing method need not even be dialogical or discursive in character. A student in mechanical engineering can be requested to construct an intricate machine, and a first glance may suffice to settle the issue on a blind date.

Whatever the testing methods, if they are dialogical in nature, the examiner must perform two distinct tasks. The first task is to adopt the appropriate role that is needed for the dialogical test to be performed, be it the role of negotiator, critical antagonist, questioner in a consistency probe, or whatever. In her capacity of negotiator, antagonist or questioner, she will put forward moves that are functional in the light of the aims of this embedded dialogue: she can make an offer, challenge an argument or ask whether the other side is prepared to concede some proposition.

In addition, during the execution of the tests, the examiner must keep track of the performance of the examinee. Thus, she is participating in the dialogue with a double intention, her second task being to measure the examinee's performance in the light of the chosen criteria. She can do so silently, without making any dialogical moves that pertain to the second task; at least, she can do so until she concludes the tests by informing the examinee about her verdict. It is quite normal, however, that the examiner feels the need to request the examinee to comment upon one of his moves in the negotiation, the persuasion dialogue or the consistency probe. For example, during a simulated negotiation, the applicant can be requested, now and then, to motivate an offer just made. In order for the examiner to be able to arrive at a thorough assessment, she may need further clarification of what the examinee has stated or an explanation of why the examinee chooses a particular move rather than another, leading to what Walton calls *the level of exegesis* (Walton 2006b, p. 765), or she may need an argument in favor of the claim that a particular move is admissible or strategically clever, which seems to be close to Walton's *level of testing* (Walton 2006b, p. 765). In such cases, the requests by the examiner and the moves requested from the examinee are not (directly) serving the purposes of the dialogue that has been chosen as a test, but rather the purpose of establishing the propositions with which the initial issue of the examination dialogue can be answered or resolved. We dub these interludes *intermediate checks*. The moves occurring during these checks can best be reconstructed as direct contributions to an inquiry, in case the initial situation is made up by an open issue, or to a persuasion dialogue, in case the examination dialogue started from a difference of opinion.

Before leaving this executive stage, the examiner must employ her special authority to decide what tests the examinee has passed or failed. The executive stage is concluded by the examiner's informing the examinee of her decisions with regard to the test results.

Example	*(continued)*
Interviewer:	How much did you earn in your previous job, Mr. X?
Mr. X:	Mostly about 2000 a month.
Interviewer:	How much would you like to earn in this job?
Mr. X:	I'd like to make twice as much.
Interviewer:	That won't work. The trade is too low to support that kind of salary.
Mr. X:	2500 will do.
Interviewer:	Why this drastic change in your demand, Mr. X?
Mr. X:	I really need the job.
Comment:	A negotiation test is proceeding. The interviewer's last question constitutes an intermediate check showing her double intention: she was not only negotiating about her salary but also testing Mr. X.

Desired outcome

An examination dialogue has a kind of concluding stage where the parties attempt to arrive at a final outcome: do the results of the tests establish the examinee as satisfying the requirements? In contrast to the test results, which are typically determined by the examiner, the examinee has a say in the final outcome, in so far as (in the case of persuasion dialogue) he sticks to or withdraws his original standpoint that he satisfies the requirements, or (in the case of inquiry) agrees or disagrees with the test results as providing a stable answer to the initial issue. But, admittedly, the distinction between the conclusion of the test and that of the entire examination dialogue is often invisible in practice.

Example	*(continued)*
Interviewer:	So, generally, it is not wise to go down so drastically.
Mr. X:	Indeed, generally not, but I had a good reason.
Interviewer:	I must say, you are really well motivated for this job. Thank you for your application. You'll hear from us by the end of the week.
Comment:	At first, Mr. X seemed not to do so well at the negotiation test. However, he did offer a good explanation for his drastic negotiation move, thereby offering a reason for his thesis that he is the man for the job. Actually, given that he absolutely did want to get the job, going down in his demands was not a bad move in the negotiation, as the interviewer may come to realize.

According to the characterization in this section, examination dialogue is to be seen as comprising two subtypes. The first subtype forms a kind of persuasion dialogue and the second a kind of inquiry. In addition, examination dialogue, of whatever brand, is a mixed dialogue type, because examination dialogues often embed dialogical testing methods of various types of dialogue.

3.6 Conclusion

Though we cannot expect everyone to endorse straightforwardly our view that the examination dialogue should not be seen as a main type of dialogue, but rather as a mixed type with two subtypes which can each be subsumed under a different main type (inquiry or persuasion dialogue), we still hope to have given some plausibility to this point of view. What became evident (to us at least) is that examination dialogues are extremely complex in their structure, mainly because of the many possibilities to have embeddings of other dialogues that are then used with a double intention: that of having this other dialogue for its own sake, and that of performing a test to be used for solving or resolving the original issue. The intermediate checks make the structure even more complicated. It also became clear that examination dialogue is a fascinating subject and that there seems to be enough material there to write a book; indeed that would be tempting. Douglas Walton must have had this experience with many other types of dialogue and argument-related activities. Fortunately for all of us, he often yielded to temptation.

Acknowledgement

We want to thank two anonymous referees for many useful suggestions.

CHAPTER 4

DIALECTICAL AND HEURISTIC ARGUMENTS: PRESUMPTIONS AND BURDEN OF PROOF

Fabrizio Macagno

In law, as in everyday conversation, presumptive reasoning is one of the most common forms of drawing conclusions from a set of premises. On Walton's view (Walton 1996b, p. 13), whereas in deduction conclusions are necessarily true if the premises are true, the conclusion of a presumptive reasoning is a simple presumption, that is, it holds in conditions of incomplete knowledge and is subject to retraction should these conditions change. These arguments are grounded on generalizations such as, 'Birds fly' or 'Americans love cars', leading to conclusions of the kind: 'Plausibly, this bird flies' or 'Bob, who is from Michigan, probably loves cars'. However, these arguments cannot be evaluated in light of standards accepted in logic (Walton 1993, p. 3), and their function is not to provide that premises necessarily imply a conclusion or to provide evidence based on statistical results. Everyday conversation is characterized by incomplete knowledge, in which only few propositions can be considered necessarily true, and only a limited number of data is collected. What is the role of plausible or presumptive arguments, if not leading to truth or statistically backed generalizations? Moreover, how can they be assessed, if the logical standards do not apply? The answer can be found shifting the paradigm from the logical concept of truth of a conclusion to the pragmatic notion of dialectical effects of a conclusion. Presumptive arguments are not true or false; they produce some effects on the dialectical setting of the interaction between interlocutors. The function of presumptive arguments is to shift the burden of proof (Walton 1989, 1988), that is, if the interlocutor is committed to the premises, he should be committed to the conclusion as well, or he has to show why the conclusion is not acceptable.

The dialectical nature of presumptive reasoning can be analyzed by inquiring into what a presumption is. We can analyze the following arguments (see (Walton 1996c, p. 17)):

1. John's hat is not on the peg. Therefore, John has left the house.

2. John has been missing for 5 years. Therefore, he can be considered as dead.

These arguments are clearly different in nature, and presuppose different types of background information. However, we can notice how they are both grounded on implicit premises shared by the interlocutors, namely that 'When John leaves the house, he wears his hat' and that 'A person not heard from in five years is presumed to be dead'. They are both presumptions, as they provide a reason to infer a conclusion from a fact, and they are matter of common knowledge, namely they are not formal logical rules, but they are part of what is commonly known. However, while the first argument can be uttered in a context in which John's habits are known, the second is based on a rule of evidence (California Evidence Code, section 667). Both arguments are reasonable and provide a probative weight in a situation in which a conclusion has to be reached from incomplete information.

The purpose of this article is to inquire into the grounds of Walton's theory of presumptive reasoning, examining the concept of presumption and its dialectical effects. In particular, the role of the presumption in shifting the burden of proof will be examined (see (Walton 2007d, Hahn and Oaksford 2007)). The role of presumption in everyday conversation will be analyzed starting from a particular type of natural dialogue, the legal discussion. Legal argumentation, conceived as a highly codified type of dialectical reasoning, can provide some general principles, which can be applied to reasoning in other types of human communication.

4.1   Presumption in argumentation

In order to describe what a presumption is, it is necessary to specify the level of our inquiry. As seen above in the introduction, presumptions are analyzed in argumentation considering factors such as the probative weight and the conclusion, or viewpoint, to be supported. However, such elements presuppose a dialogue, in which there must be necessarily subjects interacting. In such context, the first step is to describe what kind of action is performed by the subject when the presume something. Following van Eemeren and Grootendorst (1984a), we hold that when placed in a context of dialogue, every reasoning or argument is an action, or rather a complex action, carried out to accomplish a specific communicative purpose.

Walton (1992b; 1993; see also Walton and Godden 2007) describes presumption in the light of the rhetorical and logical theories as a particular speech act. Presumption can be described according to its propositional content and illocutionary effects (see (Searle 1969, chapter 3)) as "a proposition put in place as a commitment tentatively in argumentation to facilitate the goals of a dialogue" (Walton 1993, p. 138). From an illocutionary point of view, presumptions are acts, by means of which the Speaker requests the interlocutor in a dialogue to commit to a proposition, and should the Speaker fail to reject his or her commitment to the proposition, it will be taken as commitment of both parties in the subsequent dialogue (see (Walton 1992b, p. 56)). This provisional and mutual type of commitment pragmatically differentiates presumption from statements, which change the Speaker's commitments by including therein the proposition object of the speech act, and assumptions, which actually do not change the interlocutors' commitment store, as they can be freely rejected at any point in a dialogue. Walton summarizes the conditions for the speech act of presumption as follows (Walton 1992a, pp. 60-61):

I. Preparatory Conditions

    A. A context of dialogue involves two participants, a proponent and a respondent.

    B. The dialogue provides a context within which a sequence of reasoning can go forward with a proposition A as a useful assumption in the sequence.

II. Placement Conditions

    A. At some point x in the sequence of dialogue, A is brought forward by the proponent, either as a proposition the respondent is asked explicitly to accept for the sake of argument, or as a nonexplicit assumption that is part of the proponent's sequence of reasoning.

B. The respondent has an opportunity at x to reject A.

C. If the respondent fails to reject A at x, then A becomes a commitment of both parties during the subsequent sequence of dialogue.

III. Retraction Conditions

A. If, at some subsequent point y in the dialogue ($x < y$), any party wants to rebut A as a presumption, then that party can do so provided good reason for doing so can be given. Giving a good reason means showing that the circumstances of the particular case are exceptional or that new evidence has come in that falsifies the presumption.

B. Having accepted A at x, however, the respondent is obliged to let the presumption A stay in place during the dialogue for a time sufficient to allow the proponent to use it for his argumentation (unless a good reason for rebuttal under clause III. A. can be given).

IV. Burden Conditions

A. Generally, at point x, the burden of showing that A has some practical value in a sequence of argumentation is on the proponent.

B. Past point x in the dialogue, once A is in place as a working presumption (either explicitly or implicitly) the burden of proof falls to the respondent should he or she choose to rebut the presumption.

These conditions apply to all defeasible arguments (see Prakken, Reed, Walton 2005); however, an additional requirement is set in (Walton 2009), namely that A cannot be proved (or disproved) by evidence, i.e. the situation is one of incomplete knowledge. Therefore we have to add other conditions of presumption (Walton 2009b, Prakken and Sartor 2009), in particular that the inference is sufficiently strong to shift the burden of providing evidence to the other party. The effect in the dialog must be that the interlocutor must give some argument against it, or else he will risk losing the exchange.

These conditions show the dialectical effects and requisites of the speech act of presumption. Presumption is shown as simply requiring a context of dialogue, and unless rejected it shifts the burden of proving or disproving the proposition onto the interlocutor.

This account describes how presumptions work in a dialogue game; however, this account of presumptions seems to have further implications on the argumentation theory. If we examine this type of speech act using Searle's categorises, we can notice that another difference between presumptions and assertives or assumptions can be found at the level of preparatory conditions. Whereas statements and assumptions have as a propositional content a sentence (for instance, 'I state (or assume) that Bob is ill'), presumptions always require a reason, which can be stated or simply presupposed. It would be perfectly sound to state 'I state (or assume) that Bob is ill, but there are no grounds to believe that', but stating that 'I presume that Bob is ill, even though there are not reasons' would be unreasonable. While assumptions are propositions put forth to prove a further conclusion, but need to be proven later in the discussion in order to be accepted, presumptions *are* the result of a reasoning. The act of presumption requires, as a preliminary condition, the existence of some grounds supporting it. For this reason, presumptions are always the conclusion of an explicit or implicit argument.

On this perspective, presumptions become integral part of the complex act of argumentation. Van Eemeren and Grootendorst maintain that argumentation is a complex speech act "composed of elementary illocutions belonging to the category of assertives" (van Eemeren and Grootendorst 1984a, p. 34), (Walton

1992b, p. 177). However, Walton points out how the conclusions of arguments are attempts to steer the interlocutor's commitment towards a specific proposition (see Walton 1996b, p. 38), while the function of an argument is "to effect a shift of presumption towards the proponent's side" ((Walton 1992b, p. 189) for the discussion on the burden of proof, see (Prakken 2004, Prakken and Sartor 2006). Arguments, on Walton's account, are presumptive as their conclusions can be considered presumptions relative to a certain proposition, and because their purpose is to effect a shift of burden of proof relative to the thesis that the Speaker defends. On this view, we can notice how the success of an argumentative move can be evaluated according to the felicity of the act of presumption. A good presumptive argument is an argument that shifts the burden of proof onto the interlocutor. On this view, therefore, presumptions are considered as conclusions of a reasoning pursuing a specific dialogue effect, and at the same time speech acts characterized by their argumentative nature.

However, this relation between presumptions and arguments rises a crucial question: are all argument presumptive? Do all arguments shift the burden of proof in the same fashion?

4.2   Presumptions in law

A presumption in law is "an inference made about one fact from which the court is entitled to presume certain other facts without having those facts directly proven by evidence" (Hannibal and Mountford 2002, p. 464). Legal presumptions were introduced in Roman law to indicate different types of conclusions of inferences, and were divided into three categories, presumptions of fact, presumptions of law, and the so called *praesumptio iuris et de iure*, or irrefutable presumptions. When facts were deemed proved only on the grounds of logical rules, namely rules not provided by law but only shared by the common knowledge (Camp and Crowe 1909, p. 882), the presumption was classified as *praesumptio facti* or *praesumptio hominis*, namely presumptions of fact (Berger 1909, p. 646) or permissible inferences (Park et al. 1998, p. 105). Presumptions of fact are simply conclusion that the court may draw from certain facts, that is, are not mandatory (Keane 2008, p. 656). They are based on rules of inference from a previous experience of the connection between the premises and the conclusion. For instance, an example of presumption of fact may be "things once proved to exist in a particular state continue to exist in that state". If one party in a trial proves that a man was alive on a certain date, the factfinder may presume that he was alive on a subsequent date.

Presumptions of law, or true presumptions (Park et al. 1998, pp. 102-105), express legal and mandatory relationships between certain facts (called basic facts) and certain other facts (the presumed facts). They are not rules of inference drawn from everyday logic, but rules provided by law. An example of presumption of law can be the following ones (California Evidence Code, section 667; 663):

> A person not heard from in five years is presumed to be dead
> A ceremonial marriage is presumed to be valid.

For instance, if one party wants to prove that a couple is married, but has not enough evidence, he just needs to prove that a marriage ceremony had been performed. In such case, the factfinder *must* presume that the marriage is valid.

The last type of presumptions, called irrebuttable presumptions, express mandatory relationships between the basic and the presumed facts, but they cannot be rebutted (Hannibal and Mountford 2002, p. 465)). For instance, irrebuttable presumptions can be the following ones (Children and Young Persons Act 1933, s. 50; California Family Code 1994, s. 7540):

> No child under the age of 10 can be guilty of an offence.
> The child of a wife cohabiting with her husband, who is not impotent or sterile, is conclusively presumed to be a child of a marriage.

Therefore, if a child is proven to be under the age of 10, he is innocent and no evidence can rebut this conclusion. Irrebuttable presumptions are rules of substantive laws, and they are judgments, not simply instruments of evidence (Park et al. 1998, p. 106).

The three types of presumptions can be differently rebutted. Presumptions of fact can be rebutted by simply providing contrary evidence. Irrebuttable presumptions, on the other hand, do not admit any refutation. True presumptions, at last, admit different types of rebuttal strategies. The party against whom a presumption would operate has several choices (Park et al. 1998, p. 107). For instance, we can apply those possibilities to the rule of presumption of death in the following case (Miller v. Richardson, Secretary of Health, Education and Welfare, 457 F.2d 378 (1972)).

> After twenty-two years of marriage, Mrs. Miller's husband permanently left the household on May 17, 1957. Aside from a telephone call received from him several days later, Mr. Miller has not been seen nor heard from since. All efforts to find Mr. Miller were unsuccessful. Mrs. Miller commenced a proceeding in the Orphans' Court of Allegheny County, Pennsylvania to have Mr. Miller declared a presumed decedent.

The possibilities of the appellee, in this case Mr. Richardson, the Secretary of Health Education and Welfare, were the following ones:

1. Offer not evidence challenging the existence of either the foundational facts or the presumed facts. E.g. Mr. Richardson does not provide evidence. The man is considered dead.

2. Offer evidence challenging the existence of only the foundational fact. E.g. Mr. Richardson proves that the man wrote a letter three years before the trial. In this case, the presumption of law is rebutted.

3. Offer evidence challenging the existence of only the presumed facts. E.g. Richardson shows that the man left the family home shortly after a woman, whom he had been seeing, also disappeared, and that he phoned his wife several days after his disappearance to state that he intended to begin a new life in California. In this case, the presumption is rebutted as Miller's disappearance was not unexplained and implicit in his departure was an intention to continue living.

4. Offer evidence challenging the existence of both the foundational and the presumed facts. In this case, simple foundational evidence would be sufficient to rebut the presumption. For instance, Richardson proves that the man wrote a letter three yeas before the trial stating that he intended to start a new life elsewhere.

This scenario shows the general principles of persuasion. However, true presumptions (or presumptions of law) are different in strength, namely they require different rebuttals fulfilling different standards. They can be classified in three levels (Andersen 2003, p. 112), (Lempert 2000, p. 1235-47):

1. mandatory burden-of-pleading-shifting presumptions: If the party proves A, then the factfinder must find B, unless the opposing party claims B is not true

2. mandatory burden-of-production-shifting presumptions: If the party proves A, then the factfinder must find B, unless the opposing party introduces evidence sufficient to prove B is not true. Sufficient evidence may be defined as any evidence, reasonable evidence, or substantial evidence.

3. mandatory burden-of-persuasion-shifting presumptions: If the party proves A, then the factfinder must find B, unless the opposing party persuades the factfinder that B is not true. Persuasion may be defined anywhere from a preponderance to beyond a reasonable doubt.

In the first case, the opposing party has only to deny the charges. This type of presumption is set at the beginning of the pleadings, in which the defendant has only to file an answer to the summons. The second and the third case involve a crucial distinction between burden of persuasion and burden of production. The difference can be explained as follows (Murphy 2007: 71):

> The term 'burden of proof', standing alone, is ambiguous. It may refer to the obligation to prove a fact in issue to the required standard of proof, or to the obligation to adduce enough evidence to support a favourable finding on that issue.

The allocation of the burden of persuasion follows general rules relative to the type of offence. In criminal cases burden is on the prosecution to prove the facts essential to their case (see Woolmington v DPP AC 462 (1935)), whereas in civil cases "he who asserts must prove", i.e., the burden rests with the plaintiff (the party bringing the action) (Keane 2008, p. 98). Another type of burden of proof is the evidential burden of proof, which corresponds to the burden of proving all elements essential to the claim.[1] As Murphy puts it (Murphy 2007, p. 71):

> Every claim, charge or defence has certain essential elements, the proof of which is necessary to the success of the party asserting it. For example, a claimant who asserts a claim for negligence asserts: 1) that the defendant owed the claimant a duty of care; 2) that the defendant, by some act or omission, was in breach of that duty of care; and 3) that as a result of that breach, the claimant suffered injury or damage for which the law permits recovery. These elements derive, not from the law of evidence, but from the substantive law applicable to the claim, in the case the law of negligence. They are known as 'facts in issue' or 'ultimate facts'. The proof of these facts in issue depends, however, on the detailed facts of the individual case, which are referred to as 'evidential facts'. Thus, for example, in order to prove the fact in issue, negligence, the claimant might set out to prove the evidential facts that the defendant drove while drunk, too fast, on the wrong side of the road, and knocked the claimant down, breaking his leg.

If the claimant or the prosecution does not prove the essential elements (all or most of them, depending on the type of trial), the defendant is acquitted. If the elements have been supported by evidence, a prima facie case is established,

---

[1] A third type of burden of proof is the tactical burden (Walton 2008b). This type of burden can be simply defined as the procedural shifting of the burden of production in a trial, when the opponent has discharged his burden of production (Tapper 2007, p. 137).

namely should the defendant not disprove any element, the case stands or falls only by this evidence. In this case, the burden is on the defendant to provide evidence contrary to the elements and contradicting the claim. The relation between burden of persuasion and the evidential burden of proof is represented in figure 4.1 (Murphy 2007, p. 73).

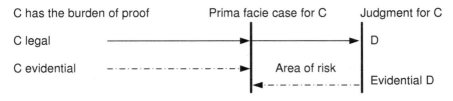

Figure 4.1: Burdens of proof

In this figure, the letters C and D stand for the Claimant (or prosecution) and the Defendant. The claimant has the burden of persuasion, or legal burden of proof; to meet such burden, he or she has to provide evidence (burden of production, or evidential). The defendant, once the prosecution or the claimant has established a prima facie case, bears the burden of providing contrary evidence (Evidential D).

A clear example can be provided by criminal proceedings. In criminal law, the prosecution has to prove beyond reasonable doubt that the defendant committed the crime he is charged with (C legal). For instance we can consider the case Mullaney v. Wilbur, 421 U.S. 684 (1975). The defendant was charged with murder, and the prosecution had to prove beyond a reasonable doubt that the defendant intentionally committed an unlawful homicide – i.e., neither justifiable nor excusable, and that he acted with malice aforethought – by providing evidence, and therefore fulfilling a burden of producing evidence (C Evidential). Circumstantial evidence showed that he fatally assaulted the victim (C provides evidence and meets the evidential burden). The defense had the burden to produce evidence contrary to the findings (D evidential), and claimed that the homicide was not unlawful since respondent lacked criminal intent, and that he lacked malice aforethought, as he acted in the heat of passion provoked by the victim's assault (D meets the burden of producing evidence). After the defense fulfilled the burden of production, the prosecution had to prove beyond reasonable doubt the absence of the heat of passion. The area of risk is the relation between the burden of production of the claimant or the prosecution (C evidential), and the burden of production of the defendant (D evidential): if the defense cannot rebut the evidence provided by the prosecution, the verdict will be reached on the basis of the prosecution's arguments.

This case is particularly interesting, because it was based on a redefinition of the crime of murder including the concept of 'malice aforethought'. Before this case, in Maine (where the judgment was rendered), like in other states of the United States, the prosecution acted on the presumption that "if the prosecution established that the homicide was both intentional and unlawful, malice aforethought was to be conclusively implied unless the defendant proved by a fair preponderance of the evidence that he acted in the heat of passion on sudden provocation". In this case, the presumption would have shifted the burden of proof onto the defendant (see Rhodes v. J Brigano 91 F.3d 803 (1996)). The essential elements of murder were the fact that the accused caused the death of the victim, and that he acted on purpose. The classification of the homicide as a murder or a voluntary manslaughter was considered a matter of qualification: in this case, the accused, already proven guilty of a crime, has to rebut the presumption that the crime has been committed in normal circumstances (presumption of sanity, see (Tapper 2007, p. 145)).

4.3   Presumptions: reasoning from lack of knowledge

The notion of presumption in law is extremely complex. Presumptions are instruments for legislatures and courts to shift burdens (Andersen 2003, p. 111); however, the reason of such a shift can be found in the notion of presumptive reasoning. Presumptive reasoning is a kind of reasoning that works in conditions of lack of knowledge. On Walton's account (see (Walton 2008b)), the notions of presumption and burden of proof are closely related to the argument from ignorance (Walton 1996a). The argument from ignorance can be described as follows (Walton 1995, p. 150); see also (Walton et al. 2008)):

MAJOR PREMISE: If A were true, then A would be known to be true.
MINOR PREMISE: It is not the case that A is known to be true.
CONCLUSION: Therefore A is not true.

This type of argument can be explained using the following example (Walton 1996a, p. 35):

> If a serious F.B.I. investigation fails to unearth any evidence that Mr. X is a communist, it would be wrong to conclude that their research has left them ignorant. It has rather been established that Mr. X is not one.

This type of argument can be considered as a particular case of a type of reasoning from oppositions, in which the relation between absence of knowledge and negation is made explicit. For instance, we can examine the following pattern of reasoning:

> This man is not dead. Therefore he is alive.

The paradigm of the possible men's conditions of existence is constituted of only two possibilities: dead or alive. If a man is not dead, he is alive. However, this type of reasoning in natural language is much more complex. The first complication is that paradigms often admit of several possibilities, like the color of the eyes; the second problem is that we do not have direct knowledge of what is negative (Sharma 2004): all we can know is that the man is not known to be dead. The more general pattern could be represented as follows:

Classification under lack of knowledge
PREMISE: If A were X, Y, Z, then A would be known to be X, Y, Z.
PREMISE: It is not the case that A is known to be X, Y, Z
PREMISE: A can be either X, Y, Z, or K. Other possibilities are not known.
CONCLUSION: Therefore A is K.

For instance, if the color of Bob's eyes is not known, and the only information available is that his eyes are not blue, not black, nor green, it could be concluded that his eyes are brown. The paradigm of the possible colors of a man's eyes is restricted to a closed paradigm. True or false, dead or alive, are special cases in which this type of reasoning from oppositions in closed paradigms applies. There are two possible strategies for rebutting this type of reasoning: showing that A is X (for instance, finding that Bob's eyes are blue), or attacking the paradigm, showing that A can be something else but X, Y, Z, and K (for instance, showing that eyes can be gray).

Classificatory presumptions in law and in everyday conversation can be analyzed as inferences from lack of knowledge following the scheme above. For instance, a person is considered to be innocent if he is not known to be guilty; a murderer is considered to be sane, if he is not known to be insane or have his mental conditions impaired, a man missing from five years is considered dead if he not known to be alive.

The concept of paradigm and reasoning from lack of knowledge can be also applied to reasoning from best explanation. For instance, we can consider the presumption that a man who killed a person using a deadly weapon did it on purpose. The pattern of reasoning can be described as follows (from (Walton 2002a, p. 44)):

Explanation under lack of knowledge
F is a finding or given set of facts.
E is a satisfactory explanation of F.
No alternative explanation E' is as satisfactory as E.
Therefore, E is plausible, as a hypothesis.

The lack of knowledge here regards the paradigm of the possible explanations. In the event that no other possible explanations of F (killing a person using a deadly weapon) are known, the conclusion will be that E (the killing was done on purpose) is the explanation. However, the paradigm of the possible explanations is not simply constituted of one possibility: it is simply not known.

The relation between paradigm and knowledge applies also to another type of presumption, namely the presumption raised by citing in trial the existence of a code of practice. Codes of practice stipulate what is the correct practice for carrying out some activities, such as the protection of personal data or ensuring a safe workplace. When the code of practice is cited in support of a prosecution, the burden on proof passes to the defendant to prove that what was done was at least as good as what prescribed in the code of practice (Ramsey 2007, p. 470). For instance, if the defendant is charged with violating the regulation requiring that an employer shall keep the workplace safe (Article 5(1) of the Order), the prosecution can cite the code of conduct provided in those situations. If the employer is shown not to have abided by one or more provisions, he is presumed to have violated the regulation (see Paul Scott v. AIB Group (UK) PLC t/a First Trust Bank, NICA 3 2003). In this case, the reasoning proceeds from the ignorance of other possible actions carried out to achieve a the wanted result, and can be represented as follows:

Cause under lack of knowledge
If actions A,B,C, are carried out, X is achieved.
Action A has not been carried out.
No other actions to achieve X are known to have been carried out.
Therefore, X has not been achieved.

Also in this case, the reasoning proceeds from the absence of knowledge of the elements of a paradigm.

The types of schemes from presumption are particular forms of reasoning from classification, effect to cause, and cause to effect in conditions of lack of knowledge. They can be considered to be heuristic forms of reasoning, aimed at providing a conclusion in a dialogue setting. The presumption is grounded on the concept of lacking evidence; for this reason, it shifts the burden to complete the missing information or incomplete paradigm to the other party. In criminal cases, the burden of persuasion never shifts in cases of presumption. The defendant is simply required to provide evidence and thereby complete the structure of reasoning. When the offense is determined, for instance the defendant has been found guilty of a crime, the issue changes, and he has the burden of proving some possible mitigations and qualifications. In civil cases, in some states in the United States, the party opposing a presumption has to disprove the presumption by preponderance of evidence. However, also in this case, the burden of persuasion regarding the issue remains on the claimant, and the party opposing the presumption has to either provide a different credible explanation or provide adequate evidence (see (Buckles 2003, p. 45)).

4.4   Presumptions as heuristic patterns of reasoning

The analysis of legal presumptions highlights a crucial distinction between two patterns of reasoning. The formal paradigm of necessarily true inferences, based on the relation between quantifiers, cannot hold if applied to human reasoning, in which quantifiers are usually omitted, and in which only few absolute truths, if any, can be the ground for necessary conclusions. Human reasoning stems from commonly accepted premises, that is, *endoxa* (see (Walton and Macagno 2006)), and proceed to conclusions through another type of accepted propositions, the ancient maxims, or principles of inference, partially included in the modern concept of argumentation schemes. For instance, the passage from some data to a classification is warranted by the scheme from verbal classification, which can be represented as follows (Walton 2006c, p. 129):

Individual Premise:	a has property F.
Classification Premise:	For all x, if x has property F, then x can be classified has having property G.
Conclusion:	a has property G

In the tradition, the classification premise was expressed as a maxim from definition, such as "what the definition is said of, the *definiendum* is said of as well" (see (Macagno and Walton 2008)). In law like in everyday conversation, it often happens that the evidence we have does not fit the definition of the concept. For instance, one of the accepted definitions of manslaughter is "the unlawful killing of a human being without malice aforethought"; however, how can we classify an event as 'unlawful', or 'without malice'? if malice is the 'intention to do injury to another party', how can an intention be established? If the law provides that 'An individual who has sustained either (1) irreversible cessation of circulatory and respiratory functions, or (2) irreversible cessation of all functions of the entire brain, including the brain stem, is dead', how can brain activities be considered to have ceased?. Furthermore, how can a man be considered dead or alive when he is missing? We can notice how the following types of reasoning are different:

Reasoning from definition	Reasoning from presumptions
— This man unlawfully killed a woman without malice aforethought — Manslaughter is the unlawful killing of a human being without malice aforethought — Therefore this man committed manslaughter	— This man stabbed a woman to death — If a wound is inflicted with a deadly weapon in a manner calculated to destroy life, malice is presumed. — This man acted with malice

The second pattern of reasoning is not grounded on a definition at all, even though it can be considered a kind of classification of malice. This type of heuristic classifications or reasoning from presumptions (see (Gigerenzer and Todd 1999)) is grounded on a different type of commonly known propositions, namely rules of thumb, stereotypes, habits (Amossy and Herschberg-Perrot 1997, Schauer 2003, Cohen 1977). They do not constitute the shared semantic system, nor the legal semantic structure (see also (Perelman 1963, 170). We can represent the different levels of commonly shared propositions as shown in figure 4.2.

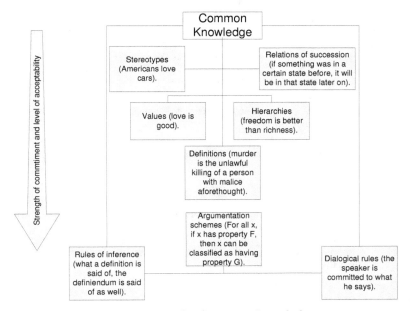

Figure 4.2: Levels of common knowledge

This figure is obviously an ideal representation of what ideally is considered as a stronger commitment. Depending on the type of discussion, and education and culture of the interlocutors, some stereotypes may even prevail over dialogue rules.

The two patterns of reasoning can be represented by two different versions of the same argument scheme. For instance, if we consider arguments from classification, we can notice that they can proceed from the normal scheme or the classification under lack of knowledge, namely from positive evidence or incomplete paradigms. Classificatory arguments can be grounded on definitions or stereotypes, they can proceed from maxims of definition or other weaker rules of inference requiring qualifications (for instance, if I define a house as four walls and a roof, and if there are four walls and a roof, then in some circumstances there can be a house as well).

The distinction between these two types of schemes, what we can call the "dialectical" schemes using the medieval name for them, and the heuristic or presumptive schemes is the basis for examining the different force of the conclusions and the possibility of rebutting them. In the first case, the interlocutor has to rebut the premises in order to rebut the conclusion. For instance, if one party argues that Bob committed a murder because he killed a woman unlawfully and with malice aforethought, the opponent can dispute the facts or the definition of 'murder', even though in law this possibility would be extremely complex. In case of heuristic reasoning, the interlocutor does not have to rebut the premises, but simply introduce new evidence. For instance, if the prosecution presumes that the homicide was committed maliciously, as the accused stabbed the victim, the defendant can simply show that it was an accident. The different role and strength of the two patterns of reasoning can be shown by the effects of the redefinition of 'rape' in Canada. In Scotland and other states, the concept of no-consent to a sexual act is stated as a presumptive, heuristic definition, imposing onto the defendant only the burden of providing evidence (Ferguson and Raitt 2006, p. 196). For instance, no-consent is presumed when the victim was asleep, or if violence or threats were used (Sexual offences act, s. 42). On the contrary, the Canadian Criminal Code includes the presumptive elements as definitions of "no consent" (section 244(3); (Temkin 2002, p. 117)):

> For the purpose of this section, no consent is obtained where the complainant submits or does not resist by reason of:
>
> (a) the application of force to the complainant or to a person other than the complainant
>
> (b) threats or fear of the application of force to the complainant or to a person other than the complainant
>
> (c) fraud
>
> (d) the exercise of authority

Whereas in the first case the defendant has only to provide evidence or a different explanation for the presumptions, in the second case he has burden of persuading that the victim was consentient. The two different treatments of the concept of 'no-consent' show how the strength of different types of human reasoning can affect the dialogue setting.

4.5 Conclusion

Presumption is a complex concept in law, affecting the dialogue setting. However, it is not clear how presumptions work in everyday argumentation, in which the concept of "plausible argumentation" seems to encompass all kinds of inferences. By analyzing the legal notion of presumption, it appears that this type of reasoning combines argument schemes with reasoning from ignorance. Presumptive reasoning can be considered a particular form of reasoning, which needs positive or negative evidence to carry a probative weight on the conclusion. For this reason, presumptions shift the burden of providing evidence or explanations onto the interlocutor. The latter can provide new information or fail to do so: whereas in the first case the new information rebuts the presumption, in the second case, the absence of information that the interlocutor could reasonably provide strengthen the conclusion of the presumptive reasoning. In both cases the result of the presumption is to strengthen the conclusion of the reasoning from lack of evidence.

As shown in the legal cases, the effect of presumption is to shift the burden of proof to the interlocutor; however, the shift a presumption effects is only the shift of the evidential burden, or the burden of completing the incomplete knowledge from which the conclusion was drawn. The burden of persuasion remains on the proponent of the presumption. On the contrary, reasoning from definition in law is a conclusive proof, and shifts to the other party the burden to prove the contrary. This crucial difference can be applied to everyday argumentation: natural arguments can be divided into dialectical and presumptive arguments, leading to conclusions materially different in strength.

CHAPTER 5

COMMITTED TO ARGUE: ON THE COGNITIVE ROOTS OF DIALOGICAL COMMITMENTS

Fabio Paglieri

Abstract

In this paper the notion of *dialogical commitment*, as defined by Walton and Krabbe, is discussed and compared with the notion of *social commitment*, as used in the field of Distributed Artificial Intelligence. The outcome of this analysis suggests that Walton's extra-psychologism on commitment is not completely warranted, albeit confusion between individual commitments and dialogical commitments should be avoided, as mandated by Walton. In contrast, it is argued that the very notion of dialogical commitment can neither be reduced to the internal mental states of the arguer, nor be understood without taking those mental states into proper account. The concept of social commitment is shown to play a pivotal role in between the individual beliefs and goals of the arguers, and the dialogical obligations they incur by effect of their exchange. As a side effect, this analytic effort reveals yet another path, partially uncharted, where Walton's work may help us building bridges across argumentation theories, cognitive science, and Artificial Intelligence.

5.1 Introduction: Walton's extra-psychologism on commitment

Commitment is a key notion in Douglas Walton's theory of argumentation: extensively elaborated in his joint monograph with Eric Krabbe, revealingly titled *Commitment in dialogue* (Walton and Krabbe 1995), the concept has remained central to his later works, without undergoing any major modification. A central theme in Walton's analysis of commitment is his extra-psychologism: it is often emphasized, in no uncertain terms, that dialogical commitments should not be understood by reference to the mental states of the arguers. In fact, one of the declared aim of Walton and Krabbe was to provide "...a treatment of commitment that is not psychological but pragmatic and critical" (Walton and Krabbe 1995, p. 12); furthermore, they insist that commitment can be considered a subjective state, i.e. a state of a given subject, only "provided we do not for *state* silently substitute *mental state*. We must rather think of a state here as a definite set of circumstances (legal, moral, social, etc.)" (Walton and Krabbe 1995, p. 21). It is worth noting that Walton and Krabbe (consciously) inherited this non-mentalistic stance from Hamblin: "A commitment is not necessarily a 'belief' of the participant who has it. We do not believe everything we say, but our saying it commits us whether we believe it or not. The purpose of postulating a commitment-store is not psychological" (Hamblin 1970, p. 264). Moreover, Walton does not seem to have reneged or attenuated this extra-psychological view of commitment in more recent writings, witness the following excerpt:

Much traditional thinking about argumentation in philosophy as well as AI has been based on a BDI (belief-desire-intention) model. But there have been many difficulties with the BDI model. Beliefs, desires and intentions are psychological states, and in trying to analyze or evaluate argumentation, trying to pin down an arguer's actual mental states can be quite a hard task. On the other hand, it is often possible to cite textual evidence to indicate what statements an arguer has committed himself to. Commitment can be seen as public. In a dialogue, a participant becomes committed to a statement in virtue of having gone "on record" by asserting it [...]. It should be clearly recognized that [...] commitment is not the same as belief. Commitment is a normative notion, meaning that the structure of the dialogue along with the arguer's recorded moves determine the arguer's commitments (Walton 2005d, pp. 63-64).

As it has been argued elsewhere (Paglieri 2007), in Walton's view dialogical commitments are *public, normative*, and based on *dialectical rules*. They are public insofar as depending only on what is explicitly stated by the speaker, plus what he[1] can be reasonably surmised to be committed to implicitly, e.g. by effect of conversational implicatures; they are normative because they bind future dialogical moves of the speaker; and they are based on dialectical rules, since the type of dialogue in which participants are engaged largely determines what kind of commitments can be expected to stem from their assertions (Walton and Krabbe 1995). Put even more succinctly, Walton's position on commitments can be reduced to the following formulation:

(dialogical moves & dialogical rules) → dialogical commitments

The extra-psychological thesis consists in maintaining that no mental ingredients are necessary to analyze the role of dialogical commitment in argumentation. This is not to say that Walton anywhere denies that commitments are ultimately based on some underlying mental states or actions: he is simply keeping that aspect of the matter out of his scope of inquiry, maintaining that this will not significantly compromise our understanding of dialogical commitments. In this essay, I want to put some pressure on such position, launching a two-pronged attack against it. First, I will try to show that the appeal to dialogical rules in Walton's definition of dialogical commitments is ultimately based on attributing a mental action to several cognitive agents — namely, acceptance of the relevant dialogical rules by the majority of the arguer's community. Hence, the practical efficacy of dialogical commitments necessarily rests on the arguers' cognitive dispositions, even when such dispositions are expunged from the definition of commitment, as Walton does. Second, I will show how dialogical commitments can be characterized as a specific case of social commitments, as they are defined in Artificial Intelligence (Castelfranchi 1995). Since the notion of social commitment relies heavily on the mental states of both the subject and the addressee of the commitment, this is tantamount to show that a socio-cognitive analysis of commitments is capable of yielding exactly the right kind of notion for argumentation theories. Moreover, such analysis suggests some insights on dialogical commitments that escaped Walton's non-mentalistic account. Finally, I will propose that the alleged opposition between dialogue-based theories and BDI models of argumentation, outlined by Walton in the quotation above, is overstated

---

[1]To simplify the exposition, henceforth I will adopt the convention of referring to the speaker (i.e. the proponent of a given argument, who undertakes certain commitments to it) using the masculine personal pronoun, whereas I will use the feminine personal pronoun to indicate the counterpart or the audience of the speaker.

and partially misleading. As a side effect, there is yet another path, partially uncharted, where Walton's work may help us building bridges across argumentation theories, cognitive science, and Artificial Intelligence.

Needless to say, my critique of Walton's extra-psychologism on commitment has no belligerent purpose, but is rather meant as a friendly skirmish, to strengthen the opponent's position by highlighting some potential ramifications of it, yet to be explored. No matter how marginal the resulting improvement may be, I still hope it will help celebrating Walton's monumental achievements in the study of argumentation.

5.2 Dialogical rules and the impossibility of mindless commitments

Dialogical rules are essential in determining the arguers' commitments. Different types of dialogue are characterized by different rules, and a key contribution offered by Walton and Krabbe consisted of elucidating how these different sets of rules affect the corresponding dynamics of dialogical commitments (1995; on the taxonomy of dialogue types with reference to commitments, see also (Walton 1998b). This has substantial repercussion on fallacies, defined as infringements of some dialogue rule: since rules change across dialogue types, a move that counts as fallacious in one dialogical context may not be fallacious in a different situation. This provides an elegant framework for dealing with fallacies, and allows to rescue some instances of good reasoning that would be otherwise mistakenly considered as fallacious (a recurrent theme in Walton's work, e.g. in his 1984; 1995; for extensive discussion of specific argumentation schemes that were often unjustly accused of being fallacious, see (Walton 1996c, 1997b, 1998a, 1999a, 2000c). The link between dialogue rules and dialogical commitments is even deeper than that: without the former, the latter could not emerge at all. Rules provide the normative framework against which commitments are established, given the dialogical moves made by each arguer. In this, Walton's approach is largely consonant with pragma-dialectics (van Eemeren and Grootendorst 1992, 2004), where again a certain set of rules determine what are the legitimate moves of the participants in a critical discussion.

The fact that Walton's notion of commitment is dependent upon some shared dialogical rules does not directly conflict with his extra-psychologism. Clearly, rules can exist and/or be postulated with no reference whatsoever to the individual mind of the arguer, and violation of a given rule, be it dialogical or otherwise, does not depend on the mental state of the violator. As the well-known jurisprudential principle reminds us, *ignorantia legis non excusat*. So, it is perfectly consistent to *define* commitments as based on dialogical rules, and yet insist that they do not refer to any particular mental state of the subject — indeed, reference to dialogical rules constitutes the necessary pivot to avoid any mentalistic ingredient in the definition of commitment.

However, Walton's purpose in defining commitments is not just positing an abstract concept, but rather proposing a key notion for the normative analysis of real-life argumentation. Dialogical commitments are normative notions in the exact sense that they describe what a rational agent should do in a dialogical exchange, without mandating that all or even most empirical agents will in fact be rational. Yet, much of Walton's work (e.g., on argumentation schemes and on dialogue types) is based on extrapolating from everyday argumentation what are perceived as instances of reasonable inference and proper dialogical conduct, *prior* to any theoretical definition of rationality — and this reliance on *prima facie* reasonableness is, to various degrees, common to most approaches in the informal logic tradition, and consistent with its deep root in pragmatics (for discussion of this point, see Johnson 2000). For instance, it would be nearly impossible to seriously defend induction and abduction as valid forms of inference, if not based on empirical evidence that they are frequently employed in everyday reasoning and, most crucially, largely successful — that is, they get things right, on average and with qualifications. Given its pragmatic foundation, argumentation theory is doomed to walk a thin line between the normative and the descriptive: it is perfectly legitimate to define and use normative concepts, but the question of their efficacy still lingers. Since the motivation for developing those concepts in the first place was the empirical observation of how they shape everyday argumentation, how do we account for that?

The same question arises also for Walton's theory of commitment: Is it possible to guarantee the *efficacy* of commitment-based argumentation, without reference to the arguers' mental states? My answer to this question is negative. More generally, I am willing to concede to Walton that it is possible to define commitment as an extra-psychological notion, i.e. with no direct reference to the arguer's mental states. But this definition is just the prologue of what commitments are expected to do in regulating interpersonal reasoning, and my contention is that, for commitments to work in practice, they need to be supported by an adequate structure of mental states. In order to see why this is the case, let me start with a little chess analogy:

> Bob and Anna are facing each other across a chess board. Suddenly, Bob takes a knight, moves it seven squares ahead of its previous position, and proceeds to remove Anna's pawn from that square. To this, Anna objects: "You can't do that!". But Bob retorts: "Sure I can — I just did!". Anna: "Yes, but the rules of chess don't allow that kind of move!". Bob: "Who said anything about playing chess?".

This narrative illustrates the kind of problems that arise when one of the participants in a given interaction is not willing to play according to the rules, or does not know how to. *Mutatis mutandis*, in argumentation the issue is whether an arguer is willing to accept[2] his (alleged) dialogical commitments: if he is not, they will not be binding for him, hence (as Walton and Krabbe noticed themselves: 1995, p. 41) they will not even count as commitments *strictu sensu*, since a commitment is something the agent is bound to do.[3] One could object that an arguer who does not accept his commitments as prescribed by dialogical rules is not playing the game of argumentation, precisely like Bob is refusing to play chess: hence, he is not dialogically committed simply because he is not genuinely involved in a dialogue to start with — or, even more radically, he is not behaving rationally. But this objection makes sense only if the dialogical rules in question are (i) defined clearly enough as to allow for unambiguous verification of their respect or violation, and (ii) agreed upon by the large majority of the speakers within a given community (in the intention of Walton and Krabbe, probably the

---

[2]The notion of acceptance used here is a technical one, based on the extensive debate in epistemology on the distinction between belief and acceptance (see for instance (van Fraassen 1980, Stalnaker 1984, Cohen 1989, Bratman 1992, Engel 1998, Tuomela 2000, Wray 2001)). On the import of this distinction for argumentation theories, see Paglieri and Castelfranchi (2007). For an updated, critical review of the literature on acceptance, see (Paglieri 2009b). Notice that framing these considerations in terms of acceptance vindicates Hamblin's observation that "we do not believe everything we say" (which is certainly true), without acceding to Walton's extra-psychologism on commitment.

[3]Notice that this is not a matter of retracting a commitment the arguer has incurred, but rather a refusal to acknowledge having that commitment in the first place. Retraction is extensively and convincingly analyzed by Walton and Krabbe (1995, pp. 9-12, 36-43), but it implies that the retracting agent has already acknowledged having a certain commitment and is now trying to get rid of it. The problem discussed here is much more basic and radical than retraction.

whole humanity), so that deviance from them can be considered as infringement of a common habit and not just a legitimate idiosyncrasy of the arguer. Both conditions can be considered established firmly enough in the case of chess, but they remain far from being proven with respect to argumentation.

What is more important here, though, is that condition (ii) makes reference to the mental states of the majority of the speakers. This means that, given a population P of potential arguers, the majority of P must accept a certain set of dialogical rules, for the resulting commitments to ensue for any given arguer x within P, irrespectively of x's own mental states. In this case, acceptance of dialogical commitments at the individual level is based on collective acceptance of a given set of dialogical rules. The only way by which the dialogical commitments of the individual can be regarded as independent from his direct acceptance of them is by reference to collective acceptance of basic dialogical rules: one way or the other, the efficacy of commitment is clearly based on the mental attitude of acceptance, be it individuated in the single arguer or in the community of the speakers, or in both.

The fact that acceptance of a given rule or proposition classifies as a mental action is unambiguously endorsed in the rich literature on the topic (for critical discussion, see Paglieri 2009a), so here I take it as an assumption of my argument. By way of example, it is interesting to consider Cohen's definition of acceptance as a mental action, as opposed to a speech-act, and with significant reference also to argumentation and commitment:

> To accept that p is to have or adopt a policy of deeming, positing, or postulating that p — that is, of going along with that proposition (either for the long term or for immediate purposes only) as a premiss in some or all contexts for one's own and others' proofs, argumentations, inferences, deliberations, etc. Whether or not one assents and whether or not one feels it to be true that p. Accepting is thus a mental act (as what was called judgement' often used to be), or a pattern, system, or policy of mental action, rather than a speech-act. What a person accepts may in practice be reflected in how he or she speaks or behaves, but it need not be. (...) Acceptance implies commitment to a policy of premising that p (Cohen 1989, p. 368).

While the mental nature of acceptance in not in question, there are two ways of conceiving its role in ensuring the efficacy of dialogue rules: either commitments are binding only when the actual arguers involved in the dialogue are accepting the relevant dialogical rules, or collective acceptance (i.e., acceptance by the majority of the speakers) of the rules is sufficient to empower commitments with their binding force, regardless of the fact that some individual arguer may not share that normative attitude. On reflection, it appears that Walton and Krabbe's notion of commitment is consistent with the latter option: it is assumed that the community of arguers tacitly agree on what the rules of the dialogical game are, so that, for instance, whenever x asserts A, x will consider himself committed to defend A under questioning, as well as whatever else A directly implies. Moreover, the parties must tacitly agree on several details of commitment-based argumentation, if this approach were to be considered a reasonable (albeit normative) model of what happens between real-life arguers. For instance, all the parties must have the capacity to distinguish between different types of dialogues and agree on the main features of each type, they must similarly apply different sanctions to different kinds of infraction (e.g., having a common understanding of the difference between a fallacy and a blunder), they have to share the same rules for retraction of commitments under different dialogical circumstances, and so on.

In the face of such a conspicuous body of mutual knowledge that the model pre-supposes, it is natural to wonder whether it is truly reasonable to assume that much, given what we know of real-life arguers and of their dialogues.

To this, I am inclined to answer with some qualified scepticism. As far as the basic tenets of commitment-based argumentation are concerned, it seems rea-sonable to assume that speakers in general are likely to tacitly conform to them. For instance, it seems relatively safe to assume that asserting A carries with it a commitment to either defend A under further questioning, or retract such a com-mitment with adequate justification (if possible). Similarly, the idea that mak-ing concessions to the counterpart entails a weak commitment not to challenge them unless something new emerges appears like a sound generalization, one that most speakers are likely to comply with in their dialogical behaviour. Other aspects of the theory, however, appear much less likely to be agreed upon, tacitly or otherwise, by the majority of the speakers — hence, they are unlikely to have any efficacy in practice. Take as a case in point the following characterization of a spurious quandary, i.e. a clash between commitments to the rules of the dialogue (C_1) and commitments to one's own priorities within the discussion (C_2):

> Suppose that the rules of a certain type of dialogue stipulate that a party X should answer a certain question in a straightforward man-ner, but that each possible answer would be very detrimental to X's position in the dialogue, so that good strategy would dictate silence on the issue. Then X should nevertheless answer the question, since, according to the assumed priority of $C_1(X)$ over $C_2(X)$, it is always better to blunder than to commit a fallacy (Walton and Krabbe 1995, p. 55).

Here the fact that blundering is to be preferred over committing a fallacy is a mere stipulation. Not only several individual arguers may vigorously disagree (this would not dull the normative force of the principle), but also the commu-nity of the speakers as a whole may feel differently. Who says that obedience to abstract principles is to be preferred over individual self-preservation? How often such self-sacrificing attitude is found or even mutually expected in human affairs, both within and outside dialogical interaction? Of course, by committing a fallacy I incur greater risks than by making a blunder, since, if I am caught in the act, I will be exposed to stronger sanctions — on this point, I tend to agree with Walton and Krabbe. But this does not imply that I should be committed to avoid taking such higher risks: on the contrary, if I am willing to play with fire, this is solely my business, and the community is in no position (and feel no inclination) to reprimand me for it. Only if my actions land me in trouble, i.e. my fallacious move is detected and criticized, then I will be sanctioned accordingly. This is especially true for dialogical interaction since this is a game without a referee — as most human interactions happen to be. There is no such thing as an "argumentative police", nor any court of appeal where to settle the matter between opposing parties. Instead, the counterpart is co-responsible in making sure that the rules, whatever they are, are respected. If I make a fallacious move and you fail to cry foul, you are implicitly conceding to it, and this makes you my accomplice in the dialogical violation. This is not to say that I am no longer responsible for breaking the rules, but rather that the social pressure is exerted by anticipating some kind of *ex post* retaliation (if the rules are violated and the cheater is caught, punishment will ensue) and not by constraining ex ante the ar-guer's strategic decision (arguers are not necessarily expected to conform to the rules, when doing so is bound to jeopardize their case).

To sum up, two aspects of dialogical rules contribute to exclude the practical possibility of "mindless commitments", i.e. commitments that exist independently from the mental states of the arguers and/or their fellow speakers. On the one hand, the very efficacy of commitments rests either on the individual arguer accepting the relevant dialogical rules, or (more typically) on the collective acceptance of such rules within the community where the argument takes place. In both cases, reference to the mental action of acceptance, be it individual or collective, is needed to ensure that commitments do not remain mere abstractions, but rather acquire full normative force on the dialogical interactions between arguers. On the other hand, some specific aspects of Walton and Krabbe's theory of dialogical commitments seem to stretch the bounds of normativity too far, so to speak, by positing dialogue rules that remain in dire need of empirical verification. In other words, it is highly debatable whether some key dialogue rules are in fact tacitly agreed upon by the majority of real-life arguers, and only empirical investigation on the approved dialogical practices within different communities of speakers will permit to settle the matter in detail. The fact that Walton and Krabbe are proposing an idealized model, one not intended to perfectly mirror reality (Walton and Krabbe 1995, p. 6), is not relevant here, since the issue is not about individual violations, occasional or systematic, of certain principles, but rather on whether the arguers as a community would subscribe to the kind of principles that the theory mandates as rational. As Johnson reminded us, "because an argument is an exercise in rationality, its status and fate in the wider culture depends on that culture's assumptions on the nature and value of rationality" (Walton 2000a, p. 11).

5.3   Committed to whom? Social commitments in dialogue

In Artificial Intelligence several notions of commitment have been used for different purposes (for a survey, see (Jennings 1993, Castelfranchi 1995)). For instance, a notion of *internal* (or *individual*) *commitment* is essential to define intentions in a BDI framework (Cohen and Levesque 1990), following Bratman's planning theory of intentions ((Bratman 1987, 1999); for some qualified criticism of that view, see Castelfranchi and Paglieri (2007)). An internal commitment (I-commitment) refers to a relation between an agent and an action. The agent has decided to do something, is determined to execute the appropriate action at the scheduled time, and the corresponding goal is a persistent one (i.e. an intention). Under these circumstances, the agent is considered committed to performing the intended action, in the sense that, to behave rationally, he should persist in pursuing his goal until he believes that such goal has been reached, or is impossible to achieve, or is no longer motivating (for further details, see (Castelfranchi and Paglieri 2007)). This type of individual commitments are very different from the dialogical commitments that Walton and Krabbe are interested to analyze, and in fact these authors pay great care in distinguishing commitments from intentions (Walton and Krabbe 1995, pp. 14-15, 26-27).

Another notion frequently discussed in AI, in particular in relation to human-computer interaction and organization theory (Fikes 1982, Winograd 1987, Gasser 1991, Rao et al. 1992), is *collective commitment*: a commitment that a whole collective subject (e.g. a firm) incurs, given the purposes of such group and the relevant background knowledge. This is the sense in which a football team can be said to be committed to victory, or a given brand can be considered to have a commitment to excellence, and so on. As noted by Castelfranchi (1995), collective commitment so defined is basically a sub-type of internal commitment, in which the subject of the commitment happens to be a group rather than an individual. Hence, this notion is also unlikely to serve the purposes of a commitment-based theory of argumentation.

A third notion of commitment analyzed in AI is *social commitment*, i.e. the commitment of an agent to *somebody else*, with respect to a given action. Social commitments (S-commitments from now on) were first analyzed by Castelfranchi (1995) as a specific form of goal adoption, and here I will follow his treatment of the notion. Given two agents (x, y) and an action A, x is committed to y to do A if and only if the following conditions obtain:

   (i) y has reason to believe that x intends to do A, i.e. that x has an internal commitment to A

  (ii) A is a goal of y (y is interested in A)

 (iii) x and y mutually know (i) and (ii), i.e. they both know that x is rationally expected to intend to do A and that A is a goal of y

 (iv) y accepts that A will be realized by or with the help of x

For instance, if Bob tells Anna that he will pay her phone bill and she accepts his offer, thereby Bob becomes S-committed to Anna with respect to the payment of her phone bill. Notice that Bob would have no S-commitment to Anna if she had no reason to expect Bob to intend to pay her phone bill (hence condition (i)), or if Bob's statement had concerned paying his own phone bill (hence condition (ii)), or if Anna had been unaware of Bob's intention (hence condition (iii)), or if Anna had just refused Bob's kind offer (hence condition (iv)). Notice also that the fact that Bob really has the intention of paying Anna's phone bill does not affect his being socially committed to her in doing so: in other words, x's I-commitment to do A is not a necessary condition for x's S-commitment to y to do A. Imagine Bob is being dishonest towards Anna, and he wants her to believe that he will pay her phone bill just to prevent her from doing so and thus causing her a financial damage. In this case, obviously Bob does not intend to pay her bill, hence he is not I-committed to it; nevertheless, by (insincerely) stating that he will pay the bill, he becomes all the same S-committed towards Anna — for instance, Anna would be entitled to protest after discovering that Bob did not pay her phone bill. This is precisely what Hamblin required of dialogical commitments: by asserting something, we become (socially, not individually) committed to it whether we believe it or not.

Let us introduce some additional terminology, to facilitate discussion of additional features of S-commitments: the *subject* of a S-commitment is the agent who becomes socially committed to somebody else with respect to action A; the *addressee* of a S-commitment is the agent to whom the subject is committed; the *object* of a S-commitment is the action or sequence of actions that the subject is committed to perform, or the state of the world the subject is committed to bring about; finally, the witness of a S-commitment is whatever agent (possibly more than one, or even collective entities) is entitled to control that S-commitments are satisfied, or take notice when they are violated (on these and other normative roles, further details are provided in Conte, Castelfranchi 1995). According to Castelfranchi (1995) (see also (Searle 1969) on this point), a key property of S-commitments is that they *generate rights in the addressee* (Anna, in our example) and *induce obligations in the subject* (Bob). In particular, as a consequence of x being S-committed to y to do A, y (the addressee) becomes entitled to: (i) control whether x does A; (ii) demand that x does A; (iii) protest with x (or with some higher authority, if any), in case x does not do A; (iv) in some cases, exact reparation for the losses occurred due to x's failure to do A (compensations, retaliations, etc.). Conversely, x (the subject), by becoming S-committed to y to do A, undertake an obligation to: (i) do what is committed to, i.e. A; (ii) not object to any of the rights that y acquired, in virtue of x's S-commitment to y (e.g. protest in case of failure). Needless to say, the subject's obligations can be violated, but not without negative consequences for the subject, if caught in the infraction and in the presence of some authority (possibly, the addressee) capable of effecting some form of punishment or penalty.

If we now try to apply S-commitments to argumentation, they appear as prime candidates to capture Walton and Krabbe's notion of dialogical commitments. Indeed, my claim is that dialogical commitments, in the sense endorsed by Walton, are just *a specific case of social commitments*. The fact that dialogical commitments are social I take to be self-evident: as the subtitle of Walton and Krabbe's book illustrates, they are "basic concepts in interpersonal reasoning", which clearly characterizes them as social entities. As for the fact that S-commitments have the property of engendering obligations in the speaker and creating rightful expectations in the addressee, this is precisely what is required from dialogical commitments in different types of argumentative interchange. So the only and crucial issue that needs to be considered concerns the *identity of the addressee*: since speech acts within a given type of dialogue create commitments in the speaker, to whom the speaker is supposed to be committed? In Castelfranchi's terminology, who is the addressee of dialogical commitments? As a case in point, let us confine the analysis to assertions, although analogous considerations could be developed for concessions and other commitment-inducing speech acts. Once a speaker has become committed to defending the proposition he asserted (the object of his S-commitment), who is the addressee of his corresponding dialogical obligations?

The most obvious suggestion would be to consider the counterpart or the audience that the speaker is facing as the addressee of his dialogical commitments. On reflection, this cannot be the case, for three reasons. First, it is unlikely that the counterpart will have as a goal that the speaker successfully discharges his dialogical obligations, e.g. that he manages to defend his position (contra condition (ii) in the definition of S-commitments): in any confrontational context (e.g. persuasion, negotiation, and quarrelling, to refer solely to Walton and Krabbe's taxonomy), the counterpart will often have the contrary goal, i.e. hoping that the speaker is incapable of defending his position under questioning. Conversely, in such circumstances the speaker would have no interest in adopting or fostering any (dialogical) goal of the counterpart, unless this helps strengthening the speaker's own position. Second, and consequently, the condition that both parties should have mutual knowledge of the counterpart's goals with respect to the speaker's commitments does not apply, thus violating clause (iii) of the definition of S-commitments. Third, there is no reason for the counterpart to necessarily expect the speaker to help fostering any of the counterpart's dialogical goals: lacking such expectation, there is nothing for the counterpart to "accept" from the speaker in terms of dialogical goal adoption, and this violates condition (iv).

The real addressee of dialogical commitments already figured prominently in the previous section: it is the *community of speakers* within which a given dialogue takes place and to which the arguers' speech acts ideally refer. Let us try to adapt Castelfranchi's definition of S-commitments to characterize dialogical commitments, assuming the speaker as subject of the commitment and the community of speakers as its addressee. Given a speaker x that asserts a proposition p within a community of speakers S, x is dialogically committed (to S) to defend P under questioning if and only if the following conditions obtain:

(i) there is reason for S to accept that x intends to defend P under questioning, i.e. that x has an internal commitment to do so

(ii) the fact that x defends P under questioning is a (collective) goal of S

(iii) x and S mutually know (i) and (ii), i.e. they both know that x is rationally
      expected to intend to defend P under questioning and that this is a goal of
      S

(iv) S accepts that the goal in (ii) will be realized by or with the help of x

Notice that, in condition (i), assuming that x intends to defend P under questioning is weaker than assuming that x believes P (hence Hamblin's proviso is satisfied): the assertion of P by x is taken to indicate that x is *ready to act as if x believes P*, whether this is true or not. In both cases, the subject still endorses the commitment to live up to his word. The fact that any specific speaker shall do so is a collective goal of the community of speakers, as required by condition (ii), since otherwise orderly discussion would be simply impossible, insofar as anyone would be free to make statements without offering any justification, as well as retracting them with no penalty attached. This is something speakers as a whole are keen to avoid, both in a Gricean perspective (Grice 1989) and from the standpoint of relevance theory (Sperber and Wilson 1986). Moreover, as stated in condition (iii), only assertions addressed to someone else entail dialogical obligations, since the community of speakers (in particular, the counterpart or the audience who witnessed the utterance of the statement) needs to be aware of what the speaker became committed to defend, as well as of the fact that the community has a stake in him satisfying such expectation: were it otherwise, we would be dialogically committed to whatever we mutter in solitude, and this is certainly not the case. Finally, as per condition (iv), the community of speakers must accept that the speaker will try to discharge his dialogical obligations, which means he should not be in any way prevented or impeded in doing so.

This analysis, albeit admittedly sketchy, suggests that we should consider the community of speakers as the addressee of the dialogical commitments incurred by an individual speaker while debating with his counterpart. This leaves open an intriguing question: if the counterpart is not the addressee of the speaker's commitments, what is her role in a commitment-based theory of argumentation? Following here Conte and Castelfranchi's taxonomy of normative roles (Conte and Castelfranchi 1995), I suggest she typically acts as the witness of the speaker's commitments: even if such commitments do not solely or directly concern her goals, she is entitled to "raise the alarm" if the speaker fails to discharge his obligations, and can protest on behalf and as a member of the community of speakers. Her right to protest, however, is not linked to her individual status as a party in the dialogue, but rather depends from her being part of the same community of speakers to which the proponent of the argument belongs. By failing to live up to his dialogical commitments, the speaker is not violating the counterpart's rights as an individual *per se*, but rather as a member of a community of speakers where certain S-commitments are endorsed by making various speech acts.

To sum up: treating dialogical commitments as S-commitments of an individual speaker towards the ideal community of his fellow speakers allows us to capture many key properties that Walton and Krabbe attribute to dialogical commitments. Hence the possibility is open to operationalize dialogical commitments by borrowing from AI the notion of social commitments and adapting it to the needs of argumentation theory. This is tantamount to demonstrate that dialogical commitments need not be considered as non-mentalistic entities, in order to exhibit the properties that Walton and Krabbe demand of them. In fact, S-commitments, including dialogical commitments understood qua S-commitments, are defined in terms of mental states and mental actions, both of the subject (knowledge in condition (iii)) and of the addressee (acceptance in conditions (i) and (iv), goal in condition (ii), knowledge in condition (iii)). This is not to say that S-commitments could be reduced to I-commitments, as discussed above and as repeatedly emphasized by Castelfranchi (1995): S-commitment is an intrinsically social concept, thus cannot be reduced to the mental states of any individual agent — and this is another point of convergence with Walton and Krabbe's desiderata for dialogical commitments. But the relational nature of dialogical commitments does not impinge on their mentalistic character: as any other S-commitment, a dialogical commitment depends on a relation *between minds*, and thus cannot be treated as a purely behaviouristic notion (contra the extra-psychologism favoured by Walton). This, in turn, opens new perspectives on the relationship between BDI systems and commitment-based argumentation in AI, as discussed in the next section.

5.4   BDI systems and commitment-based argumentation: friends or foes?

In light of what was discussed so far, it is now time to get back to Walton's initial quotation on the alleged opposition between BDI models and commitment-based argumentation in AI. BDI is an acronym that stands for beliefs, desires, and intentions: BDI models analyze the behaviour of rational agents as being dependent on these classes of internal representations — more precisely, as being directed towards and governed by the realization of the agent's intentions in light of the agent's beliefs (Rao and Georgeff 1991, Wooldridge 2000), where intentions are usually described as a sub-set of desires that happen to meet certain conditions (they are feasible, they have been chosen for execution among competing options, etc.). With reference to human agents, BDI theories describe rational behaviour as being based on the psychological dispositions of believing, desiring, and intending. The same dispositions need not be understood as psychological when referred to artificial agents, but rather as *functional concepts* that allow intelligent anticipatory regulation of the agent's behaviour. Depending on the domain of application, BDI models have been enriched with additional notions (e.g. obligations, norms, expectations, etc.), sometimes defined as molecular entities composed of a certain combination of the three basic primitives, sometimes introduced as additional primitives (for a discussion on the import of this distinction, see Castelfranchi, Paglieri 2007; Paglieri 2009a).

The alleged opposition between BDI models and dialogue-based systems in AI is based on Walton's extra-psychologism on commitment: insofar as we have reason to consider dialogical commitments as extra-psychological entities, then indeed there seems to be a radical difference with BDI approaches, where commitments, be they individual, collective, or social, are analyzed in terms of the agents' internal states. However, the purely non-mentalistic character of dialogical commitments have been questioned in section 2, whereas in section 3 we have seen how (a certain type of) social commitments exhibit many of the desired properties for dialogical commitments. Hence there is reason to regard Walton's contraposition of BDI models and dialogue-based systems as overstated, and perhaps even potentially misleading. To be sure, there is much I still agree with in Walton's approach. For one thing, it is certainly true that commitments should not be confused with either beliefs or intentions. Moreover, commitments cannot be reduced to the internal states of the speaker, whereas they can be reduced to (i.e. defined in terms of) a specific *relation* between the internal states of both arguers and those of the community of speakers they belong to. On the other hand, there is a specific point where Walton's position needs amendment, in light of the considerations developed so far: it is *not* the case that, "in trying to analyze or evaluate argumentation, trying to pin down an arguer's actual mental states can be quite a hard task" (Walton 2005d, p. 63). This would be true only if we were looking for the wrong kind of mental states, e.g. beliefs corresponding to the arguer's assertions (contra Hamblin's recommendation) or individual intentions of vindicating the arguer's position (contra Walton and Krabbe's distinction). But if we confine our attention to the right kind of mental states, i.e. those implied in the definition of dialogical commitments as S-commitments of the arguer to the relevant community of speakers, there is no difficulty whatsoever in "pinning them down". In fact, dialogue rules can be reinterpreted as collective norms regulating what kind of mental attitudes (collective and individual) are implied by different speech acts across various dialogical contexts.

This is not to say that formalizing the notion of dialogical commitments within BDI systems is going to be a straightforward affair. On the one hand, BDI models themselves suffer from their own weaknesses (for some motivated criticisms, see (Castelfranchi 1998, Castelfranchi and Paglieri 2007)), albeit of a different sort from what Walton suggested, and the right set of primitives needs to be defined, for these models to be able to capture the internal states implied by dialogical commitments. On the other hand, dialogical commitments *qua* S-commitments make reference to the community of speakers as the addressee of the commitment, and this presents two non-trivial difficulties: first, there is need to define more precisely the collective subject labelled as "relevant community of speakers", and to explore the implications of its definition (for instance, it remains to be seen whether it makes dialogical commitments culturally dependent, and whether this is a desirable result); second, we shall have to provide an operational notion of certain collective mental states (i.e. collective acceptances and collective goals), and this is notoriously a problematic endeavour, witness the rich debate on the topic (see for instance (Gilbert 1989, Searle 1995, Tuomela 1995, 2002, Velleman 1997, Bratman 1999, Tollefsen 2002, Tummolini and Castelfranchi 2006)).

In spite of these difficulties, I hope to have convincingly argued that there is room for integration between BDI models and dialogue-based approaches, and that the notion of commitment, far from being the seed of discord, can act as a bridge between these two different ways of looking at the agent's dialogical obligations. Incidentally, this indicates another source of inspiration for AI models in Walton's rich theory of argumentation, that has already proven to be often seminal in computer science (Walton 2000b, 2005a, Reed and Norman 2003, Reed and Walton 2005). In the case of dialogical commitments, it would be extremely fruitful to use Walton and Krabbe's functional analysis (dispensing with its non-mentalistic twist) to determine a set of constraints that any BDI-based analysis of dialogical commitment would have to comply with. In other words, it is possible to define dialogical commitments by making reference to the agent's internal states and their social relation, very much in the fashion of BDI models, but the resulting definition must ensure that dialogical commitments have the key functional properties individuated by Walton and Krabbe — possibly, along the lines sketched in section 3.

5.5   Conclusions

In this paper I took issue with a secondary but important aspect of Walton's theory of dialogical commitments: the view that dialogical commitments could be profitably analyzed without reference to the arguers' mental states and actions. While accepting the basic tenets of the dialogical framework developed by Walton and Krabbe, I endeavoured to show that their notion of dialogical commitments is fully compatible with, and even necessarily related to, a certain relational structure of mental states, pertaining not only to the arguers, but also to the community of speakers where their arguments take place. This indicates that dialogical commitments have deep cognitive roots that Walton's analysis had neglected to uncover, and that are likely to generate some surprising ramifications in the development of a commitment-based theory of argumentation. Among such implications there is the possibility of reconciling BDI models and dialogue-based approaches,[4] instead of considering them as opposed, as Walton sometimes seems to suggest.

Against this view, one may object that the mentalistic perspective has nothing to add to the standard extra-psychological position, where dialogue rules and commitments are conceived as abstractions to explain the admissibility of certain facts — namely dialogues and dialogue moves which are considered to be acceptable or not. Even agreeing to look at dialogical commitments *qua* S-commitments, what difference does it make for the sake of argument analysis? I already hinted at the answer to this question in section 2: the mentalistic perspective is essential to explain the *efficacy* of dialogue rules and dialogical commitments in binding the arguers' moves, and argumentation theories need to address also that efficacy, since they are ultimately based on a pre-theoretical appreciation of what is effective (without being deductive) in everyday reasoning and argumentation. "Efficacy" here should not be confused with rhetorical value, and certainly does not suggest that whatever form of argument is valid, insofar as it sways the audience in the intended direction. More modestly, as per its dictionary definition, efficacy means the power to produce an effect, and the effect we are concerned with here is the normative pressure exerted on arguers by dialogical rules and commitments. My point is that argumentation theories could only assume but not explain that power, without reference to the mental states and actions of the arguers.

As a final remark, I would like to propose that Douglas Walton, in his impressive explorations of natural and computational models of argument, has been somehow overcautious in advocating the integration of psychological results and notions within argumentation theories, whereas he has championed similar forms of integration with respect to other disciplines, e.g. computer science and legal studies. If I may venture a suggestion, I believe this attitude should be reconsidered: psychological research can offer much to the study of argumentation, not only in terms of empirical results (on this point, see Hample 2005), but also providing conceptual models of the "cognitive side" of arguments (an issue I have belabored elsewhere, alone and with others, see for instance (Paglieri and Castelfranchi 2006, Paglieri 2007, Paglieri and Castelfranchi 2010, Paglieri 2007, Paglieri and Woods 2009, Paglieri 2009a); see also (Gabbay and Woods 2003) for a broader approach to the matter). Walton's keen analytical insight could add great momentum to current efforts of finding the middle ground between psychological concerns and dialogical constraints, and his contribution in this respect would be very much welcome.

---

[4]Another interesting perspective that BDI models could open up for argumentation theories concerns the relation between desires and intentions, which would contribute to explain arguers' actions and values within different dialogue types. However, since this connection was already discussed in previous work, I decided not to dwell on it here: the interested reader will find more on the relationship between desires and intentions in Castelfranchi and Paglieri (2007), whereas the implications of looking at argumentation as a goal-oriented practice are explored in (Paglieri and Woods 2009) and (Paglieri and Castelfranchi 2010).

CHAPTER 6

WALTON'S THEORY OF ARGUMENT AND ITS IMPACT ON COMPUTATIONAL MODELS

Chris Reed

6.1 Introduction

This paper focuses upon two areas of Walton's work. The first is his development of a theory of argumentation schemes. Though rooted in earlier research such as Hastings (1963) and, to an extent, Perelman and Olbrechts-Tyteca (1969), Walton's account of argumentation schemes (Walton 1996b), (Walton et al. 2008) is a highly significant and novel development of what is essentially an Aristotelian idea. Walton argues that inference – and in particular, the sort of everyday, presumptive inference that is commonly employed by arguers – cannot be accounted for by some single, over-arching abstraction, but rather, that it is a varied collection of structured topoi or mechanisms by which arguments can be constructed or analysed. One of the great strengths of Walton's approach is that it lays out a methodology for describing these argumentation schemes. Associated with a scheme's premises and conclusions are a set of critical questions that help to uncover the implicit components to which appeal is being made or the exceptions which would undercut the inference's applicability. So although the approach is in some ways quite loose and informal, it manages to impose order and structure on that looseness and informality. This balance between, on the one hand, pragmatism borne of empirical study, and on the other, orderliness borne of philosophical rigour, has yielded a theory which manages to provide common ground between the theoreticians and the empiricisits. It has been as successful in teaching critical thinking (see, for example, its natural and appealing presentation in Walton's textbook (2006c)), and in developing large-scale analyses (Reed 2005), as it has in furthering the debate on the nature of inference (Walton and Reed 2005). But most importantly for this paper is the effect that the approach has had on computer science, and on artificial intelligence in particular. Here, Walton's theory of argumentation schemes has provided a framework for many researchers working with particular, and sometimes very specific, forms of argument. From models of practical reasoning (Atkinson et al. 2006b) to systems designed to support the process of identifying suitable organs for transplant (Tolchinsky et al. 2006), Waltonian argumentation schemes have provided exactly the right balance between theoretical consistency (fitting well with AI models of defeasible reasoning for example), and practical utility (being applicable to very specific scenarios that provide the context for many AI research projects).

Argumentation schemes are not, however, the only example of Walton's work having significant impact computationally. The second area, and the one which has had the longest influence on AI, has been Walton's work on structures of dialogue, running from his early collaborations with Woods (Woods and Walton

1978) and later Krabbe (Walton and Krabbe 1995) through to more expansive accounts such as (Walton 1998b). Again, Walton's approach was not developed in isolation (see (Wells 2007) for a historical overview), but particularly the account in (Walton and Krabbe 1995) has been very influential in computer science for many of the same reasons as his argumentation schemes work. The balance between clarity in the theoretical development and practical utility in application has led to commitment-based dialogue game specification emerging as one of the predominant tools for working with computational models of dialogue, particularly in the area of the design, implementation and evaluation of protocols for multi-agent systems.

More recently, Walton and colleagues have been developing an account of how these two aspects of his work intersect. With models of dialogue explicitly supporting presumptive argument, and critical questions serving to accentuate the dialogical side of argumentation schemes, it is reasonable to hope that argumentation schemes can be connected to commitment based models of dialogue. Reed and Walton (2007) and Walton et al. (2008) have taken some preliminary steps in this direction, and this presents a specific challenge to computational users of Waltonian theory. Whilst models of commitment based dialogue are well developed in multi-agent systems, and models of argumentation schemes are well developed in AI reasoning systems, there is no work that brings the two together into a coherent whole. Building on some initial steps in (Reed et al. 2008), this paper aims to explore how this unification can be achieved. Specifically, the paper first summarises how much of Walton's work can be characterised computationally in terms of models for the representation of argument, and then summarises one existing approach to such representation, and the extensions to that approach that tie together Walton's models of dialgoue with those of argumentation schemes. Finally, the paper explores a particular challenge that is faced by this approach (the problem of calculated properties) and links that challenge back to one of the coernerstones of Waltonian argumentation theory: commitment.

6.2   The Need for Representation

Artificial Intelligence has long been an idiosyncratic hybrid of pure theory and pragmatic engineering. Nowhere is this more true than in computational models of argument. The mathematical theories of argument which originate in works such as (Dung 1995) have been enormously influential in theoretical models of reasoning in AI, because they provide the machinery for handling issues such as defeasibility and inconsistency in ways that traditional classical logics are not able to support. These same mathematical theories are, however, barely recognisable as theories of argumentation as the philosophical and communication scholarly communities would know them – they serve rather as 'calculi of opposition'[1].

At the same time, AI is also home to applications of theories of informal logic (Gordon et al. 2007), of pedagogic critical thinking (Reed and Rowe 2004), of rhetoric (Crosswhite et al. 2003) and of legal argumentation (Walton 2005a): these applications are all rooted squarely in the tradition of argumentation theory as a discipline, and diverge from it in ways that are typically incremental and driven by pragmatic necessity.

Whilst the fecundity of the research area has been clear (see, e.g., (Rahwan and Simari 2009) for a representative set of papers), the diversity and sheer number

---

[1] The phrase 'calculus of opposition' is due to Henry Prakken: he suggested it in a private discussion, but it strikes me as a particularly apt description not only of what is supported by Dung's concept of acceptability, but also very much the spirit behind it.

of different systems has led, inevitably, to fragmentation. It was this problem that led in 2005 to a workshop to explore possible means of harmonisation between approaches and systems. The remit of the meeting was avowedly practical: to try to find ways that these systems might start to work together. But practical, engineering issues turn very quickly to deep and open philosophical issues: What consitutes an enthymeme, a fallacy or an inference? What differentiates presumptions and assumptions in argument? How can linguistic and psychological conceptions of argument be reconciled? Are propositions the right atoms from which to construct argumentation complexes? What is the character of the rules that govern argument dialogues? And so on.

Clearly, it is impractical to hope that these questions might be resolved once and for all, so the approach is in two parts. In the first, computational developments are fixed, quite pragmatically, upon what is currently the best understanding of the various issues. Walton's work figures large here. But the work has also tapped into pragma-dialectics, into speech act theory, and into the work of theorists such as Brockriede, Freeman, Goodwin, Groarke, Hitchcock, Johnson, Kienpointner, Krabbe, O'Keefe and Perelman amongst many others.

The second part of the approach is to tackle as little as possible at the first iteration – whilst still achieving something significant. For this minimal possible goal, the focus was upon representing arguments. Whilst there are many AI systems that reason with arguments, present arguments, render arguments in natural language, try to understand natural arguments, visualize arguments, navigate arguments, critique arguments, support the construction of arguments, mediate arguments, and so on, we cannot hope to solve problems special to each. It seems reasonable to assume, however, that all of these systems might want to store arguments in some structured format. This, then, is the focus. If we want to set out to try to support harmonisation between systems, and to do so in a way that is as closely tied as possible to current models from the theory of argumentation, then we start with a simple task that is common across most AI systems of argument: representation.

6.3 The Argument Interchange Format

The Argument Interchange Format (AIF) (Chesñevar et al. 2006) was developed with the aim of developing a means of expressing argument that would provide a flexible – yet semantically rich – way of representing argumentation structures. The AIF was put together to try to harmonise the strong formal tradition initiated to a large degree by Dung (1995), the natural language research described at CMNA workshops since 2000[2], and the multi-agent argumentation work that has emerged from the philosophy of Walton and Krabbe (1995), amongst others.

The AIF can be seen as a representation scheme constructed in three layers. At the most abstract layer, the AIF provides a hierarchy of concepts which can be used to talk about argument structure. This hierarchy describes an argument by conceiving of it as a network of connected nodes that are of two types: information nodes that capture data (such as datum and claim nodes in a Toulmin analysis, or premises and conclusions in a traditional analysis), and scheme nodes that describe passage between information nodes (similar to the application of warrants or rules of inference). Scheme nodes in turn come in several different guises, including scheme nodes that correspond to support or inference (or rule application nodes), scheme nodes that correspond to conflict or refutation (or conflict application nodes), and scheme nodes that correspond to value judgements or preference orderings (or preference application nodes). At this

---

[2]http://www.cmna.info

topmost layer, there are various constraints on how components interact: information nodes, for example, can only be connected to other information nodes via scheme nodes of one sort or another. Scheme nodes, on the other hand, can be connected to other scheme nodes directly (in cases, for example, of arguments that have inferential components as conclusions, e.g. in patterns such as Kienpointner's (1992b) "warrant-establishing arguments"). The AIF also provides, in the extensions developed for the Argument Web (Rahwan et al. 2007), the concept of a "Form" (as distinct from the "Content" of information and scheme nodes). Forms allow the AIF to represent uninstantiated definitions of schemes (this has practical advantages in allowing different schemes to be represented explicitly – such as the very rich taxonomies of Walton et al. (2008), Perelman and Olbrechts-Tyteca (1969), Grennan (1997), etc. – and is also important in law, where arguing about inference patterns can become important).

A second, intermediate layer provides a set of specific argumentation schemes (and value hierarchies, and conflict patterns). Thus, the uppermost layer in the AIF ontology lays out that presumptive argumentation schemes are types of rule application nodes, but it is the intermediate layer that cashes those presumptive argumentation schemes out into Argument from Consequences, Argument from Cause to Effect and so on. At this layer, the form of specific argumentation schemes is defined: each will have a conclusion description (such as "A may plausibly be taken to be true") and one or more premise descriptions (such as "E is an expert in domain D"). Walton's schemes (Walton 1996b), (Walton et al. 2008) have been developed in full for the AIF.

It is also at this layer that, as Rahwan et al. (2007) have shown, the AIF supports a sophisticated representation of schemes and their critical questions. In addition to descriptions of premises and conclusions, each presumptive inference scheme also specifies descriptions of its presumptions and exceptions. Presumptions are represented explicitly as information nodes, but, as some schemes have premise descriptions that entail certain presumptions, the scheme definitions also support entailment relations between premises and presumptions. The AIF has here largely followed the lead of a collaboration between Walton and two AI researchers, (Gordon et al. 2007).

Finally the third and most concrete level supports the integration of actual fragments of argument, with individual argument components (such as strings of text) instantiating elements of the layer above. At this third layer an instance of a given scheme is represented as a rule application node the terminology now becomes clearer. This rule application node is said to fulfill one of the presumptive argumentation scheme descriptors at the level above. As a result of this fulfillment relation, premises of the rule application node fulfill the premise descriptors, the conclusion fulfils the conclusion descriptor, presumptions can fulfill presumption descriptors, and conflicts can be instantiated via instances of conflict schemes that fulfill the conflict scheme descriptors at the level above. Again, all the constraints at the intermediate layer are inherited, and new constraints are introduced by virtue of the structure of the argument at hand. Fig 6.1 shows diagrammatically how all of these pieces fit together (but it should be borne in mind that Fig 6.1 aims to diagram how the AIF works - it is a poor diagram of the argument that is represented). The "fulfils" relationships are indicated by dotted lines, and inference between object layer components (i.e. the arguments themselves) by thick lines. Remaining lines show "is-a" relationships. Note that many details have been omitted for clarity, including the way in which scheme descriptions are constructed from forms.

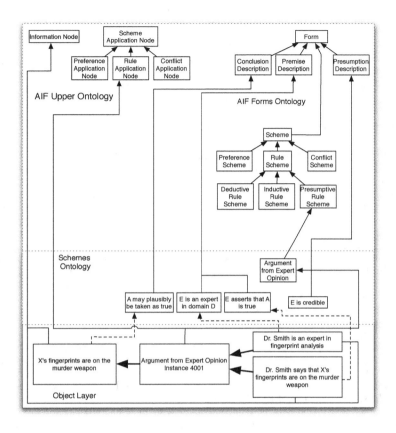

Figure 6.1: Three levels of AIF representation.

6.4   Extensions to Handle Dialogue

The next step is to allow the representation of dialogue. Several preliminary steps in this direction have been taken in (Reed 2006), (Modgil and McGinnis 2008) and (Reed et al. 2008). The motivation for this work can be summarised through O'Keefe's distinction between $argument_1$ and $argument_2$: $argument_1$ is an arrangement of claims into a monological structure, whilst $argument_2$ is a dialogue between participants - as O'Keefe (1977)[p122] puts it, "The distinction here is evidenced in everyday talk by ... the difference between the sentences 'I was $arguing_1$ that P' and 'we were $arguing_2$ about Q.'" Clearly there are links between $argument_1$ and $argument_2$ in that the steps and moves in the latter are constrained by the dynamic, distributed and inter-connected availability of the former, and further in that valid or acceptable instances of the former can come about through sets of the latter. An understanding of these intricate links which result from protocols and argument-based knowledge demands a representation that handles both $argument_1$ and $argument_2$ coherently. It is this that the dialogic extensions to the AIF set out to provide.

The fundamental building blocks of dialogues are the individual locutions. In the context of the AIF, Modgil and McGinnis (2008) have proposed modelling locutions as information nodes. We follow this approach primarily because statements about locution events are propositions that could be used in arguments. So for example, the proposition, *Chris says, 'ISSA will be in Amsterdam'* could be referring to something that happened in a dialogue (and later we shall see how we might therefore wish to reason about the proposition, *ISSA will be in Amsterdam*) – but it might also play a role in another, monologic argument (say, an argument from expert opinion, or just an argument about Chris' communicative abilities).

Associating locutions exactly with information nodes, however, is insufficient. There are several features that are unique to locutions, and that do not make sense for propositional information in general. Foremost amongst these features is that locutions often have propositional content. The relationship between a locution and the proposition it employs is, as Searle (1969) argues, constant – i.e. "propositional content" is a property of (some) locutions. Whilst other propositions, such as might be expressed in other information nodes, may also relate to further propositions, (e.g. the proposition, *It might be the case that it will rain*) there is no such constant relationship of propositional content. On these grounds, we should allow representation of locutions to have propositional content, but not demand it for information nodes in general – and therefore the representation of locutions should form a subclass of information nodes in general. We call this subclass, "locution nodes". There are further reasons for distinguishing locution nodes as a special case of information nodes, such as the identification of which dialogue(s) a locution is part of. (There are also some features which one might expect to be unique to locutions, but on reflection are features of information nodes in general. Consider, for example, a time index - we may wish to note that Chris said, 'ISSA will be in Amsterdam' at 10am exactly on the 1st March 2010. Such specification, however, is common to all propositions. Strictly speaking, *It might be the case that it will rain* is only a proposition if we spell out where and when it holds. In other words, a time index could be a property of information nodes in general, though it might be rarely used for information nodes and often used in locution nodes).

Given that locutions are a (subclass of) information nodes, they can, like other information nodes, only be connected through scheme nodes. There is a direct analogy between the way in which two information nodes are inferentially related when linked by a rule application, and the way in which two locution nodes are related when one responds to another by the rules of a dialogue. Imagine, for example, a dialogue in which Chris says, 'ISSA will be in Amsterdam' and Simon responds by asking, 'Why is that so?'. In trying to understand what has happened, one could ask, 'Why did Simon ask his question?' Now although there may be many motivational or intentional aspects to an answer to this question, there is at least one answer we could give purely as a result of the dialogue protocol, namely, 'Because Chris had made a statement'. That is to say, there is plausibly an inferential relationship between the proposition, *'Chris says ISSA will be in Amsterdam'* and the proposition, *'Simon asks why it is that ISSA will be in Amsterdam'*. That inferential relationship is similar to a conventional inferential relationship, as captured by a rule application. Clearly, though, the grounds of such inference lie not in a scheme definition, but in the protocol definition. Specifically, the inference between two locutions is governed by a *transition*, so a given inference is a specific application of a transition. Hence we call such nodes transition application nodes, and define them as a subclass of rule application nodes. (Transition applications bear strong resemblance to applications of schemes of reasoning based on causal relations: this resemblance is yet to be further explored, but further emphasises the connection between inference and transition).

So, in just the same way that a rule application fulfils a rule of inference scheme form, and the premises of that rule application fulfil the premise descriptions of the scheme form, so too, a transition application fulfils a transitional inference scheme form, and the locutions connected by that transition application fulfil the locution descriptions of the scheme form. The result is that all of the machinery for connecting the normative, prescriptive definitions in schemes with the actual, descriptive material of a monologic argument is re-used to connect the normative, prescriptive definitions of protocols with the actual, descriptive material of a dialogic argument.

With these introductions, the upper level of this extended AIF is almost complete. For both information (I-) nodes and rule application (RA-) nodes, we need to distinguish between the old AIF class and the new subclass which contains all the old I-nodes and RA-nodes excluding locution (L-) nodes and transition application (TA-) nodes (respectively). As the various strands and implementations of AIF continue, we will want to continue talking about I-nodes and RA-nodes and in almost all cases, it is the superclass that will be intended. We therefore keep the original names for the superclasses (I-node and RA-node), and introduce the new subclasses I' and RA' for the sets I-nodes\L-nodes and RA-nodes\TA-nodes respectively. The upper level of the AIF is thus as in Figure 6.2. This Figure shows all the *isa* links in the ontology, but misses out many of the *fulfills* and *has* edges which have been described elsewhere (Reed et al. 2008), (Rahwan et al. 2007). The Figure also shows a number of schemes to show how they slot into the overall hierary (but these are the schemes themselves, rather that instances of the schemes).

One final interesting question is how, exactly, L-nodes are connected to I-nodes. So for example, what is the relationship between a proposition p and the proposition 'X asserted p'? According to the original specification of the AIF, direct I-node to I-node links are prohibited (and with good reason: to do so would introduce the necessity for edge typing – obviating this requirement is one of the advantages of the AIF approach). The answer to this question is already available in the work of Searle (1969) and later with Vanderveken (Searle and Vanderveken 1985). The link between a locution (or, more precisely, a proposition that reports a locution) and the proposition (or propositions) to which the locution refers is one of illocution. The illocutionary force of an utterance can be of a number of types (Searle and Vanderveken (1985) explore this typology and its logical basis in some detail) and can involve various presumptions and exceptions of its own. In this way, it bears more than a passing resemblance to scheme structure. These schemes are not capturing the passage of a specific inferential force, but rather the passage of a specific illocutionary force. As a result, we refer to this schemes as *illocutionary schemes* or Y schemes. Specific applications of these schemes are then, following the now familiar pattern, illocutionary applications or YA nodes. These too are shown in Figure 6.2, with the naming convention that illocutionary schemes are referred to with gerunds (asserting, promising, etc.), whilst transitional inference schemess are referred to with nouns (response, statement, etc.), which both ensures clarity in nomenclature, and is also true to the original spririt and many of the examples in both the Speech Act and Dialogue Theory literatures.

6.5 Calculated Properties

The AIF cannot handle arguments based on the results of computation over AIF structures. Arguments based on counting other arguments, weighing other argu-

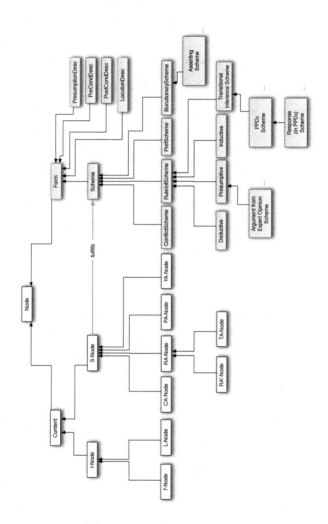

Figure 6.2: Upper levels of AIF extended to handle dialogue

ments, comparing other arguments and evaluating other arguments all involve processes (counting, weighing, comparing, evaluating) that cannot be captured in the AIF itself (and nor should they be, for otherwise the AIF would swell to some general purpose programming language). These various processes might collectively be thought of as ways of calculating properties about the arguments that the AIF represents. It is not that such arguments cannot be represented at all. But rather, if arguments are based on these *calculated properties* – arguments such as "the prosecution has not provided sufficient evidence for a conviction, so the accused is released" – then they can only be represented in the same way as normal propositions. The AIF has no way of capturing the link between such a statement and, say, the existence or non-existence of a set of other nodes. For monologic arguments this is a relatively small problem, but excludes, as the previous example demonstrates, some relatively common forms of legal argument. It may be that most or all of these cases can be handled with preference applications. But for dialogue, the matter is more serious. Protocol rules are very often defined on the basis of calculated properties of dialogue histories: the existence or non-existence of particular claims, the current status of claims, and most significantly, commitments.

Our current approach to tackling this problem is pragmatic. Arguments based upon calculated properties do occur, so we must represent them. To be more accurate, such arguments do occur naturally, and therefore the AIF *will* be employed by its users to represent them. The connection between these propositions and the AIF structures from which their truth values can be determined are no more represented in the AIF than are the connections between other, more conventional, propositions and the real world phenomena that establish their truth values (if such phenomena exist). It is not the job of the AIF to represent or calculate some objective truth (or even multiple subjective truthes). It is possible, in contrast, to imagine a process that runs over a given fragment of AIF, and can determine whether or not certain calculated properties hold. There might be any number of such processes – one to calculate what counts as sufficient evidence for a prosecution case in a given court system, another to determine which arguments are justified according to a given acceptability semantics (Dung 1995), etc. The results of these computational processes may then be compared either automatically or manually with claims about calculated properties that are represented explicitly in the AIF. But the job of ensuring that all and only true calculated properties are represented is not and cannot be the responsibility of the AIF alone.

With calculated properties representable as conventional I-nodes, their frequency in dialogue protocols becomes slightly less of a problem. So, for example, requiring that an opponent is committed to a proposition p before challenging p might require the fact that 'opponent is committed to p' is represented explicitly as an I-node. Commitment is, from an argumentation theory point of view, the most important calculated property, as it forms a core part of many dialogue games. Statements about (that is, propositions which refer to) commitment are I-nodes like any other. This is reminiscent of Locution nodes, or L-nodes, which are also a special type of I-node. Though L-nodes have a particular role to play in their most common application in representing dialogue, they can also be used as propositions simpliciter: the fact that Bob said q might be used as a premise in an argument for Bob's sentience, for example. So too I-nodes that correspond to statements about commitment can be used in contexts beyond the dialogue game at hand. A very common example of such commitment-based arguments is in circumstantial *ad hominem* arguments, where a speaker's commitments (or in some cases, the locutions themselves) are used to form an argument against the speaker's current claim. As with the rest of the AIF, claims about commitment (that a particular individual is committed to a particular claim, possibly in a particular context) occur just once in the AIF structure, and can be connected into multiple arguments. These commitment-oriented propostions seem entirely reasonable to add in to the AIF structure as a dialogue proceeds. Unfortunately, there are many other types of calculated property that one can imagine being used by a dialogue protocol that are much less appealing. One would not want a proliferation of nodes such as 'the opponent has not spoken in seven turns,' or 'the last-but-two move was not an assertion,' despite the fact that such propositions may very well be required for application of a transition scheme (that is, for the application of a structural rule in a protocol). The solution to this problem is actually already built in to the AIF. As described in (Rahwan et al. 2007), scheme nodes (there, rule of inference schemes in particular) have within them definitions of both implicit premises and exceptions. They are defined by the critical questions associated with a scheme and provide stereotypical growth points that allow an argument to be extended through challenge or questioning. Implicit premises, in particular are important to us here. In rule of inference schemes, implicit premises are often included in the scheme but rarely questioned. If they are not made explicit, then they do not occur in the AIF graph – they simply remain an un-exploited potential growth point in the scheme. The same thing happens for the calculated properties that are so often used in protocol definitions. If the fact that 'the opponent has not spoken in seven turns' is not disputed (by either the original participants in the dialogue or by any subsequent analyst or contributor) then it remains an implicit part of the transition scheme, and never becomes an explicit part of the AIF graph. In this way, the AIF provides a mechanism for representing these calculated properties when necessary, but does not clutter analyses with enormous numbers of implicit premises.

It is not difficult to imagine dialogic computational processes, akin to the monologic processes above, that perform various tasks on an AIF structure: verifying that a given AIF fragment conforms to the rules of a protocol; listing alternative ways in which a given interlocutor could legally respond at a given point in a dialogue, etc. But again, these computational process are run on top of AIF; they are not a part of it. Even the process of having a dialogue itself is separate from the AIF. Such a process (which could well involve an interface which allows a set of users to contribute to a dialogue according to the rules laid out in the protocol) would be responsible for updating AIF structure, most certainly, but at each stage in the process that static AIF structure records the history of the dialogue to that point, with the connections between the locution nodes and the I-nodes to which they refer.

In both the monological and dialogical settings, this approach to handling calculated properties is consistent with Finocchiaro's (monological) work on meta-arguments (Finocchiaro 2007) and Krabbe's earlier work on metadialogue (Krabbe 2003), in that the meta-moves are explicitly representable. What the AIF does not do however (or at least not in its current incarnation) is develop an infinitely extensible language of meta- and meta-meta-arguments as has been advocated elsewhere in the AI literature (Wooldridge et al. 2005), as there is not yet any convincing need in realistic scenarios. This is certainly an avenue for further research, however.

Finally, an interesting difference between inference schemes in monological argument structures and transitions schemes in dialogical argument structures is that the former will rarely constrain their premises to be specific types of calculated properties. There will be few rules of inference, for example, that are based upon whether one argument defeats another. In contrast, transition schemes will very often be based upon calculated properties such as commitments. There are exceptions, however. We have seen how some legal argumentation makes stereotypical use of calculated properties; such usage may very well be suited to being captured in argumentation schemes. And similarly, transition schemes will sometimes make use of purely structural (i.e. non-calculated) properties – 'always assert after a challenge,' for example. When a dialogue protocol is made up of transition schemes that are exclusively of this type, the protocol is Markovian[3] (in the general case, protocols, and especially those expressed as dialogue games, are not intrinsically Markovian).

6.6   Conclusions

This paper has aimed to review just enough of the technical machinery of the Argument Interchange Format to allow the reader to get a clear understanding of the philosophical masts to which it is nailing its colours. In particular, the AIF is subscribing wholeheartedly to Walton's theory of argumentation schemes, because it offers a way of dealing with the diversity of argument forms that strikes a balance between theoretical elegance and practical utility. In addition, the extensions to the AIF that support dialogue lean heavily on commitment-based models of dialogue developed by Walton, Woods, Mackenzie and others. It is slightly ironic, in this context, that commitment itself ends up not as a first class data object like the premises, claims, inferences and locutions of arguments, but rather, as a somewhat more elusive calculated property. Though this is surprising, it is perhaps a natural consequence of the deterministic way in which commitment update is defined in dialogic models.

There have been many examples of generalised machine-representable dialogue protocols and dialogue histories, e.g. (Robertson 2005), (W3C 2002), but these approaches do not make it easy to identify how the interactions between dialogue moves have effects on structures of argument (i.e. $argument_1$), nor how those structures constrain dialogue moves during argument (i.e. $argument_2$). Though there are still challenges that the AIF faces in its expressivity and flexibility, we have shown that representing complex protocols and rich argument structures in a common framework is now straightforward, and that the ways in which those protocols govern or describe dialogue histories is directly analogous to the ways in which schemes govern or describe argument instances. These strong analogies provide representational parsimony and simplify implementation. This is important because AIF representations are far too detailed to create by hand, and the range of software systems will stretch from corpus analysis to agent protocol specification, from Contract Net (Smith 1980) through agent ludens games (McBurney and Parsons 2002) to PPD (Walton and Krabbe 1995). The success of AIF will be measured in terms of how well such disparate systems work together.

As the argument web starts to become a reality, the diversity of applications and users will place enormous strain on the representational adequacy of the infrastructure, so it is vital that that infrastructure rests upon a solid foundation. Walton's research forms a central part of that foundation, and has smoothed the road between the philosophy of argument and the engineering of large scale argument technology.

---

[3]i.e. the legitimate future moves depend only upon the current move and not upon other features of the dialogue history.

CHAPTER 7

ANALYZING AND EVALUATING COMPLEX ARGUMENTATION IN AN ECONOMIC-FINANCIAL CONTEXT

Rudi Palmieri and Eddo Rigotti[1]

7.1 Introduction

In his extraordinarily intense and fertile research activity, reflected by an impressive number of publications, Douglas Walton has not only studied in depth numerous theoretical crucial concepts in argumentation theory but has also focused on several contexts in which argumentation occurs, considering different fields, like law (e.g. (Walton 2002a, 2005a, 2008c)), health care (e.g. (Walton 1985b, 2009c)), media (e.g. (Walton 2007c)), public discourse (e.g. (Walton 2000a, 2001b, 2005g)).

In its research and education activities our group in Lugano (ILS)[2] has devoted, in the last years, significant efforts to financial communication (cf. (Rigotti and Morasso 2007, Rocci and Palmieri 2007, Rigotti and Palmieri 2008, Palmieri 2008b,a, 2009, Rocci 2009)) a field whose relevance has strongly increased today, in connection with the global financial crisis. Now, Douglas Walton has also considered economic-financial argumentation by focusing in particular on what he calls "the argument from waste" or "sunk costs fallacy" (Walton 2002b), but, in

[1]The authors wish to thank particularly the two reviewers for the precious comments on a earlier version of the paper and Giovanni Barone-Adesi for the advices given on specific economic/financial aspects of the paper. They all have contributed significantly to the improvement of the paper. However, only the authors are to be held responsible for any error or inaccuracy in the paper.

[2]The Institute of Linguistics and Semiotics (ILS) of the University of Lugano focuses, both in its research and educational activities, on the relationship between argumentation and the context in which it occurs. As one of the considered contexts is finance, in 2004, this Institute has set up in collaboration with the Institute of Finance (Faculty of Economics) a Master Program in Financial Communication, whose main objective is to provide students with the competencies in financial economics and banking integrated with juridical and communicative-argumentative skills strongly required by the professional context. Four research projects on economic/financial argumentation, supported by the Swiss National Science Foundation, are currently ongoing at the ILS. Eddo Rigotti leads a project studying argumentation as a tool for resolving conflicts between managers and shareholders of publicly listed stock corporations (Grant: PDFMP1-123093). The authors of this paper are also conducting a recently approved project studying the argumentative practices adopted by Swiss banks in order to comply with Anti Money Laundering and Counter Terrorism Finance rules while at the same time preserving the fiduciary relationship with the suspected client (Grant: CR11T1_130652/1). Eddo Rigotti carried out the present development of the AMT (Argumentum Model of Topics) for the analysis of financial argumentation as part of his contribution to the research project entitled "Modality in argumentation. A semantic-argumentative study of predictions in Italian economic-financial newspapers." The project is supported by the Swiss National Science Foundation (Grant: 100012-120740/1) and directed by Andrea Rocci. Finally, Eddo Rigotti is actively involved in another project led by Andrea Rocci which investigates the argumentative function and rhetorical exploitation of keywords corporate reporting discourse (Grant: PDFMP1_124845). At the educational level, ILS is the leading house of Argupolis (www.argupolis.net), a doctoral school devoted to the study of argumentation practices in context, financed by the Swiss National Science Foundation (Grant: PDAMP1-123089).

our view, his major contribution to the study of economic/financial argumentation is related to his wide and deep investigation on practical reasoning (Walton
1990a, 2005g, 2007b, 2009d), a typically decision-oriented argumentative activity.

Indeed, argumentation in financial interaction proves its relevance most of
all in relation to the decision-making processes from which financial activities
arise, like investing in a certain company; assigning a rating to a firm; advising
clients in the banking context; shareholders' voting of a certain proposal made by
corporate managers — like a merger, a capital increase, a new board of director
—, etc.

Finance can be defined as the art of creating resources — rather than simply
dividing them — by matching the capital with an original idea able to increase
it (cf. (Lazonick 1991, Snehota 2004, Rigotti and Morasso 2007)). In this view,
the main actors of the financial interaction are *entrepreneurs* — people with ideas
but no capital for realizing them (cf. (Kirzner 1973)) — and investors/savers —
people with excess of capital but without projects for increasing or keeping its
value over time (Healy & Palepu 2001).

The relevance of argumentation for financial and economical decisions is proven by the fact that a key value like trust, that is typically exchanged in argumentative interaction (Rigotti 1998), also constitutes the substance of financial
reality. Not coincidentally, *trustworthiness* is a fundamental value at stake in financial decisions, especially when the interaction involves as decision-makers
laypeople who, not being experts of financial concepts and mechanisms, must,
for this reason, entrust decisions about their wealth to expert financial managers,
both institutional intermediaries and private advisors (cf. (Cottier and Palmieri
2008)). For all these experts, argumentation represents an instrument for building, maintaining and also restoring their trustworthiness in front of the market
participants (investors, clients, regulators, etc...) and the society in general, including political authority as well as citizens and households, who, by the way,
are also savers and investors. The current crisis, not by chance often described as
a crisis of trust (cf. (De la Dehesa 2008, Bernake 2009, Palmieri 2009)), is showing
very clearly how crucial reputation, reliability and trustworthiness are for entrepreneurs, and companies in general, in order to convince investors to finance
their business activities and for intermediaries to show the reliability of their financial strategies.

However, trust does not represent the only concern of financial argumentation: beyond the construction of trust or the assessment of trustworthiness, financial decision-making involves numerous other argumentative strategies, like
the feasibility of a project in terms of internal constitutive rationality, or in terms
of compatibility with conditions established by the context, and even compliance
with juridical norms and moral values; the expediency of a proposal in terms of
realization of the interests and goals of the stakeholders; the comparison with
alternative economic activities.

In the present paper we address financial argumentation by focusing on a significant example, a complex argumentation recently advanced by an outstanding
German entrepreneur in a particular circumstance that, in several aspects, is typically bound to the current financial and economical crisis. It is, in fact, an intervention related to the automobile industry, one of the sectors most harmed by the
current crisis, which discusses a measure — integration through merger of previously independent companies — that is frequently adopted in order to improve
the conditions of troubled companies, which, by remaining alone, would not ensure their competiveness or even compromise their existence. Now, because of
the competitive nature of markets, a measure that might represent an opportunity for one, is expected to represent a threat for the other. In this connection,
counter-measures may be considered or neglected by competing companies in

order to defend their position within the market.

In what follows we work out the reconstruction and evaluation of this argumentation. The argument schemes (or *loci*) emerging from this reconstruction are analyzed and some of them are evaluated, The evaluation is accomplished by verifying the respect of the applicability conditions imposed by the maxim, which is the inferential connection implicitly linking the argument and the standpoint. Indeed, this kind of assessment represents a useful instrument for determining the quality of the reasoning processes grounding financial decisions.

Our analysis makes use of the *Argumentum Model of Topics* (AMT), set up in our Institute of Linguistics and Semiotics, University of Lugano, in particular by Eddo Rigotti and Sara Greco-Morasso (see (Rigotti 2006, 2008, 2009, 2006, Rigotti and Morasso (to appear)). After accomplishing an analytical overview (van Eemeren et al. 1993a), we propose, following AMT, an integrated representation of the argument structure, which makes explicit all the premises necessary for justifying the passage from the argument to the standpoint.

7.2 The story

On the 12th of May 2009, during the presentation of the new VW Polo in Olbia (Italy), Ferdinand Piëch[3], Chairman of the VW's Supervisory Board, was asked for his opinion in relation to the possible combination of its competitor Opel with Fiat and Chrysler and its possible implications for VW. Piëch's reply was reported by different sources. Here we report a passage taken from the Manager Magazin Online:[4]

> Piëch äußerte sich gelassen zu einem mglichen Bündnis von Fiat, Opel und Chrysler. Schon bei Volkswagen und Audi habe es etwa 15 Jahre gebraucht, aus den beiden Unternehmen einen integrierten Konzern zu schmieden, sagte Piëch. Daher mache er sich keine Sorgen über die sich anbahnende Allianz. "Zwei Kranke in einem Doppelbett oder gar drei geben noch keinen Gesunden. Ich bin sicher, dass die, die im Moment über Zusammenschlüsse nachdenken, keine 15 Jahre Zeit haben". [...] Piëch warnte, die unterschiedlichen Unternehmenskulturen knnten ein Hindernis für eine erfolgreiche Allianz sein.
>
> *Translation: Piëch replied relaxed to a possible alliance between Fiat, Opel and Chrysler. Already [in the combination of] Volkswagen and Audi about 15 years were necessary in order to forge the two companies into an integrated group, Piëch said. Therefore, he is not worried about the impending alliance: "Two sick people in a double bed, or even three ones, don't make one healthy person. I am sure, that those who are now thinking to merge, have not 15 years at their disposal". The different corporate cultures may represent an obstacle to a successful alliance, Piëch warned.*

A passage of this text presented a particular communicative strength and not surprisingly was reported by numerous press agencies in numerous languages. We start, in our analysis from it, as reported below:

---

[3]Ferdinand Karl Piëch is Chairman and former CEO of Volkswagen and is one of the largest shareholders of Porsche. He is the grandson of Ferdinand Porsche, founder of the homonymous automobile company and designer of the popular VW Beetle. Piech won the award of Car Executive of the Century in 1999. In 2009, he succeeded in combining VW and Porsche through merger, after defeating the opposition by Porsche top managers, concluding in this way a long controversy that has strongly involved the German public opinion.

[4]"Piëch brüskiert Wiedeking", URL: www.manager-magazin.de/unternehmen/artikel/0, 2828,624192,00.html. Last visit: August 4, 2009.

> *"Two sick people in a double bed, or even three, don't make one healthy person"*

The reasoning developed by Piëch typically belongs to action-related argumentation, or practical reasoning (Walton 1990, 2005b, 2007a, 2009a; Rigotti 2008), and more specifically to prudential reasoning. The standpoint can be formulated as "The project of combining Fiat-Opel-Chrysler is imprudent". The following inferential procedure is activated:

A. If the project of an action cannot bring about the desired outcome, it is imprudent (major premise or *maxim*)

B. The project of combining Fiat-Opel-Chrysler cannot bring about the desired outcome (minor premise)

C. The project of combining Fiat-Opel-Chrysler is imprudent (conclusion)

Walton (1990a, 2007b) distinguishes sufficient and the necessary conditions for practical reasoning. It is clear that the capacity for an action to realize a certain goal is a necessary, though not sufficient, condition for its expediency. Therefore, being this condition absent, the action would be imprudent.

The passage reported above ("two sick people...") is intended to justify the minor premise B: impossibility for the project to be realized. Piëch shows to adopt an argument from analogy even though the inferential procedure looks rather obscure and is clearly questionable. In order to evaluate analogical reasoning, for example, by means of critical questions (cf. (Walton 2005e, Christopher-Guerra 2008)), the first question that should be made concerns the real similarity of the two compared realities, namely physical sickness and corporate financial troubles.

Indeed, very often, financial/economic crisis as well as political and social crisis, are read in terms of illness, of disease, where the patients correspond to the concerned contexts in trouble — ranging from the market, to the country, from the industrial sector to single firms — and the failure of these organizations is frequently identified with their death.[5] It is worthwhile here to quote an article dealing with well-known financial failures, written by one of the leading scholars in financial economics:

> "In finance, as in pathology, we can learn more from failure than from success. This lecture examines three famous financial failures, Metallgesellschaft's oil futures business, LTCM and related hedge fund failures, and the current travails of ENRON, and performs a post mortem on each to see what can be learned. Not surprisingly, the cause of death was similar in each case, or, to put it more familiarly, history always repeats itself" (Ross 2002, pp. 9-27).

It is indeed a case of partial and not total similarity (which could occur only in cases of proper isomorphism); in such cases, there are some aspects (properties) for which the analogical principle does work and some other aspects for which it

---

[5]Rare manifestations of disagreement about the large use the quasi-medical reading of social crises in terms of illness appear also in commentaries made by illustrious columnists of the financial press. For example, consider the opinion presented by Paul Ingrassia — specialist in the car industry and winner, in 1993, of the Pulitzer Prize for Beat Reporting — about the trouble of Detroit's big three automakers, in which the identification of the companies' possible bankruptcy with their death was strongly criticized (as bankruptcy would have given the opportunity to restructure the companies in a substantial way). In this case it is clear that human death would not be totally analogous to corporate failure (see "The Auto Makers Are Already Bankrupt", WSJE, Nov. 21, 2008).

does not work[6]. For example, if one says that the European Union is like a family, then we can reasonably conclude that, as in family brothers are expected to help each other, in the European Union too member States are expected to help each other. But, from the fact that in family, along the time, all members get old and die, we could not come to the conclusion that the member states of the European Union are also expected to get old and die.

But Piëch's reasoning suffers from another, more trivial, logical vice: in exploiting the physical illness/financial trouble comparison, he refers to a wholly improbable scenario presupposing that, if two or three healthy people are in the same bed, they would make a healthy individual. In other words, the state of affairs of the pathological context from which Piëch wants to infer conclusions regarding the actual financial context is in itself not capable of making sense.

7.3 Activating the principle of charity for a maximally argumentative reconstruction

In a more charitable reconstruction, Piëch's reasoning is to be reinterpreted as a metaphorical presentation of an argument from the parts to the whole directly concerning the economic-financial context.

Thus, we can translate this reasoning as follows:

> Two sick people in a double bed, or even three ones, don't make one healthy person =
> **The combination of two or three troubled companies cannot produce but a troubled company.**

Let us represent the essential ingredients of this argument with the Y-structure (see below) proposed by Rigotti within a recent re-elaboration of the ancient doctrine of Topics (see (Rigotti 2006, 2009, Rigotti and Morasso 2009, Rigotti et al. 2006, Rigotti and Morasso (to appear)).

The reasoning procedure underlying the argumentation is developed on the right line, which starts from a maxim, i.e. an inferential connection generated by an ontological relation (locus), that activates a logical scheme leading, through a minor premise, to the final conclusion, the latter corresponding to the standpoint. On the left line, we find the argument's material component, i.e. premises that must be shared by the co-arguers in order to ascertain the minor premise of the procedural line, and justify the final conclusion. Of course, only if the conditions

---

[6]Walton and Macagno (2009, p. 158) discuss the argument from analogy by remarking that "two subjects can be considered analogous even though they are completely different, provided that the two subjects can be subsumed under a common functional genus". In our opinion, we can speak of functional genus when the two concerned subjects, though for many aspects thoroughly divergent, feature the same properties in relation to the issue at stake. In this respect, another argumentation developed in US automaker crisis represents indeed a case in point. An analogy between the Katrina hurricane in New Orleans and the automakers' crisis in Detroit was exploited in order to justify the Federal Government's aid for the latter. Both events should be interpreted as calamities (or "acts of God": events not imputable to human responsibility). The following inferential connection (maxim) is activated in such type of reasoning: "If X presents a set of features also present in Y and justifying Z for Y, then Z is justified for X too". In relation to this argumentation, if non-imputableness to human responsibility, seriousness and need for major support reasonably justified the Federal Government's aid for Katrina, then they wholly justify the Government's aid for Detroit automakers' crisis. Not by chance, the arguer gathers within the same functional genus Katrina and the automakers' crisis, as different species of hurricanes, distinguishing natural and economic hurricanes. This reasoning can assume a presumptive nature in relation to the questionable nature of the endoxon implementing the maxim: indeed, it is far from being ascertained that the automakers' crisis is not imputable to any human responsibility. Following a critical commentator like Paul Ingrassia, "Hurricane Katrina was an act of God. The car crisis is an act of man" (see "The Auto Makers Are Already Bankrupt", WSJE, Nov. 21, 2008).

imposed by the maxim are correctly applied by the material premises and these premises are true, the conclusion can be correctly drawn.

The minor premise is the result of a syllogistic procedure whose major and minor premises correspond to two specific types of material premises: the major premise represents an endoxon, i.e. is a statement based on an opinion generally accepted within a certain field of interaction (the financial system, a university, a country, etc...); the minor premise (not to be confused with the minor premise of the procedural component on the right side) is a datum, i.e. a factual assertion presented as specifically related to the particular situation of discussion.

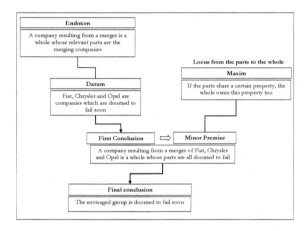

Figure 7.1: Argument analysis 1

Piëch's argument deals with the ontological relation of whole and parts: it involves the relation between a reality in its whole and its constituents. This ontological relation is called *locus from the parts to the whole* (*locus a partibus ad totum*).[7] The maxim exploited by Piëch sounds as "if the parts share a certain property, the whole owns this property too". In order conclude that the parts of the group Fiat-Chrysler-Opel will fail very soon, we need to know that in the business context a merger entails the consolidation of two or more companies (parts) into a single entity (whole) and recognize the fact — the datum — that the parts of the envisaged group, namely Fiat, Chrysler and Opel, are doomed to fail soon.

The Y-structure (so named because its form recalls that of the letter "y") can be used not only for the reconstruction of the argument but also for its analysis and

---

[7] Peter of Spain (Summulae Logicales, 5.7;5.14-5.23; in particular 5.14-5.18) distinguishes between totum universale and totum integrale; the totum concerned in the latter case is conceived as the result of an operation of integration in which the components, though pre-existing, do not necessarily maintain their original functional position. Consider, for example, the components of a car and, in general, of a machine, or the member states towards EU.

evaluation. Each box may be submitted to critical questions (cf. (Walton 2005e, Christopher-Guerra 2008)). The Y-structure allows a critical assessment of the argument's whole soundness, inclusive of logical validity as well as relatedness to context and common ground information.

We should evaluate the validity of the inference at work by considering under what conditions the property of the parts can be transferred to the whole and vice versa. In this respect, van Eemeren and Grootendorst (1999) showed that the possibility of transferring properties is limited to structure-independent and non-relative properties, as the table 7.1 summarizes.

Transferable (+) and non-transferable properties	structure-independent properties (2a)	structure-dependent properties (2b)
Absolute properties (1a)	red, white, blue, glass, iron, wooden (+)	round, rectangular, edible, poisonous (-)
Relative properties (1b)	heavy, small, light, big, fat, slim (-)	good, expansive, strong, poor (-)

Table 7.1: (source: (van Eemeren and Grootendorst 1999))

It is clear that relative properties cannot be transferred from the parts to the whole. For example, the fact that the parts of a hay bale (the blades of hay) are extremely light does not entail that the hay bale is light. In this case, the maxim stated above could not be applied and whoever invokes it, by saying that since the parts (the blades of hay) are light, the whole (the hay bale) is light too, would commit the so-called *fallacy of composition*. Analogously, the opposite reasoning is also invalid (*fallacy of division*). The evidence that a big dictionary is heavy does not entail that each single page of it is heavy too. Therefore, our maxim holds only for absolute properties, though not all of them allow a correct use of this logical connection. The property dealt with must also be structure-independent in order to ensure to our maxim its validity. Let us consider this figure (cf. (van Eemeren and Garssen 2009)):

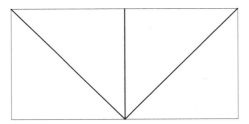

The figure as a whole is doubtlessly a rectangle but it is composed of four right triangles. It is clear that the properties of the parts (being triangle) are not transferred to the whole, which is a rectangle.

In conclusion, only absolute/structure-independent properties can be transferred from the parts to the whole, making in this way our maxim valid. Examples of this type of properties are colors ("if the parts are red, the whole is red") or materials (if the legs and the board are wooden, the table is wooden) .

Now, what is the property invoked by Piëch when he states the imprudence of the combination of Fiat, Opel and Chrysler? According to VW's CEO, the financial sickness, the trouble, characterizing the three automakers will negatively determine the group resulting from the combination, which, still in Piëch's view, is condemned to suffer. Thus, the property that would be transferred from the three automakers (the parts) to the combined group (the whole) is *being in trouble*, or, following Piëch's medical metaphor, financial disease (unhealthiness).

At a first sight, it seems clear that if the members of a group are all in trouble, the whole will necessarily suffer. How can one person help another one to solve a problem if she suffers from exactly the same problem? In more concrete terms, if John and Jim can lift maximum 50kg each and this object weights 60kg, John cannot lift the object on behalf of Jim. However, John can certainly help Jim in lifting the object if the two combine their strengths and lift together the object. In this case, the physical weakness of the single men (the parts) is not transferred to the couple (the whole).

More interestingly, suppose that John can lift even 200 kg and he wants to move a table weighting 60 kg from his bedroom to the living room. Suppose also that the path from the bedroom to the living room is very twisted and requires manual abilities that John does not hold (to quote a well-known advertisement by Pirelli: "power is nothing without control") but Jane does. Jane, in her turn, cannot lift more than 40kg. It is immediately evident that John and Jane can strategically combine their respective skills (power and manual ability) in a synergic way.

Also, two blind individuals could hardly help each other in moving from one place to another but, a blind person and a deaf one might manage to do numerous things together, like Richard Pryor and Gene Wilder did in the funny movie "See no evil, hear no evil".[8]

This situation is analogous to the world of corporate mergers. Mergers, in fact, are intended to create synergies and more in general to exploit the particular strengths of the companies involved so that the resulting new company is more valuable and profitable than the sum of the companies' standalone value (2+2=5). In this case only the combination of single components can produce an efficient whole. But, of course, every business combination turns the risk of destroying value rather than creating it (cf. (Bruner 2005)).

Therefore, it should be established whether the Fiat-Opel/Chrysler combination is potentially able to create synergy or not. The Fiat-Chrysler alliance has been motivated by this existence of this potential. The idea behind the combination is that Chrysler, which without the strategic alliance with Fiat would have gone bankruptcy,[9] might improve the quality of its products by exploiting Fiat's technology, while Fiat, beside getting part of the US bailout injection, might have the opportunity to expand its business overseas, in the USA. However, we don't have information for establishing where the expected synergies from the alliance with Opel would come from. Indeed, this is exactly the point that Piëch seems to insist on when he says that:

> "die unterschiedlichen Unternehmenskulturen könnten ein Hindernis für eine erfolgreiche Allianz sein".

---

[8] Another similar example comes from the 11 Oscar-awarded movie Ben-Hur. When Judah Ben-Hur makes his return from the galleys, he finds Simonides, his old loyal steward, that who was imprisoned and beaten so much that he lost the use of his legs. Simonides lives with Malluch, a dumb man met in jail. Seeing Ben-Hur worried for his conditions, Simonides ensures him:
I am twice the man I was [...] I and Mullach were released the same day. Since then I have been his tongue and he has been my legs. Together we make a considerable man!

[9] At the beginning of 2009 Chrysler was near to bankruptcy and desperately needed a government bailout in order to survive. The US government decided to grant the financing at the condition that Chrysler accepted the deal with Fiat. The deal foresaw a common strategic plan and the immediate purchase by Fiat of a 35% stake in Chrysler, which in the future could increase toward a controlling ownership.

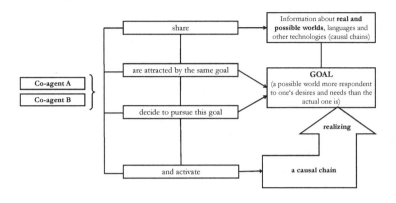

Figure 7.2: Ontology of cooperation

("the different corporate cultures may represent an obstacle to a successful alliance")

The statement by Piëch refers indeed to a typical and recurrent problem in all mergers, especially when decisions concern cross-border mergers. More particularly, past experience shows a peculiar relevance of this issue in the German context.[10]

The argument relying on differences in the corporate culture, though not having absolute validity, is by the way relevant. The ontology of cooperation, with the conditions for the construction of cooperation, is referred to (see figure below): the effectiveness of the combination of different systems of knowledge and technology regarding a reasonable management of the causal chain and, more importantly, the compatibility (entailing non competitiveness) of goals (vision and mission).[11]

In his argumentation supporting the standpoint of imprudence of this alliance bound to the impossibility of success, Piëch eventually develops another complex argument: VW's experience of merger with Audi (*locus from example*) shows that a long time (15 years) is taken in order to realize a successful merger. Now, according to Piëch, the involved companies have not this time at their disposal because their diseases and their cultures' incompatibility will kill them out

---

[10]It is worthy to recall, in this relation, the controversy that accompanied the Vodafone-Mannesmann's takeover [1999-2000], bound to the typically German principle of co-determination (Mitbestimmung) (cf. (Nowak 2001, Höpner and Jackson 2001, 2006)).

[11]The activation of a causal chain brings about the transition of a possible world into a real one. As one of the reviewers remarks, the co-agents activate in the causal chain those items of shared information and those shared technologies that are requested for the realization of the pursued goal

sooner (*locus from time*). In other words, their unhealthiness and their intrinsic differences are an insuperable obstacle to endurance, which is necessary for realizing a successful integration leading to a competitive company.

7.4   Integrating the argument scheme into the argument structure

We see that numerous arguments are advanced by Piëch. It is thus expedient at this point to make a analytical overview of the argumentative interaction in order to clarify the function of each of the advanced arguments and how these arguments are organized for supporting the arguer's standpoint. According to van Eemeren, et al. (1993), in order to reconstruct a text as an argumentative discussion, the analyst has to identify:

- the issue under discussion,

- the parties involved in the critical discussion — in terms of protagonist and antagonist — and their standpoint in relation to the issue,

- the arguments advanced in support of the standpoint,

- the structure in which the arguments are organized.

In the case discussed by this paper, the issue concerns the potential threat that the possible Fiat-Chrysler-Opel alliance would represent for VW (and Piëch). Piëch is clearly the protagonist of the standpoint that a Fiat-Chrysler-Opel alliance would not be a threat for VW, which is equivalent to say that *Fiat-Chrysler-Opel will not become competitive*. We could consider the journalists, and the public opinion more generally, as antagonists who virtually cast doubts about this view. By accepting to reply the question, Piëch is acknowledging the existence of the issue and assumes the burden of proving his specific position. Following the method suggested by Pragma-Dialectics (cf. (van Eemeren et al. 2002)), the following figure represents the argument structure showing which arguments support the standpoint and how these arguments are organized and combined in order to accomplish this task (see also (Henkemans 1997)).

As previously mentioned, the main reason given in support of this claim is that the envisaged group will not endure enough time to become competitive. Thus, a *locus from time* grounds the standpoint at the first layer. This argument, however, presupposes that:

1. a period of 15 years is required for creating a competitive group from a merger

2. Fiat-Chrysler-Opel have not this time at their disposal because they will fail soon.

The acceptability of the two statements is not taken for granted by Piëch who advances further reasons in order to prove each of them. Thus, as the figure shows, a second layer of argumentation must be introduced.

The argument belonging to the *locus from time*, directly subordinated to the standpoint, is reconstructed in figure 7.4 through the Y-structure proposed in the AMT.

As said, VW's Chairman did not limit himself to this move but committed himself to ground both the endoxon and the datum. Therefore, it is not properly adequate to speak of endoxon — which, being by definition already shared, at

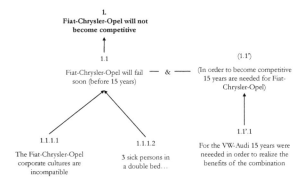

Figure 7.3: Argument structure

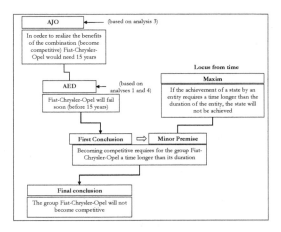

Figure 7.4: Argument analysis 2

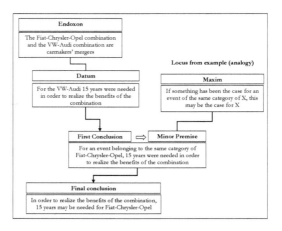

Figure 7.5: Argument analysis 3

least in the arguer's views, needs not to be proved — and of datum — which is by definition a piece of evidence. They instead represent an *argumentatively justified opinion* (AJO) and an *argumentatively established fact* (AED) respectively. In this way, they become sub-standpoints, which require further arguments in order to be accepted.

The major premise, stating that a period of 15 years is needed in order to create a competitive combined company, is justified through an example of the VW-Audi combination, whose Y-reconstruction is reported below (argument analysis 3, figure 7.5). However, the reference to a single example is capable of showing that something is possible, not that something will necessarily take place.

The minor premise is certainly the most critical point. Piëch, in fact, proposes a multiple argumentation intending to show that the combined group will fail before 15 years. We speak of multiple argumentation (van Eemeren et al. 2002, Henkemans 1997, 2000) when more than one argument is put forth in support of the standpoint and these arguments are independent from each other, meaning that each of them could individually support the standpoint. This does not mean that the individual argument is sound but only that it works regardless of the others, i.e. there is no strategic combination but only an addition of reasons. We speak of coordinative argumentation when the *conjunction* of two or more arguments supports the standpoint[12] (For instance if I say that investors must prefer stock A to stock B because A has a higher expected return and because A is less risky than B. In fact, looking at the higher expected return is not a sufficient criterion for a rational investment decision as the level of uncertainty (risk) bound

---

[12]For a systematic treatment on the difference between multiple and coordinative argumentation, see (Henkemans 1997)

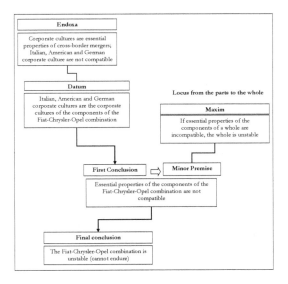

Figure 7.6: Argument analysis 4

to this return must be considered as well).

In our case, the two arguments concern the bad state of the three companies that would be transferred to the whole combined group (argument analysis 1, figure 7.1) and the incompatibility of the respective corporate cultures (argument analysis 4, figure 7.6). Both invoke the locus from the parts to the whole, though activating two different maxims generated by it.

The argument based on the difference of corporate cultures is not at all trivial, even though it presupposes an *endoxon* — a sort of absolute incompatibility between Italian, American and German corporate cultures — that might be questioned. A huge literature studying post-merger integration argues that the differences in corporate culture, if not adequately tackled, can strongly prevent the realization of the prospected benefits. These considerations highlight the importance, beyond shareholders and managers, of other corporate stakeholders, such as employees and customers (Bastien 1992, Cartwright and Cooper 1993).[13]

The four y-structured arguments can be intersected, producing the following complex figure that summarizes the whole argumentative organization of Piëch's speech. This figure recovers the argument structure, maintaining the explicit formulation of the single arguments and all their relevant premises:

---

[13]The role of stakeholders in corporate mergers varies across countries. In Europe, and in particular in Germany (see note 11), where companies are governed by the principle of co-determination, the influence of stakeholders seems to play a more relevant role (cf. (Degeorge and Maug 2008) for a review of European corporate finance).

Figure 7.7: combined Y-structure

7.5   Concluding remarks

This paper has illustrated the reconstruction and evaluation of a complex argumentation developed within a specific social context.

Figure 7.7 represents a first attempt to integrate the reconstruction of the argument structure with the reconstruction of the argument scheme proposed by AMT. This relation should be specified more in detail; in particular we should verify the hypothesis that the argument structure only indicates the conjunction of two material starting points, which the AMT indentifies as endoxon and datum, whereas the underlying maxim is in general left implicit, though representing an essential component of the inferential structure of argument.

It should be verified whether the *loci* identified by the analysis of this specific case study — or at least some of them — are characteristic of other similar argumentations belonging to the economic-financial context. The *locus from analogy*, for example, seems to be frequent in financial argumentation (but also the *appeal to expert opinion*, especially in the banking context). The typical decision-making concerning a certain investment is often based, among several factors, on the performance of *similar* investment strategies, or on the past performance of the *same* investment strategy. To what extent information on the past can help predicting the future is object of strong debates in financial theory, dividing in particular those who support the efficient market hypothesis from those who question this view and propose a behavioural approach to explain financial decision-making.[14] In any case, the analysis of this kind of argument also in its verbal manifestation

---

[14] The Efficient Market Hypothesis (Fama 1967, 1970) argues that securities' market prices fully reflect all available information. If this hypothesis is true, using information on past events for making profit would be useless because, in an efficient market, the price of the concerned asset would have already included this information. The EMH is based on some strong assumptions, stating in particular that investors are fully rational and that information at their disposal is complete, perfect and symmetric. Since 1970s, the EMH started to be questioned by the behavioural finance approach, which proposes

(and not only in its mathematical modelling) can contribute in understanding how financial reasoning typically works and, when fallacies emerge from the study of financial failures[15], may explain more precisely the causes of such unsuccessful — and often damaging — business activities.

The realm of finance is often wrongly reduced to figures and charts, overlooking the fact that economics is a social science dealing with a typically human activity, and that financial decisions are taken by people who do not process formal data like machines, but reason through concepts belonging to ordinary language. Dramatic events, like the current crisis, always remind us this aspect and — why not — suggest that the understanding of finance and the realization of financial activities may improve through the integration of argumentation both at the theoretical and practical levels. Of course, the path toward this objective is still very long.

---

models for financial decision-making in which the EMH's extreme assumptions are relaxed (cf. (Shleifer 2000). The controversy within the financial literature is still open and today there is not yet definitive agreement about market (in)efficiency.

[15]Some initial indications in this direction have emerged from the study of real cases like the bankruptcy of Enron (Palmieri 2007); the sale of illegal financial products by an Italian bank (Palmieri 2006); and the well-known failure of LTCM (see Rigotti & Palmieri, Financial argumentation 2009. Argumentum elearning module. www.argumentum.ch). A lack of reasonableness, understood in particular as an approach to reality that exceeds mere rationality and takes into account all the relevant factors and the concreteness of the situation (Rigotti 2009, 2006), is typical of these financial debacles. Behind this seems to lie a dangerous view of human rationality, reduced to maximization of own wealth and blind application of abstract models for resolving problems (Martini et al. 2008).

Part II

Schemes and Fallacies

CHAPTER 8

ARGUMENTATION SCHEMES: FROM INFORMAL LOGIC TO COMPUTATIONAL MODELS

Trevor Bench-Capon and Katie Atkinson

8.1   Introduction

Two of Douglas Walton's ideas have had a particular influence on modelling argumentation in Artificial Intelligence. One is the notion of dialogue types, developed with Erik Krabbe (Walton and Krabbe 1995), and the other, which will be the topic of this paper, is the notion of argumentation schemes as presumptive justifications, subject to critique using a number of questions characteristic of the scheme (Walton 1996b) (Walton et al. 2008).

Walton's notion of argumentation schemes developed out of his long standing interest in fallacies. In particular there is a need to account for the fact that many of the well known fallacies often seem to be used quite properly to support positions in everyday argumentation. Thus, although fallacies such as the Argument from Ignorance, Argument from Expert Opinion and various forms of abductive argument, are strictly speaking logical fallacies, they also seem, in the right circumstances, to be accepted as justifying their conclusions. Thus the fallacy can be seen not so much in the form of the argument, but rather in the improper use of the argument. So, the notion of argumentation schemes was developed in order to explain the proper use of such arguments: argumentation schemes represent stereotypical patterns of reasoning which can presumptively support conclusions when used properly, but which have also the possibility of being fallacious when improperly used.

For proper use, first it has to be recognised that these arguments justify their conclusions only presumptively: anyone using such an argument must, when challenged, be prepared to offer further justification or else withdraw the conclusion. Second it has to be accepted that the conclusion depends on a number of assumptions, characteristic of the scheme. While these assumptions can legitimately be made, they need to be justified if questioned in the context in which the scheme is deployed. For example, if an argument from expert opinion is used, it is assumed that the expert was making a sincere, unbiased pronouncement on a topic within his field of expertise. The conclusion must therefore be withdrawn if reasons to think that the expert is biased can be produced: for example that his research was sponsored by a manufacturer with a vested interest in his conclusions.

The primary use for these schemes was for the analysis of naturally occurring arguments. Given a piece of prose setting out an argument, the text could be broken up into individual arguments. Where these represented instantiations of argumentation schemes, they could then be considered, using the critical questions characteristic of the particular schemes, to see whether they were acceptable, or whether they had failed to address the critical questions sufficiently to discharge the burden of proof. In this form they proved a useful tool for informal logic. With regard to AI, the first use of argumentation schemes was in argument diagramming systems. Argument diagrams had originally used Toulmin's scheme (Toulmin 1958), as in e.g. (Marshall 1989), but the richer range of schemes proposed by Walton gave a greater scope which could be exploited in more general argument visualisation tools such as Araucaria (Reed and Rowe 2004). Computational modelling of argument is not, however, limited to analysis of argument – there is also a desire to generate arguments from some underlying knowledge model. Since natural argumentation is not restricted to logical deduction, the use of argumentation schemes gave promise that a richer repertoire of arguments based on these schemes could be generated, extending the scope of computational argument. For example, the sufficient condition scheme for practical reasoning was used in (Atkinson et al. 2006b) to provide a computational model of justifications for actions: this particular line of work will be further discussed in section 3. More generally, the Carneades system (Gordon et al. 2007) explicitly uses argumentation schemes as the main driver for its inference mechanism.

In this paper we will consider the use of argumentation schemes for the computational generation of arguments. In section 2 we will step back a little and consider some practices in AI developed independently of the informal logic work on argumentation schemes, but which share some of the motivations and characteristics. In section 3 we will discuss the adaptation of a particular scheme for computational use, and indicate some of the clarifications and decisions that are required if this is to be done successfully. In section 4 we will look at some other schemes, those related to Position to Know, and see how far they meet, or can be made to meet, the requirements for computational use. In section 5 we offer some concluding remarks.

8.2 AI Before Argumentation Schemes

Just as informal logic recognised that sound deductive reasoning was insufficient to analyse the whole range of legitimate arguments that can be used, so too AI found that it was necessary to go beyond deduction and make use of similar reasoning patterns in order to be able to derive the kind of conclusions it desired.

In classical Logic Programming, the need to use Horn Clause logic to make the computation tractable posed some difficulties for representing negation, since the heads of such clauses were positive literals, meaning that we could never conclude that something was *not* the case. The solution was to extend Horn Clauses with negation as failure, so that if a particular fact could not be demonstrated from the program, it was taken to be false. This is, of course, simply a form of the Argument from Ignorance:

I do not know that X is the case, so X is not the case

and is in general fallacious. In order to be a legitimate justification for X, the assumption that *if X were the case, then X would be known*, has to be satisfied. In Logic Programming this was addressed through the Closed World Assumption, which effectively assumes that everything, or at least everything of concern to the program, is known to the program. While this assumption is not generally true, the assumption can be made under circumstances formalised in (Clark 1977). What Clark proposes is the *completion of the database*, arguing that negation as failure may be used soundly for some relation R if the procedure for R is complete (a procedure for a relation is the set of clauses with the relation as head, and completeness means that no other clauses for that relation exist: i.e. the clauses constitute individually sufficient and collectively necessary conditions). If the procedure is complete, then the clauses of the procedure can be seen as individually sufficient *and collectively necessary* conditions, so that if none of the clauses are satisfied, the negation of the head is logically justified.

This aspect of logic programming bears striking similarities to Walton's notion of argumentation schemes. A style of reasoning which is potentially fallacious is used, but conditions under which it can be regarded as legitimate are provided. We could see the Closed World Assumption as a critical question: *can we be sure that the clauses of this procedure collectively supply a necessary condition for the truth of the predicate?*

Another common form of system dating from the 1980s is the expert system which operates by using backward chaining on a set of observations to find an explanation. This explanation is only *the* solution if the system has all possible explanations in the knowledge base. This can be seen as using an argumentation scheme, Argument from the Best Explanation, but relies for its legitimacy on there being no other explanations. Again therefore, this requires the Closed World Assumption to hold, and so is again subject to the critical question proposed above. Note here that the role of the Closed World Assumption is different in that it is required even if the system makes no use of negation.

For a third argumentation scheme relating to the logic programs and expert systems of the 1980s, consider the use of Query the User (Sergot 1984). Often an expert system will gather facts from the user, and accept these facts unquestioningly. The justification of these facts, therefore, is simply that the user stated them. This might be seen as use of an argumentation scheme akin to Argument from Position to Know, since the user is assumed to be able to answer the questions correctly. It does not, however, satisfy the critical questions for such a scheme: there is no reason to believe that the assumptions of that scheme are met, and no assurance that the user will be sincere. It is perhaps better to regard the justification as being an Argument from Prior Commitment, *you should believe this because you told it to me*, which at least should have persuasive force for the user who supplied the information.

We can see therefore that these early programs were effectively making use of argumentation schemes – exploiting potentially fallacious patterns of reasoning in order to produce conclusions which could be justified even though not logically inevitable, subject to certain conditions being satisfied. Note, however, that these conditions – the analogue of critical questions – are couched in terms which are very specific to the knowledge model being used. This is essential if they are to be given a precise and formal statement suitable for use within a computational model. The informally expressed critical questions associated with the schemes are not, however, completely avoided. For example, with respect to the Argument from Ignorance, we can pose critical questions such as *Has an appropriate effort to find the information been made?* and *Is this the kind of knowledge which can be said to be complete?* These questions are properly directed at the designer of the program: the program is acceptable only if the claim that the procedure is complete can be made, and this requires proper care on the part of the designer, and that the Closed World Assumption be applied only to predicates for which complete knowledge is possible. Thus a user will only accept the conclusions using these argumentation schemes if there is confidence that the critical questions have been considered and resolved in the design of the system.

The lessons from this are first that it is very natural to use argumentation schemes in computer systems: similar problems were being encountered and similar solutions proposed independently in both AI and informal logic. The contribution of Walton's notion of argumentation schemes is that it gives a better rationale for the common AI practices. The second lesson is that informal analysis of argument is not required – and because it is designed to operate on natural language cannot reasonably be required – to give very precise and rigorous definitions of the schemes and the questions. In contrast, in computational contexts, because the schemes are operating on a well defined knowledge model, it is both possible and necessary (if they are to be applied by the program) to provide precise definitions in terms of the underlying logical model. This is clearly shown by a comparison of the various formulations of Argument from Ignorance and the specification of the Closed World Assumption for logic programs. In the next section we will give an extended case study of the adaptation of an informal argumentation scheme for computational use.

8.3   Development of the Sufficient Condition Scheme for Practical Reasoning

Argumentation schemes can be used both for reasoning about beliefs and for reasoning about actions, known as *practical reasoning*. Philosophical work on practical reasoning has centered on the practical syllogism introduced by Aristotle, but is regarded as somewhat problematic because it is essentially abductive and thus potentially fallacious. Walton's notion of argumentation schemes offers a solution to these difficulties, and in (Walton 1996b) he offers two argumentation schemes for practical reasoning: the sufficient and the necessary condition schemes for reasoning about action. These schemes treat the practical syllogism not as deduction, but as a presumptive argumentation scheme. Our discussions concentrate on the sufficient condition scheme which is as follows:

> W1: G is a goal for agent a
> Doing action A is sufficient for agent a to carry out goal G
> Therefore agent a ought to do action A.

Walton associates with this scheme four critical questions:

> CQ1: Are there alternative ways of realising goal G?
> CQ2: Is it possible to do action A?
> CQ3: Does agent a have goals other than G which should be taken into account?
> CQ4: Are there other consequences?

This scheme is perhaps sufficiently precise for analysing informal arguments, where the goal can be interpreted as the context demands. But, as we have previously argued in (Atkinson et al. 2006b), the notion of a goal here is ambiguous, potentially referring indifferently to any direct results of the action, the consequences of those results, and the reasons why those consequences are desired.

Consider someone who wishes to travel to London to meet a friend, John, who will shortly be leaving the country. Using W1 any of the following three arguments could justify his action:

> W1a: I wish to be in London
> Going to London is sufficient for me to be in London
> Therefore I ought to go to London.

> W1b: I wish to see John
> Going to London is sufficient for me to see John
> Therefore I ought to go to London.

W1c: I wish to maintain my friendship with John
  Going to London and meeting John is sufficient for me to maintain my
    friendship with John
  Therefore I ought to go to London (and meet with John).

That the distinction is important can be seen by the objections that can be
made against the different arguments. 'John is in Manchester' is an objection
to W1b and W1c but not W1a. Similarly telephoning John is an alternative for
W1c but not for the other two. Moreover we need to establish which of these we
mean for the purposes of mapping into our knowledge model: if we want W1c,
then we must explicitly represent meeting John in our state, whereas for W1a it
is sufficient to allow it to be a consequence of the propositions true in the state.
The representation of the goal of W1b also has implications for the knowledge
model.

We therefore see uses of W1a-c as enthymemes for a compete argument along
the lines of *I want to be in London to meet John in order to maintain our friendship:
going to London is sufficient to achieve this, so I ought to go to London.* Accordingly,
in (Atkinson et al. 2006b) Walton's scheme is developed into the more elaborated
scheme:

AS1 In the circumstances R
  we should perform action A
  to achieve new circumstances S
  which will realise some goal G
  which will promote some value V.

What this scheme does in particular is to distinguish three aspects which are
conflated into the notion of goal in Walton's scheme. These aspects are: the *state
of affairs* which will result from the action; the *goal*, which is those aspects of the
new state of affairs for the sake of which the action is performed; and the *value*,
which is the reason why the agent desires the goal. As indicated by the example
above, making these distinctions opens up several distinct types of alternative
to the recommended action. We may perform a different action to realise the
same state of affairs; we may act so as to bring a different state of affairs which
realises the same goal; or we may realise a different goal which promotes the
same value. Alternatively, since the state of affairs potentially realises several
goals, we can justify the action in terms of promoting a different value. In coming
to agreement this last possibility may be of particular importance: we may want
to promote different values, and so agree to perform the action on the basis of
different arguments. Furthermore, different agents may have their own different
values that they want to promote leading each to propose different actions to
achieve a goal.

For an example suppose that Trevor and Katie need to travel to Paris for a
conference. Trevor offers the argument "we should travel by plane because it is
quickest". Katie replies with the argument "we should travel by train because
it is much pleasanter". Trevor and Katie may continue to disagree as to how
to travel, but they cannot deny each other's arguments. The conclusion will be
something like "we should travel by train because it is much pleasanter, even
though travelling by plane is quicker". Because two people may have different
preferences, values, interests and aspirations, people may rationally choose different options: if Katie prefers comfort to speed she will rationally choose the
train, but this does not mean that Trevor cannot rationally choose the plane if he
prefers speed to comfort. To relate this back to the AS1 scheme, we can through
this example see the distinction between a goal and a value. The goal is to be in

Paris for the conference, and this is not in dispute: the dispute is how that goal should be realised and turns on the values promoted by the different methods of travel. What is important is not the state reached, but the way in which the transition is made from the initial state to the goal state.

Starting from the definition of the AS1 scheme, we can make a systematic effort to identify critical questions. We can question the various elements of the scheme, for example, whether the action is possible, or whether the value is something worthy of promotion. Also we can question the connections between elements: does the action bring about the state; does the the state realise the goal, etc. Finally we can consider alternatives in a more articulated way: we can realise the goal by reaching a different state, or promote the value by realising a different goal. In this way we can hope to determine a complete list of critical questions. As given in (Atkinson et al. 2006b), AS1 has associated with it sixteen critical questions:

CQ1: Are the believed circumstances true?
CQ2: Assuming the circumstances, does the action have the stated consequences?
CQ3: Assuming the circumstances and that the action has the stated consequences, will the action bring about the desired goal?
CQ4: Does the goal realise the value stated?
CQ5: Are there alternative ways of realising the same consequences?
CQ6: Are there alternative ways of realising the same goal?
CQ7: Are there alternative ways of promoting the same value?
CQ8: Does doing the action have a side effect which demotes the value?
CQ9: Does doing the action have a side effect which demotes some other value?
CQ10: Does doing the action promote some other value?
CQ11: Does doing the action preclude some other action which would promote some other value?
CQ12: Are the circumstances as described possible?
CQ13: Is the action possible?
CQ14: Are the consequences as described possible?
CQ15: Can the desired goal be realised?
CQ16: Is the value indeed a legitimate value?

Thus far we have elaborated the scheme so as to give a specification of what we require. This is, however, not yet enough for computational use: we need to relate the scheme and the critical questions to an underlying knowledge model. One way of doing this would be to use a BDI model, as in (Atkinson et al. 2005), but a more natural approach is to use a state transition system. In (Atkinson and Bench-Capon 2007), AS1 and CQ1-16 were defined in terms of an Action-based Alternating Transition System (AATS), a structure originally defined in (Wooldridge and van der Hoek 2005).

For an AATS we begin with a finite set Q of possible *states*, with $q_0 \in Q$ designated as the *initial state*. Systems are populated by a set Ag of *agents*. Each agent $i \in Ag$ is associated with a set Ac_i of possible actions, and it is assumed that these sets of actions are pairwise disjoint (i.e., actions are unique to agents). The set of actions associated with the set of agents Ag is denoted by Ac_{Ag}, so $Ac_{Ag} = \bigcup_{i \in Ag} Ac_i$.

A joint action j_C for a set of agents Ag is a tuple $\langle \alpha_1,...,\alpha_k \rangle$, where for each α_j (where $j \leq k$) there is some $i \in Ag$ such that $\alpha_j \in Ac_i$. Moreover, there are no two different actions α_j and $\alpha_{j'}$ in j_{Ag} that belong to the same Ac_i. The set of all joint actions for a set of agents Ag is denoted by J_{Ag}, so $J_{Ag} = \prod_{i \in Ag} Ac_i$. Given an element j of J_{Ag} and an agent $i \in Ag$, i's action in j is denoted by j^i.

As given in (Atkinson and Bench-Capon 2007), this allows an AATS to be defined as follows:

An *Action-based Alternating Transition System* (AATS) is an $(n + 7)$-tuple $S = \langle Q,$
$q_0, Ag, Ac_1, \dots , Ac_n, \rho, \tau, \Phi, \pi \rangle$, where:

- Q is a finite, non-empty set of *states*;

- $q_0 \in Q$ is the *initial state*;

- $Ag = \{1,...,n\}$ is a finite, non-empty set of *agents*;

- Ac_i is a finite, non-empty set of actions, for each $i \in Ag$ where $Ac_i \cap Ac_j = \emptyset$ for all $i \neq j \in Ag$;

- $\rho : Ac_{Ag} \to 2^Q$ is an *action precondition function*, which for each action $\alpha \in Ac_{Ag}$ defines the set of states $\rho(\alpha)$ from which α may be executed;

- $\tau : Q \times J_{Ag} \to Q$ is a partial *system transition function*, which defines the state $\tau(q, j)$ that would result by the performance of j from state q. Note that, as this function is partial, not all joint actions are possible in all states (cf. the precondition function above);

- Φ is a finite, non-empty set of *atomic propositions*; and

- $\pi : Q \to 2^\Phi$ is an *interpretation function*, which gives the set of primitive propositions satisfied in each state: if $p \in \pi(q)$, then this means that the propositional variable p is satisfied (equivalently, true) in state q.

In order to express preferences between states, the AATS was extended in (Atkinson and Bench-Capon 2007) to include a notion of values. The idea is that a value may be promoted or demoted (or neither) by a transition between two states.

- Av_i is a finite, non-empty set of values $Av_i \subseteq V$, for each $i \in Ag$. The set of all values for a set of agents Ag is denoted by Av_{Ag}.

- $\delta : Q \times Q \times Av_{Ag} \to \{+, -, =\}$ is a *valuation function* which defines the status (respectively, promoted, demoted or neutral) of a value $v \in Av_{Ag}$ ascribed to the transition between two states: $\delta(q_i, q_j, v)$ labels the transition between q_i and q_j with one of $\{+, -, =\}$ with respect to the value $v \in Av_{Ag}$.

Given this knowledge model we can define AS1 and its critical questions in terms of it.

AS2   The initial state $q_0 = q_x \in Q$,
Agent $i \in Ag$ should participate in joint action $j_n \in J_{Ag}$ where $j_n{}^i = \alpha_i$,
Such that $\tau(q_x, j_n)$ is q_y,
Such that $p_a \in \pi(q_y)$ and $p_a \notin \pi(q_x)$, or $p_a \notin \pi(q_y)$ and $p_a \in \pi(q_x)$,
Such that for some $v_u \in Av_i$, $\delta(q_x, q_y, v_u)$ is +.

CQ1: $q_0 \neq q_x$ and $q_0 \notin \rho(\alpha_i)$.

CQ2: $\tau(q_x, j_n)$ is not q_y.

CQ3: $p_a \notin \pi(q_y)$.

CQ4: $\delta(q_x, q_y, v_u)$ is not +.

CQ5: Agent $i \in Ag$ can participate in joint action $j_m \in J_{Ag}$, where $j_n \neq j_m$, such that $\tau(q_x, j_m)$ is q_y.

CQ6: Agent $i \in Ag$ can participate in joint action $j_m \in J_{Ag}$, where $j_n \neq j_m$, such that $\tau(q_x, j_m)$ is q_z, such that $p_a \in \pi(q_z)$ and $p_a \notin \pi(q_x)$ or $p_a \notin \pi(q_z)$ and $p_a \in \pi(q_x)$.

CQ7: Agent $i \in Ag$ can participate in joint action $j_m \in J_{Ag}$, where $j_n \neq j_m$, such that $\tau(q_x, j_m)$ is q_z, such that $\delta(q_x, q_z, v_u)$ is +.

CQ8: In the initial state $q_x \in Q$, if agent $i \in Ag$ participates in joint action $j_n \in J_{Ag}$, then $\tau(q_x, j_n)$ is q_y, such that $p_b \in \pi(q_y)$, where $p_a \neq p_b$, such that $\delta(q_x, q_y, v_u)$ is –.

CQ9: In the initial state $q_x \in Q$, if agent $i \in Ag$ participates in joint action $j_n \in J_{Ag}$, then $\tau(q_x, j_n)$ is q_y, such that $\delta(q_x, q_y, v_u)$ is –, where $v_u \neq v_w$.

CQ10: In the initial state $q_x \in Q$, if agent $i \in Ag$ participates in joint action $j_n \in J_{Ag}$, then $\tau(q_x, j_n)$ is q_y, such that $\delta(q_x, q_y, v_w)$ is +, where $v_u \neq v_w$.

CQ11: In the initial state $q_x \in Q$, if agent $i \in Ag$ participates in joint action $j_n \in J_{Ag}$, then $\tau(q_x, j_n)$ is q_y and $\delta(q_x, q_y, v_u)$ is +. There is some other joint action $j_m \in J_{Ag}$, where $j_n \neq j_m$, such that $\tau(q_x, j_m)$ is q_z, such that $\delta(q_x, q_z, v_w)$ is +, where $v_u \neq v_w$.

CQ12: $q_x \notin Q$.

CQ13: $j_n \notin J_{Ag}$.

CQ14: $\tau(q_x, j_n) \notin Q$.

CQ15: $p_a \notin \pi(q)$ for any $q \in Q$.

CQ16: $v_u \notin V$.

Only now – having come a very long way from W1 – do we have an argumentation scheme capable of use for the computational generation of arguments. Of course, in producing these definitions, some interpretation of the original questions in terms of the target knowledge model was necessary[1]. Also, of course, it would have been possible to use some other underlying model. What is essential, however, is that we have a well specified model of the knowledge we will use to generate the arguments, and a well specified description of the scheme and the critical questions in terms of that model. Whereas in the analysis of informal argumentation the scheme and the questions can be interpreted (and re-interpreted) in terms appropriate to the particular context, for computational use all this needs to be fixed in advance.

Thus we can see that there are a number of steps that need to be undertaken when developing an argumentation scheme for computational use:

1. Any ambiguous terms in the argumentation scheme must be made precise. This may involve replacing one term by several, since the scope of the term appropriate to the context cannot be determined at run time;

---

[1] In (Atkinson and Bench-Capon 2007) a seventeenth critical question was added to the list to distinguish cases where an action failed due to the intervention of another agent, as permitted through the use of joint actions in the underlying AATS structure. This question could be covered by CQ2, but the use of the AATS allows us to make the useful distinction between actions which fail because of the actions of another agent and those that fail from natural causes, and so the additional question better exploits the knowledge model we are using.

2. Given the elaborated scheme, all critical questions must be identified in a systematic manner;

3. A suitable knowledge model must be selected;

4. The scheme and the critical questions must be restated in terms of the model; this may involve some further interpretation.

So far this full process has been performed for very few schemes. In the next section we will look at the use of a family of related schemes in a computational context.

8.4 Position to Know in a Computational Context

In this section we will consider a family of argumentation schemes which can be termed Position to Know schemes. Essentially all of them have the form *X is in a position to know whether P: X says that P, so P*. The different schemes arise from what puts X in a position to know that P. Some examples are:

- a period of study which has made X an expert (Argument from Expert Opinion);

- being present at a particular place and time (Argument from Witness Testimony);

- being a particular person, as when we say *I should know whether that hurts or not* (Argument from Privileged Access);

- some kind of public standing or community acceptance (Argument from Authority).

Some of these, particularly the first two, have received attention in the AI and argumentation literature (see e.g. (Gordon et al. 2007) for Expert Opinion and (Bex et al. 2003) for Witness Testimony). However, these accounts do not necessarily have the precise analysis necessary for computer generation of arguments. In (Bex et al. 2003), no specific critical questions are given; instead there is a discussion:

> The role of such critical questions has been discussed extensively in the legal literature on witness testimony and examination. Schum (1994, p. 325) has identified three requirements of the credibility of the testimony of a witness that can be questioned: (1) veracity, or whether the witness believes what she said, (2) objectivity, or whether what was reported corresponds to the event believed, and (3) observational sensitivity, or observations of linkages between events. Bromby and Hall (2002) devised a system to advise on the credibility of witness testimony by citing factors of (1) competency, (2) compellability, including the connection between the witness and the accused and any immunity the witness may have, and (3) reliability, which includes position to know factors. There remain many fine points to be clarified.

The kind of considerations mentioned here are quite appropriate for human use on an informal instance (and for the kind of analysis undertaken in (Bex et al. 2003)), but would be less suitable for computer realisation. This is perhaps unsurprising, since determining whether someone is telling the truth is an advanced social skill, and one in which there are few very reliable performers.

Expert Opinion may be more tractable. In (Gordon et al. 2007), Walton's definition is first cited:

> *Major Premise.* Source E is an expert in the subject domain S containing proposition A.
>
> *Minor Premise.* E asserts that proposition A in domain S is true.
>
> *Conclusion.* A may plausibly be taken as true.

Six critical questions are given:

1. How credible is E as an expert source?

2. Is E an expert in the field that A is in?

3. Does E's testimony imply A?

4. Is E reliable?

5. Is A consistent with the testimony of other experts?

6. Is A supported by evidence?

Later in (Gordon et al. 2007), when considering the representation of this scheme for use in the Carneades system, the critical questions are recast as:

- Premise. E is an expert in the subject domain S containing the proposition A.

- Premise. E asserts A.

- Assumption. E is a credible expert.

- Exception. E is not reliable.

- Exception. A is not consistent with the testimony of other experts.

- Assumption. A is based on evidence. Conclusion. A.

The authors of (Gordon et al. 2007) do not specify any form of the knowledge base, but one must assume that either there is a knowledge base which allows the required propositions to be derived, or there is some means of querying the user to determine, for example, whether E is reliable. To generate arguments in a specific application, however, we believe that there is rather more work to be done, so that notions such as reliability can be specified with sufficient precision in terms of the knowledge model to be used by the application.

Considered as a general problem, addressing the above issue would be a daunting task. In particular applications, however, things may in fact be easier. Firstly a computer application is likely to use specific resources, rather than trawling a large number of unspecified resources to find support. Suppose I want to argue that Peter Lever played more cricket matches for England than Paul Allot. I can do so on the basis that Lever played 17 times for England and Allot only 13. To justify these claims I can cite the expert opinion of www.cricinfo.com, which is endorsed by Wisden and considered completely decisive for cricket matters. This site is credible, reliable, based on evidence, and if anyone disagrees with it, *tant pis* for them. Effectively I can use the Argument from Expert Opinion here because I can be confident that the critical questions are not problematic. Similar definitive resources exist in other areas, such as www.imdb.com for facts about films and actors. If, however, the application is to use some less authoritative source, such as Wikipedia, the same confidence cannot be assumed. Here we may well wish to make some effort to answer at least the question of consistency with other sources, and find some corroborative information, or make some effort to discover conflicting opinion.

Suppose then we wish to use specific Internet resources to solve questions posed to us. In order to do so we will need to rely on some analogue of Argument from Expert Opinion, call it Argument from Internet Resource. Use of such an argument is sound only if the chosen source has the right credentials, that is, if the designer who has selected this resource is satisfied with respect to a set of critical questions similar to those posed against the Argument from Expert Opinion. It is essential that the chosen source be reliable, credible, based on evidence, and be believed preferentially in cases of disagreement. The justification, in terms of the diligence of the designer, is thus very similar to the justification of the use of the Argument from Ignorance in logic programs as discussed in section 2.

In Multi-Agent Systems, where an agent will be told things by other agents, we would need to consider justifying acceptance of information received in this way using a scheme akin to that of Witness Testimony. In many current proposals, the problem is circumvented by the assumption of benevolence, in which agents can be assumed to be sincere and reliable. This is relatively plausible for closed systems, but less so for open systems, where it is dangerous to assume anything about other agents. This is a real problem for open multi-agent systems, and one which would doubtless benefit from hard thinking about how the insights from the Witness Testimony scheme and its critical questions can be adapted for use in this context.

8.5   Concluding Remarks

In this paper we have considered one of the strands of Walton's work which has had a significant impact in AI, namely his notion of argumentation schemes as providing presumptive justification of a conclusion subject to critical questioning. The underlying insight, that several forms of reasoning traditionally classified as fallacious are in fact desirable and necessary in informal argumentation, and can be legitimately used if certain considerations are observed, has a mirror in the practice of AI in logic programs and expert systems. They too made use of possibly fallacious forms of reasoning, subject to constraints, such as the Close World Assumption, which were able to legitimise their use in particular contexts. Argumentation schemes provide an interesting rationale for these practices. If we wish to use the schemes for purposes other than human analysis of pre-existing arguments, for example to generate arguments, the schemes as stated in informal logic cannot be used immediately. Rather it is necessary to rethink them in terms of their intended computational use and supporting knowledge model, so that all need for contextual interpretation is removed. A case study of the process was given for one particular scheme, and some considerations relating to other schemes were advanced.

Douglas Walton's influential argumentation schemes remain an important insight, one which needs to be embraced by computational modelling of argument. His work comes from informal logic and was designed for manual analysis of argumentation. For computational purposes much detailed work remains to be done before the use of argumentation schemes can be as standard and well understood as logical deduction.

CHAPTER 9

CONTEXTUAL CONSIDERATIONS IN THE EVALUATION OF ARGUMENTATION

Frans van Eemeren, Peter Houtlosser, Constanza Ihnen & Marcin Lewiński

9.1 Introduction

Calling Professor Douglas Walton one of the most renowned argumentation scholars is certainly more than a polite hyperbole. Walton's career as an argumentation theorist spans almost four decades and in his work he has treated virtually every topic in the discipline. No other argumentation scholar has been that prolific and no other scholar has focussed so consistently on problems central to the discipline.

One of Walton's focal points is the contextual dependency of the quality of ordinary argumentation. His basic idea is that the assessment of whether an argumentative move is reasonable or fallacious is always conditional upon the background against which the argumentation develops. In Walton's approach – developed further with Krabbe and others – the context of argumentative moves is specified by defining the *type of dialogue* the moves are part of. He distinguishes between six general dialogue types: persuasion dialogue, negotiation, inquiry, deliberation, information-seeking dialogue, and eristic dialogue. Walton (1998b) presents them as 'conversational contexts of argument' that function as normative models against which everyday arguments can be properly judged as sound or fallacious. His fundamental assumption is that each of the six types has specific rules for (acceptable, reasonable, relevant) argumentative behaviour and therefore different criteria for what makes up good argumentation.

Walton argues correctly that his concept of dialogue types revives in fact, as so often happens in the study of argumentation, a classical Aristotelian idea, viz., that fallaciousness depends not just on form, but on the context of dialogue (Walton 1992d, p. 143).[1] One may add that Aristotle developed a rhetorically minded conceptualisation of the contexts of argumentation with his division of the deliberative, the forensic and the epideictic genre.[2]

Another discipline that has undertaken detailed study of contexts of discourse is linguistic pragmatics. Inspired by Malinowski's early ethnographic attempts at a pragmatic description of language use in a 'context of situation,' this line

---

[1] For a more elaborated comparison of Aristotle's five types of argument and Walton's types of dialogue, see (Walton 1998b, pp. 11-14, 237-244).

[2] Contemporary rhetoricians, such as Bitzer (1968), pursued the idea of 'the rhetorical situation' as the broad 'context in which speakers or writers create rhetorical discourse.' See also van Eemeren (2008). The concept of the rhetorical situation has been used by Jacobs to emphasize the contextuality of critical fallacy judgments. According to Jacobs (2002), good (or reasonable) arguments are those that 'make the best of the situation,' introducing adjustments to or of the particular rhetorical situation.

of research has been developed in Hymes' (Hymes 1962, 1972) ethnography of communication with its central notion of a 'speech event,' and in pragmatics by researchers such as Levinson (1992) who saw important analytical advantages in conceptualizing various 'activity types' in which language users are involved. The aim of these socio- and pragma-linguistic studies was to describe some typical contexts of communication in agreement with the way in which they are perceived by the language users in order to obtain a clearer picture of how speech exchanges function in certain conventional circumstances.[3]

Wittgenstein's notion of 'language games' – inspirational too to linguistic pragmatics – has stimulated philosophical and logical research into the contextual conditions of reasoned discourse. When it comes to the study of argument, various theories of dialogue games have been proposed. Their goal generally is to construct different models of logical validity using concepts of formal dialectics and game theory (e.g., (Barth and Krabbe 1982)). Such normative theorising is posited on the assumption that in different (constructed) games of dialogue different arguments are valid, and thus acceptable.

This broad and formal research tradition is very important to a successful scrutiny of the contextual dependency of argumentation. It is no coincidence that both Walton in his approach to argumentation and the researchers working within the pragma-dialectical tradition either explicitly or implicitly draw on the concepts just referred to. The similarity that we notice between Walton's approach and the pragma-dialectical approach to argumentation is that they both aim to combine a pragmatic perspective on argumentation with dialectical standards for its assessment. In particular, Walton and pragma-dialectics alike see the concept of types of contexts as at the same time an indispensable tool for a description of what is happening in a given argumentative situation and a locus of application of certain standards of assessment (Walton and Krabbe 1995, van Eemeren and Houtlosser 2005). All the same, we note that, while sharing some basic background as well as goals and problems, the two theoretical approaches differ as to the solutions they propose and the form these solutions take. This, we hope to make clear in the analysis presented in this paper.

The basic goal of our paper is to provide an outline of a well-justified and consistent answer to the question of how to incorporate contextual considerations in the evaluation of argumentation. To reach this goal we first briefly summarise, in section 2, Walton's view on the issue of the contextuality of fallacies, focusing on the concepts of dialogue types, mixed dialogues and dialectical shifts. In section 3 we discuss some theoretical problems that we think Walton's conceptualisations entail. Our major concern here is the question of the exact theoretical status of the concept of dialogue types. In section 4 we point to practical difficulties in the actual analysis and evaluation of ordinary argumentation that may result from the fact that Walton's theoretical approach is, about this point, unclear. In the section following our discussion of Walton's approach we present the pragma-dialectical view of the contextuality of fallacies. By such a juxtaposition of two theories so similar in general aims, yet distinct in solutions, we hope to refine our understanding of the way in which contextual factors can enter into the analysis and evaluation of argumentation.

9.2   Walton's dialogue types

Walton's approach to the contextuality of fallacies centres on the concept of dialogue types and the connected notions of mixed dialogues and dialectical shifts. The way in which Walton introduces them points to their theoretical provenance,

---

[3]For a recent overview of various discourse-analytic approaches to context, see: (Blum-Kulka 2005).

that is, to Hamblin's fundamental assumption that 'dialectic is the study of contexts of use in which arguments are put forward by one party in a rule-governed,
orderly verbal exchange with another party' (Walton 1998b, p. 6). Indeed, the
roots of Walton's concept of dialogue types as 'conversational contexts of argument' can be traced back to the 'systems of dialogue rules' developed by formal
dialecticians, which pertain to game-like and self-contained dialogues defined
by the goal of the game and regulated by strict logical rules determining how
this goal can be reached in an acceptable way. As Walton and Krabbe observe,
the use of such systems was highly specialised, as they had to provide solutions
to problems of logic and mathematics; 'each system of dialogue (or dialectical
system) defines its own concept of logical consequence (its concept of validity,
its "logic")' (Walton and Krabbe 1995, p. 4). In other words, each system pre-
scribes how one can win a particular game of dialogue in a valid way. As Walton
and Krabbe note, an important step from these abstract constructs to actual arguments has been made by Hamblin, who argued 'that such dialectical systems
could be used to model contexts of dialogue in which argumentation takes place
in everyday conversation of various kinds' (Walton and Krabbe 1995, p. 5). Despite this move towards argumentative reality, the goal of such modelling was
decidedly normative – dialectical systems were to help the analysts to 'justify
critical judgments that an argument in a real case is fallacious or nonfallacious'
(Walton and Krabbe 1995, p. 6).

This theoretical backdrop enables us to get a clearer picture of how Walton's
'new dialectic' seeks to use the types of dialogue to extend the formal approach
to the rich context of ordinary argumentation:

> The key here is that the concept of dialogue has to be seen as a con-
> text or enveloping framework into which arguments are fitted so they
> can be judged as appropriate or not in that context. So the concept of
> dialogue needs to be normative: it needs to prescribe how an argu-
> ment ought to be used in order to fit in as appropriate, and to be used
> rightly or correctly (as well as incorrectly, in some cases). But the con-
> cept of dialogue also needs to fit the typical conversational settings
> in which such arguments are conventionally used to make a point in
> everyday argumentative verbal exchanges.
>
> The concept of a dialogue [...] is that of a conventionalized, purposive
> joint activity between two parties (in the simplest case), where the
> parties act as speech partners. It is meant by this that the two parties
> exchange verbal messages or so-called speech acts that take the form
> of moves in a game-like sequence of exchanges (Walton 1998b, p. 29).

Such a definition, which attributes both 'normative' and 'conventionalized'
qualities to dialogue types, gives this theoretical concept a double function: not
only should dialogue types prescribe which argumentative behaviour is correct,
or reasonable, within the bounds of a well-delineated language game, but they
also have to mirror in their structure 'the typical conversational settings' or – as
they are called by many – 'speech events' that are characteristic of a given community (Hymes 1962, 1972). As used by Walton, the notion of context is still
limited here to dialogue types understood as rule-governed and generic conversational entities.[4]

---

[4]Recently, Walton and Macagno introduced a notion of 'dialogue context' referring to 'a broader
notion of dialogue,' which includes, among other things, 'common ground,' 'interpersonal relationship,'
and 'social constraints' between arguers (Walton and Macagno 2007, p. 110). This extends the contextual
considerations pertinent to argumentation analysis and evaluation beyond the goal-directed and rule-
governed structure of the dialogue types, bringing Walton's theoretical framework closer to being a
rhetorical perspective.

Walton and Krabbe's basic assumption is 'that the critical discussion [...] is the most fundamental context of dialogue needed as a normative structure in which fallacies and other errors of reasoning can be analyzed and evaluated' (Walton and Krabbe 1995, p. 7). Yet, they also observe a wide variety of other argumentative dialogues and organize this plurality of dialogues in a typology of six 'general types' (Walton and Krabbe 1995, p. 66): persuasion dialogue (critical discussion), negotiation, inquiry, deliberation, information-seeking dialogue, and eristics. These types are primarily distinguished through their main goal: 'resolution of conflicts by verbal means' (critical discussion), 'making a deal' (negotiation), 'reaching a (provisional) accommodation in a relationship' (eristics), etc. Next, the six basic types differ as regards the initial situation, participant's aims (not to be confused with the goal of a dialogue as such), and side benefits.

According to Walton, the usefulness of the concept of dialogue types in argumentation theory lies in its capacity to account systematically for the difficulties related to the identification of fallacies in different contexts. As is clear from the discussion above, Walton's dialogue types are in the fist place supposed to fulfil a normative function that is central to argumentation theory. In the simplest formulation this context-dependent normativety amounts to a claim that 'a good argument is one that contributes to a goal of the type of dialogue in which that argument was put forward' (Walton and Krabbe 1995, p. 2). Conversely, a fallacy 'is an infraction of some dialogue rule' (Walton and Krabbe 1995, p. 25). All the terms Walton and Krabbe use to explain their views in these matters – 'good strategy,' 'bad strategy (blunder),' 'fallacy' – acquire their (varied) meanings in the context of (the rules of) a specific type of dialogue. Walton and Krabbe explain this dependence in a radically pluralistic way: 'what constitutes a fallacy in one game of dialogue does not need to constitute a fallacy in another (it could be just a blunder, or even be entirely all right)' (Walton and Krabbe 1995, p. 25). Examples illustrating this principle are plenty: personal attacks (especially of the *abusive ad hominem* type) are fallacious in critical discussion, but entirely fine in eristic dialogues; appeals to threat (or *ad baculum* arguments) are, again, unreasonable in critical discussion, but 'quite typical and characteristic,' as Walton (1992d, p. 141) puts it, in the context of negotiation.

Walton's view of the contextuality of fallacies is, in sum, that each dialogue type constitutes a separate normative model of argumentation, with its own specific rules prescribing what good (and fallacious) argumentation is.[5] For such a theoretical framework to be practically useful in the analysis and evaluation of actual argumentation, two interrelated issues need to be dealt with: (1) the relation of the six normative (general) dialogue types to the plethora of types of communicative contexts actually perceived by ordinary language users, and (2) the exact ways in which fallacies occur in different types of dialogue.

An effort is made to deal with the first issue through the concept of *mixed dialogues*. Walton and Krabbe are well aware that their six basic types of dialogue cannot cover all ordinary speech events (Walton and Krabbe 1995, p. 82). Therefore, they assume that such speech events are composites of more of the six dialogue types. They take it, in other words, that there is a multiplicity of various normative types of dialogue constituting together this particular speech event. A political debate, for instance, as we know it in Western democracies, escapes any easy one-speech-event-to-one-dialogue-type classification. Walton (1998b, p. 223) regards a political debate as a type of context for argumentation

---

[5]This 'postmodern and relativistic standard of rationality,' as Walton calls it (Walton 1998b, p. 30), is most clearly reflected in his concept of dialectical relevance (Walton 1998b, p. 35): 'This postmodernistic pluralism yields an approach to dialectical relevance [...] that offers six distinctively different ways of explaining the nature of relevance or irrelevance of an argument in a given case, depending on the type of dialogue exchange one was supposed to have been engaged in, relative to the given case.'

that involves a mixture of no less than five (out of six) types of dialogues: it is partly information-seeking dialogue, partly deliberation, partly eristic dialogue, partly negotiation, and partly critical discussion. In such a complex case some obvious problems of evaluation arise. By which standards associated with the six basic types of dialogue should we, for instance, judge the arguer's performance in the 'mixed' speech event of political debate? Walton's (easy) solution is that 'it is conditionally permissible to evaluate a political debate [...] from a point of view of a critical discussion' (Walton 1998b, p. 224).

The second issue, concerning the fallacies, is dealt with by viewing it as a problem of a diachronic multiplicity of dialogues. The conceptual tool to solve this problem is the notion of *dialectical shifts*. Walton's central observation is that discussions that emerge and develop are liable to take turns that – in his theoretical framework – can be perceived as shifts from one type of dialogue to another. The central distinction between such shifts is the normative division between licit shifts and illicit shifts. Licit shifts are overt and mutually agreed moves away from the dialogue the participants were originally supposed to carry out to another type of dialogue that still serves the goals of the original dialogue (Walton 1992d, pp. 138-139). Illicit shifts are, by contrast, covert and unilateral attempts to change the type of dialogue that is going on into one that is wrongly presented as being in line with the exchange taking place in the original dialogue. It is the illicit type of shift that is typically 'associated,' as Walton puts it, with the informal fallacies. *Ad hominem* attacks, for example, typically signify an illicit shift from other types of dialogue to a quarrel, while *ad baculum* attacks are connected with illicit shifts from a different type of dialogue to a negotiation dialogue.

9.3 Theoretical complications arising from Walton's definition and classification of dialogue types

In our view, any theory of argumentative contexts derives its value from its functionality in analysing and evaluating real-life argumentation. The fundamental idea behind Walton's consideration of types of contexts seems indeed to be that this should lead to a significant improvement of the analysis and evaluation of actual arguments. Having a better understanding of how arguers typically behave in a given type of context, and what norms they are expected, or even required, to follow, must allow for a more differentiated appraisal of everyday argumentation than would result from a straightforward application of context-independent normative standards, such as the logical validity norms or the pragma-dialectical rules for critical discussion. When adopting such a pragmatic touchstone in assessing the value of Walton's theory we encounter some problems that his account may present to an analyst of actual argumentative discourse.

To start with, it is not precisely clear what the theoretical status is of the concept of a dialogue type. The notion of a dialogue type lays claim to both the normative and descriptive: dialogue types are at the same time defined as 'normative ideal models' and as 'conventionalized activities.' Here arises a problem. On the one hand, normative concerns are given priority, which is made explicit when Walton and Krabbe emphasize that dialogue types as 'normative models that represent ideals of how one ought to participate in a certain type of conversation if one is being reasonable and cooperative' should not be confused with 'an account of how participants in argumentation really behave in instances of real dialogue that take place [...] in a speech event' (Walton and Krabbe 1995, p. 67). On the other hand, however, the concept of dialogue types that Walton and Krabbe propagate has unmistakably an empirical flavour, as is particularly evident in Walton's characterization of the various types of dialogue. When he makes his case for the context-dependent fallaciousness of *ad baculum* arguments, for instance, Walton supports his position by observing that 'during a negotiation type of dialogue, threats and appeals to force or sanctions are quite typical and characteristic' (Walton 1992d, p. 141). In this case, and in many more cases adduced by Walton, it is the observation of an empirical regularity – describable in quantitative terms such as 'often,' or quantifiable terms such as 'typically' and 'characteristically' – that creates the normative basis for giving a fallacy judgment.[6]

This conflation of normative and descriptive perspectives makes us think that Walton's account ignores a philosophical, but also methodological, distinction regarding the concept of rule-governed behaviour that seems pertinent to us. When it comes to the study of language, it is important to distinguish clearly between: (1) observable behavioural regularities, or patterns of language use; (2) the norms underlying some of these regularities, which language users may have understood and internalized during their socialization; (3) external norms for judging language use that are analytically stipulated by the theorist because they are relevant from an external and generally theoretical perspective.

Traditionally, the study of behavioural regularities in argumentative discourse has been the domain of discourse analysts and other pragma-linguists, who have aimed to describe – with careful attention to every detail – the sequential and other patterns of dialogical arguments, deliberately avoiding any normative theorizing. Even in such a purely empirical perspective, argument can be viewed as a rule-governed verbal activity aimed at managing disagreement and 'organized by conventions of language use in which two cooperative speakers jointly produce the conventional structure' (Jackson and Jacobs 1980, p. 251). The important dilemma facing the analyst in such empirical research is, as Jackson observes with regards to the order of turn taking in natural conversations, 'whether these patterns represent rules or regularities' (Jackson 1986, p. 142).

A problem normatively oriented scholars are confronted with in relation to this dilemma is how to distinguish between a pattern that is infrequent but correct and a pattern that is both infrequent and incorrect.[7] When it comes to language use, in order to be able to solve this problem they need to have some further insight into the norms of correctness prevailing in a given communicative community in a certain context. In principle, these norms of language use can – pace the methodological difficulty just pointed out – be abstracted from the behaviour of the language users, be formulated by the language users themselves, or even be explicitly institutionalised in a written code, such as the rules of *netiquette* for Internet communication. A systematic description of such existing norms yields descriptive models of actual 'empirical' normativety. Using Pike's (Pike 1967) 'emic'-'etic' distinction, in the case of such descriptive models we can speak 'emically' about reconstructing the 'internal' normativety that is regarded sound in practice. Doing so is fundamentally different from speaking 'etically' in the study of argumentation about an 'external' normativety based on analytic considerations concerning the goals the discourse has to serve in view of the ideal critical set and defined by the theorist and the procedures it has to comply with to achieve these goals. It is the stipulation of a certain set of norms for achieving

---

[6]One of the reasons why it does not really become clear that Walton's types of dialogue are in fact contrary to his normative aims empirical categories is that in Walton's descriptions the norms pertaining to the various dialogue types are not unequivocally related to the goals of the activity types concerned.

[7]Most interesting (and most worrisome) are, of course, the cases of frequent and incorrect patterns of language use. The notion of 'an often committed fallacy' belongs to this category. Such examples clearly point to the need for distinguishing between descriptive and normative considerations in the study of language use and, more in particular, argumentation.

a certain critical goal by an argumentation theorist that leads to the construction of an ideal model of external normativety. Both the formal dialecticians and the pragma-dialecticians have taken this kind of approach. Despite their theoretical differences, they even have similar general aims: to develop ideal models of argumentative dialogues that point out what optimally reasonable argumentative behaviour in perfect conditions amounts to, and to characterize argumentative behaviour that falls short of this ideal as the commitment of a specific kind of fallacy.

Bearing these distinctions in mind, we cannot cope with the question to which of the three categories we mentioned Walton's dialogue types belong. Even Walton and Krabbe's explicit treatment of the main point is not of much help:

> It would be nice if the answer were clear-cut and if each logic system had a tag on it, saying whether it is to be taken descriptively or normatively. But that is not the way things have worked out. All serious logic systems seem to have descriptive and normative uses, but to different extents. The point is that descriptive accuracy and normative content are both important and, moreover, interdependent. Purely descriptive systems cannot be used as instruments of evaluative criticism. A purely normative system that is too far removed from what actually goes on cannot be applied to what actually goes on (Walton and Krabbe 1995, p. 175).

Imagine, for the sake of illustration, a context of argumentation in which the discussants 'typically and characteristically' recognise certain syllogistic forms as valid, although every student who has completed an elementary class in logic would easily see that they are fallacious. Let us say that in this context the categorical syllogism 'Some A are B, Some B are C: Some A are C,' as described by Jackson (1995: 264), is 'exceedingly likely to pass as valid.' Will such a persistent acceptance of this form of syllogism render it reasonable in this type of context? Granted that 'conventional validity' based on intersubjective agreement is indeed a precondition for reaching a conclusive judgment concerning the acceptability of argumentative moves, we would like to emphasise that determining their 'problem-solving validity' should come first. This means that, while recognising the interdependence of problem-solving validity and conventional validity, and consequently of normative and empirical models of argumentation if they are to have any real significance, we do not agree with keeping it vague, and even doing so deliberately, whether the primary source for coming to a judgment concerning the reasonableness of argumentation is an external normative source or an internal normative source. The problem is, that within the theoretical framework of Walton's dialogue types one cannot answer the question if such an acceptance by ordinary language users can be the basis of a context-sensitive judgment of the reasonableness or fallaciousness of a syllogistic form of reasoning like the one we mentioned.[8]

Related to the problem of the source of Walton's normativety, and thus of the rules of particular types of dialogue, is another issue that concerns the goals of the various types. Again, the question is: are they based on empirical analyses or stipulated on the basis of theoretical considerations? In other words, are these

---

[8]We are, of course, aware of the fact that this example may be rather extreme, since Walton seems to treat formal fallacies as context-independent (for instance in the fourth step of his 'dialectical method of evaluating arguments' (Walton 1998b, p. 251)). However, many informal fallacies related to an incorrect use of argumentation schemes are treated in Walton's method in exactly the same way. If this is intentional, we wonder about Walton's views concerning the universality of these formal and informal argument forms and how such universality fits into Walton's postulated 'postmodern and relativistic standard of rationality' (Walton 1998b, p. 30).

goals familiar, or at least reflectively recognizable, to the discussants concerned or are they formulated by some theorist, in this case Walton himself?

In addition, the enormous diversity of the goals Walton assigns to the various dialogue types raises the question of which of these dialogue types are really argumentative. Can, for example, 'reaching a (provisional) accommodation in a relationship' be seen as an argumentative goal? If such goals – and, consequently, the types of dialogue concerned – are not inherently argumentative but may simply have an argumentative component, many more types of dialogue can be distinguished, whether 'basic' or 'mixed,' that are similarly argumentative, which presents a serious problem to a theoretical framework that is meant to provide context-based argumentative standards of normativety. If, on the other hand, all these types of dialogue are fundamentally argumentative, one would like to know what definition of 'being argumentative' is used. In particular the information-seeking dialogue and the inquiry, basic types of dialogue that – according to Walton's descriptions – do not necessarily involve any conflict or confrontation, as is commonly required for a discourse to be considered argumentative, are then problematic.[9] Whichever view Walton may have intended to enforce, the main point that has to be made seems to us that there must be a theoretical rationale for considering discourses or verbal moves to be argumentative that is independent of the specific empirical environment – or type of dialogue – in which they occur.

Because the status of the goals of dialogue types has an immediate bearing on the typology of dialogues, and the extent to which these dialogues are argumentative is problematic, two interrelated questions are pertinent regarding the typology. First, what exactly is the function of the typology? In other words, what is in the first place the purpose of dividing contexts of argumentation? Second, what is the principle of classifying types of dialogue in precisely this way? This second question enquires about the kind of underlying theoretical framework that justifies this classification. Or is the classification just based on categories recognized by ordinary language users?

9.4   Practical problems in identifying dialogue types and evaluating argumentation

A context-sensitive evaluation of argumentation requires a proper identification of the type of dialogue the arguers are actually involved in. More specifically, since Walton associates fallacies with illicit shifts from the one type of dialogue to another type of dialogue, judging the functionality of the relation between the new context and the old context (a basic criterion for the (il)legality of a shift) requires a good grasp of the context before the shift and the context after the shift. Whether the new dialogue is functionally relevant depends, according to Walton, after all on 'retrospective evaluation' of the goals of the original dialogue:

> Some cases of retrospective evaluation are based on an explicit agreement by the two parties to take part in a certain type of dialogue. In other cases, conventions and institutions enable us to clearly make this determination. But in many cases of argumentation in everyday conversation, such agreements are implicit and can be determined only by making a contextual judgment of the expectations of the participants, by looking at the evidence given through their verbal exchanges, as the sequence of arguments proceed.

---

[9]Interestingly, Walton and Krabbe themselves point out that an 'argumentative' type of dialogue is 'conflict-centered' (Walton and Krabbe 1995, p. 105).

> In some cases, it is not clear to the participants themselves, or to any-
> one else, exactly what type of dialogue they were supposed to be
> engaging in when they had an argumentative exchange, and it is pre-
> cisely this indeterminacy that is often exploited by deceitful and falla-
> cious arguments. In evaluating such arguments, the best we can do is
> make a conditional judgment that such and such an argument would
> be fallacious if evaluated from the point of view and standards of
> argumentation appropriate for a particular type of dialogue (Walton
> 1998b, pp. 205-206).

According to Walton, the participants in a dialogue are supposed to be engaged in a certain type of dialogue, but may not be able to tell exactly what type
of dialogue. In our view, this is also due to the fact that the types of dialogues
Walton distinguishes are theoretical categories. The same people would probably
have no major difficulties in classifying their actual verbal activity as, for example, heart-to-heart-talk, sales talk, talk man-to-man, women's talk, bull session,
chat, polite conversation, chatter (of a team), chew him out, give him the low-
down, get it off his chest, or griping, because all these labels represent, according
to Hymes (1962, p. 24), 'well known classes of speech events in our culture.' Walton's problem, then, is to 'translate' such indigenous categories of speech events
in a justifiable way into his six-fold taxonomy.

To facilitate this task, Walton will make use of the concept of 'mixed discourse': 'certain conversational types of speech exchanges which we tend to
think of as univocal or homogeneous really turn out to be composite from the
point of view of our typology' (Walton 1998b, p. 218). Classifying actual speech
events as cases of mixed discourse, the way we see it, is only one of the solutions
an argumentation analyst employing Walton's framework may resort to. An analyst can also decide that one type of dialogue is clearly dominant, while mere
'admixtures from other types of dialogue' come only into play as 'flavours' or
'overtones' of this primary type (Walton and Krabbe 1995, p. 70). However, as
many examples presented by Walton show, in case of conversational departures
from the type of dialogue that is carried out, the analyst can also decide that a
shift has taken place. If a number of such shifts follow one another, this can create a 'cascading effect,' rather than a mix of concurrent dialogues (Walton and
Krabbe 1995, pp. 106-107). Another option – not really entertained by Walton
and possibly outside the scope of his perspective – is that a given speech event
is a poorly executed version of one of his six ideal types. The fundamental analytic problem is that Walton offers no clear principle and – as a consequence – no
applicable practical criteria for deciding which option an argumentation analyst
ought to choose from the various options that have been distinguished.

Should, for example, a heated scholarly debate be identified and evaluated –
in accordance with general institutional expectations – as an instance of a 'critical
discussion' (with a flavour of a quarrel)? Or is it perhaps better – to account for
the behaviour of the enraged professors – to regard it as an 'eristic discussion,'
in this case a subtype of eristics that is 'closer to the critical discussion than the
quarrel is' (Walton and Krabbe 1995, p. 79)? Or is it rather to be considered as a
'mixed' type of dialogue consisting of elements from a critical discussion, from
eristics, and, possibly, from inquiry (which would bring it, going by Walton and
Krabbe (1995, p. 85), close to the 'Socratic dialogue')? Or should one favour a
sequential analysis and notice a cascading effect – not unlike the one Walton and
Krabbe (1995, pp. 106-107) described –from inquiry, to persuasion dialogue (critical discussion), to debate, degrading finally in a quarrel? Each such decision
would result in a different evaluation of the actual discourse, since the evaluations would be based on dialogue-specific standards of argumentative rationality

that are different for a critical discussion, an inquiry, or an eristic dialogue.[10]

However, evaluating arguments relative to the context of the ideal dialogue the arguers are supposed to be involved in is only a part of Walton's understanding of the problem of contextuality of fallacies. At the centre of his approach lies the idea of dialectical shifts from one type of dialogue to another. Therefore, if an analyst opts for disclosing dialectical shifts in actual speech events,[11] reconstructing different combinations of pre- and post-shift dialogues would lead to a different assessment of their mutual functionality and thus to different fallacy judgments. Why, we wonder, does it not suffice to say that the fallacious move itself is inappropriate – in the sense of dysfunctional – in the dialogue the parties are engaged in? One does not get the impression that the fact that there is a dialectical shift plays a decisive role in the analysis. Still, Walton and Krabbe argue that the added value of the concept of dialectical shifts lies in its role as an explanatory device in solving the important issue of why a fallacious argumentation technique can be persuasive:

> The use of this particular technique might be quite inappropriate and incorrect in one type of dialogue, running quite contrary to the goals of the dialogue, but if the context has shifted to another type of dialogue, the use of this same technique may now be quite appropriate. If the shift has not been perceived, or if it was a covert or unilateral shift, not made out in the open, or agreed to by both parties, the incorrect argument may appear correct on the surface to the uncritical respondent or observer (Walton and Krabbe 1995, p. 115).

Briefly put, illicit dialectical shifts are not only dysfunctional to the achievement of the goals of the primary dialogue type – which makes them fallacious – but they are also covert and unilateral, which makes them deceitful and persuasive.

Apart from the problem of the identification of types of dialogue we already mentioned, we notice several more difficulties with an account like this. Above all, it is not at all clear how these criteria of (il)licitness should be applied. First, how can an analyst determine whether the new dialogue does not inhibit the goals of the original dialogue? The problem here is that, apart from a detailed, semi-formal stipulation of systems of rules for permissive and rigorous persuasion dialogues (Walton and Krabbe 1995, pp. 123-172), Walton offers no models of dialogue types in terms of sets of rules for securing the functionality of moves relative to the main goal of a given dialogue.[12] Therefore, judging the functional relevance of argumentative moves – both *within* one dialogue type and *across* dialogues of different types, as in the case of dialectical shifts – remains simply a matter of common sense.[13] And this is not the strongest brand of rational assessment for a serious critic of argumentation to work with.

---

[10]Let us mention just two examples of such evaluative differences. According to Walton, an argument to the person is perfectly fine in an eristic dialogue, but in the context of both a critical discussion (persuasion dialogue) and an inquiry it easily becomes a fallacious argumentum ad hominem. Next, retraction of previously incurred commitments is certainly allowable in eristics ('I said so before, now I don't, and so what!'), conditionally allowable in a (permissive) persuasion dialogue, but unacceptable in inquiry defined by 'the property of cumulativeness' (which means, according to Walton (1998b, p. 70) in an inquiry 'that the participants can't change their minds.').

[11]One of the problems here is that we are talking about an option rather than a necessary step in a well-justified analysis.

[12]In other words, Walton offers no check of the problem-solving validity of any of the dialogue types other than the persuasion dialogue.

[13]Walton limits himself to semi-theoretical and semi-empirical observations: 'Negotiation dialogues may profit both from inquiries and from persuasion dialogues as subdialogues, but not the other way around. A shift from a persuasion dialogue or from an inquiry to negotiation seems always to obstruct

A second analytic problem is that it is hard to understand how any 'dialectical' shift can be 'unilateral.' The general pragmatic line followed by the authors of *Commitment in Dialogue* makes it impossible to construe shifts as mental states of one of the arguers. Instead, a shift must be an externalized argumentative move that is manifested in text. Therefore, it seems to us that a 'unilateral shift' must signify a monologual attempt of one of the arguers that is either accepted or rejected by the interlocutor, so that it becomes a 'dialectical' (or dialogical) shift.

Directly related to the second analytic problem is the third, the problem of dialectical shifts being 'covert.' It is a crucial element of Walton's explanation of the treacherous character of fallacies that they are deceitful because they are reasonable in the context of the new type of dialogue in which they are situated after a covert dialectical shift has taken place whereas they are in fact fallacious if judged properly by the standards applying to the original dialogue. Paradoxically, the relative acceptability of such fallacious moves is based on them being perceived as part of a new type of dialogue that is introduced covertly. The addressees are, in other words, deceived by the fallacy, because they see it as a 'quite appropriate' contribution to the new dialogue, while – due to the fact that this new dialogue was introduced by means of a shift that was unilateral and covert – they have 'not perceived' that a new dialogue is actually taking place.[14] How, we wonder, can the condition that an argument is perceived as reasonable in the new context of dialogue in the explanation of a persuasive character of fallacious arguments be reconciled with the condition that this new context has not been perceived?

The recognition of these three problems leads us to believe that the concepts of mixed dialogues and dialectical shifts are not very functional as analytic tools. Instead of furthering a differentiated and balanced analysis and evaluation of actual argumentative discourse, they generate complications that we are unable to dissolve within the theoretical framework developed by Walton.[15] This is why we turn now to a different theoretical framework.

9.5   An alternative view of the contextuality of fallacies

The pragma-dialectical approach to fallacies

We emphatically agree with Walton (and Krabbe) that a theory of fallacies is incomplete as long as it is unclear how the norms that are proposed as standards for reasonable argumentation can be brought to bear on actual instances of ordinary argumentative discourse. We also agree that the best way of fulfilling this requirement is to combine normative insights expressed in dialectical models of regimented argumentative exchanges and descriptive insights in the conduct of argumentative language use (van Eemeren and Grootendorst 1984b, van Eemeren et al. 1993b, Walton and Krabbe 1995, van Eemeren and Houtlosser 2007b). The way in which we aim to reconcile the normative and the descriptive dimensions, however, differs significantly.

In our view, the theorizing about fallacies has to start from a general and coherent perspective of what it means to act reasonably in argumentative discourse, and this perspective must be the expression of a clearly defined ideal of reasonableness. Philosophically, the pragma-dialectical theory starts from a critical view of reasonableness, which is given a theoretical shape in the ideal model of

---

or at least to impair the original dialogue' (Walton and Krabbe 1995, p. 73). So far no consistent classification of such (dys)functional relations between different types of dialogue has been proposed.

[14] At times Walton even talks about fallacies 'causing the argumentation to shift' (Walton 1992d, p. 140). In such a 'reverse' account, however, one cannot speak of fallacies being the result of dialectical shifts, let alone of unilateral and covert shifts.

[15] This does not mean, of course, that it is impossible to deal with these complications.

a 'critical discussion.' This model of a critical discussion specifies the stages and the types of speech acts that are instrumental in resolving a difference of opinion by critically testing the tenability of the standpoints at issue. The procedural rules applying to the speech acts performed by the parties in the discussion in each of the stages of the critical discussion express the norms that are instrumental to resolving a difference of opinion on the merits (van Eemeren and Grootendorst 1984b, 2004).

The theoretical device of a critical discussion has, on the one hand, a dialectical dimension and, on the other hand, a pragmatic dimension. The model is dialectical because it is premised on two parties who try to resolve a difference of opinion by means of a methodical critical exchange of discussion moves. The model is pragmatic because, rather than purely in formal terms, the discussion moves are treated as speech acts (van Eemeren and Grootendorst 2004). Due to its integration of 'pragmatic' and 'dialectical' features, the critical discussion model can serve heuristic and evaluative purposes. With the help of the model, the argumentative moves made in empirical reality can be reconstructed in terms of speech acts that play a part in a critical procedure for resolving a difference of opinion. This methodical reconstruction of argumentative discourse, which is to result in an 'analytic overview' of the resolution process, is an indispensable step to making real-life utterances accessible to critical assessment by dialectical norms of reasonableness. Once the reconstruction has been carried out, the various speech acts represented in the analytic overview can be evaluated according to the rules applying to the stage in which they are performed. Any move constituting an infringement of one of the rules for critical discussion is a possible threat to concluding the difference in a reasonable way and must therefore (and in this particular sense) be regarded as fallacious.

The notions of 'strategic manoeuvring' and 'argumentative activity type' build on the pragmatic and descriptive dimension of argumentative discourse and are brought in to cast light on real-life argumentative discourse from the viewpoint of normative concerns. In developing the concept of strategic manoeuvring, we started from the observation that in empirical reality discussants do not just aim at performing speech acts that will be considered reasonable by their fellow discussants; they also orient their contributions towards gaining success, that is, to achieving the 'perlocutionary' effect of acceptance. From our point of view, this means that each discussion move, at every stage of the resolution process, is assumed to be simultaneously oriented towards the 'dialectical' aim of the stage concerned and the 'rhetorical' analogue of getting things one's way. 'Strategic manoeuvring' is the term we use to refer to the arguers' attempts to reconcile – and diminish the potential tension between – the simultaneous pursuit of these two aims. If a party allows its commitments to a critical exchange of argumentative moves to be overruled by the aim of persuading the opponent, the strategic manoeuvring 'derails.' Because derailments of strategic manoeuvring always involve violating a rule for critical discussion, they are on a par with the obstructive moves we designated as fallacies (van Eemeren and Houtlosser 2002, 2003).

As we have argued in earlier publications, insights into the strategic design of the discourse gained by making use of the pragmatic notion of strategic manoeuvring can strengthen the pragma-dialectical reconstruction of argumentative discourse, in particular its justification (van Eemeren and Houtlosser 2002, 2007b). In addition, starting from a general classification of modes of strategic manoeuvring, soundness conditions for each mode of manoeuvring can be developed that make it possible to decide whether a mode of manoeuvring under analysis satisfies or violates a discussion rule. Such soundness conditions must stipulate which generic and empirical circumstances must obtain for a particular mode of manoeuvring to remain sound (van Eemeren and Houtlosser 2003).[16] The pragma-dialectical theory of argumentation as developed by van Eemeren and Grootendorst already provides general norms to distinguish between reasonable and unreasonable argumentative moves as well as analytic procedures for reconstructing verbal moves as specific speech acts playing a part in any of the stages of the resolution process, but specific criteria are still lacking for deciding whether or not a certain speech act agrees with the norms relevant to the particular stage to which it belongs. Formulating soundness conditions for the various modes of strategic manoeuvring is a way of rendering the critical discussion rules ever more applicable to the evaluation of ordinary argumentative practice.

The notion of 'argumentative activity type,' as introduced in pragma-dialectics by van Eemeren and Houtlosser (2005), refers to argumentative practices and routines in the Gumperz (1972) sense, that are more or less institutionalised in empirical reality and are characterised by specific goals, rules, and other kinds of conventional preconditions for argumentative discourse.[17] Insight into the conventional preconditions – practical constraints and opportunities – for strategic manoeuvring holding for the various activity types can be of help in defining the strategic function of the argumentative moves that are made, thus strengthening our methods for reconstruction and analysis, and in determining whether or not the soundness conditions for particular modes of strategic manoeuvring have been fulfilled. For that reason, paying careful attention to the peculiarities of the activity type in which the discourse occurs has become part of our procedure for giving an accurate justification for the application of our critical norms to the speech acts that we evaluate. Within the pragma-dialectical framework, the difference between legitimate and fallacious manifestations of strategic manoeuvring is demarcated by establishing whether the soundness conditions applying to a particular mode of strategic manoeuvring in a particular argumentative activity type have been met.[18]

Differences between Walton's approach and the pragma-dialectical approach

Although we agree with Walton that fallacy judgments are always contextual judgments, the pragma-dialectical approach and Walton's approach differ in at least three crucial respects: (1) Walton (and Krabbe's) dialogue types are neither equivalent to our notion of a critical discussion nor to our notion of an argumentative activity type; (2) our rules for critical discussion express general norms for sound argumentation that are not limited to a particular argumentative activity type or a specific dialogue type; (3) we consider fallacy judgments to be context-dependent because the criteria for determining whether an argumentative move complies with a critical discussion rule may vary from one context to another – and this is not so because the norms for reasonable argumentation per se are

---

[16]For an example of soundness conditions, see the conditions formulated for strategic manoeuvring by making an appeal to authority in (van Eemeren and Houtlosser 2003).

[17]We consider activity types as argumentative if argumentation plays an important, if not essential, role in achieving the overall goal of the activity type concerned, as is, for instance, the case in a great deal of the activity types belonging to the general clusters of adjudication, mediation, negotiation, problem-solving, and public debate. 'Love letters' and 'news reports,' on the other hand, we do not consider as argumentative activity types, although they may contain elaborated pieces of argumentation. This is because the overall goals of these activity types can be achieved just as well without them containing any argumentative element. Precisely because argumentation is vital to reaching a reasonable outcome of argumentative activity types, it is necessary to maintain critical standards in evaluating its soundness.

[18]For a more elaborate treatement of argumentative activity types in the pragma-dialectical theory, see (van Eemeren 2010), especially Chapter 5.

relative to a particular argumentative activity type or dialogue type. In elaborating on each of these differences we shall explain in more detail how contextual features are taken into account in the pragma-dialectical method for analysis and evaluation.

The model for a critical discussion and argumentative activity types

The pragma-dialectical analysis and evaluation of argumentative discourse can benefit from an enquiry into the characteristics of the various argumentative activity types in which such a discourse may occur, but only if the empirically-based activity types are clearly distinguished from the theoretical ideal of a critical discussion. Argumentative activity types are generally recognized empirical entities consisting of observable communicative practices that share certain basic goals and conventions. Parliamentary debates, legal indictments, and Internet forums are all specific examples of argumentative activity types. Our notion of a critical discussion does not refer to such empirical phenomena; in argumentative reality no 'tokens' of a critical discussion can be found. A 'critical discussion' is by no means an activity type, not even an activity type to which a privileged status has been assigned. The theoretical construct of a critical discussion represents a normative ideal of reasonable argumentative discourse while argumentative activity types are descriptive categories referring to empirically-based communicative practices. Unfortunately, the crucial difference between the two is frequently overlooked.

Due to the descriptive status of the concept of argumentative activity types, the observation of regularities and the identification of explicit and implicit norms participants accept and expect others to observe are crucial to the characterisation of an activity type. A critical discussion, by contrast, is worked out analytically in view of a certain ideal of reasonableness. The pragma-dialectical procedure for conducting a critical discussion is not intended to be a reflection of existing patterns of argumentative behaviour or empirically shared argumentative rules, but constitutes a theoretically based model for resolving differences of opinion on the merits.

The task of an analyst who examines real-life argumentative activity types is to describe the norms that ordinary actors actually apply (but not necessarily correctly) in the activity types under consideration. Whichever methodology is used to determine the norms of a certain activity type, the empirical problem that is to be resolved can be summarized as follows: 'Which norms determine in the argumentative practice concerned what are allowable argumentative moves in this particular activity type?' By contrast, the job of a theorist proposing a normative model for conducting argumentation is to formulate the rules for having argumentative exchanges on the basis of a well-considered philosophical ideal of reasonableness. This means that in this case the theorist is concerned with an entirely different question: 'Which norms should guide in all cases the conduct of argumentative discourse if the discourse is to be reasonable?' In answering this question argumentation theorists may take different routes. Pragma-dialecticians believe that the best way of responding is to develop a general view of the immanent goal of argumentative discourse that is in line with a well-considered philosophy of reasonableness and then develop a system of rules for achieving this goal that is problem-valid and has the potential of becoming conventionally valid as well.

A general normative framework for all argumentative activity types

Although it is in fact not really clear to us whether Walton's dialogue types have a normative or an empirical status (or represent some combination of the two), it is certain that each dialogue type is claimed to be a normative model (Walton and Krabbe 1995).[19] This leads us to the second crucial difference between Walton's approach and our own. Walton promotes the pluralistic view that argumentation needs to be judged as correct or incorrect in relation to a multiplicity of 'models of reasoned dialogue.' When it comes down to deciding which dialogue type should be the standard in the evaluation of a given case of argumentative discourse this view is bound to cause practical problems. Such problems are alien to pragma-dialectics, because in this approach argumentative discourse is in all cases judged by resorting to the model for a critical discussion.[20]

The norms expressed in the rules for critical discussion are intended to be general norms for sound argumentation that can be used to evaluate argumentative discourse regardless of the context of activity type in which the discourse takes place. The generality we claim for these norms reflects not only our critical – as opposed to anthropological – philosophy of reasonableness, but also our meta-theoretical starting point that argumentation is in principle aimed at convincing a – meaning any, not just some particular – rational critic who judges reasonably of the acceptability of the standpoint at issue (van Eemeren and Grootendorst 1984b, 2004). Other specific purposes argumentation may serve in the context of a specific argumentative activity type, like stimulating the acceptance of a certain public policy, scrutinising government's performance, or reaching a compromise in a conflict of interests, are disregarded in the normative framework of a critical discussion. The ideal model of a critical discussion proposes a set of procedural standards for determining when argumentation is to be regarded acceptable by a rational critic who judges reasonably.[21] As a consequence of our critical philosophy of reasonableness, the standards for determining the acceptability of argumentation are situated theoretically in the dialectical context of an ideal model for conducting a critical discussion designed to resolve differences of opinion 'on the merits.'

The fact that we regard the model for a critical discussion applicable to the full plurality of argumentative activity types that can be discerned, does not imply that we ascribe to all these argumentative activity types, let alone to non-argumentative activity types, the purpose of resolving a difference of opinion on the merits by conducting a critical discussion. This purpose we only ascribe analytically to the argumentative exchanges occurring within these activities. We firmly believe, however, that recognizing the particular goals that characterize each of the various argumentative activity types, so that the one activity type can be clearly distinguished from the other, is a prerequisite for explaining the specific way in which argumentation in disciplined in each particular activity type and plays a vital role in the identification of fallacies. To establish a systematic link between the various argumentative activity types and a critical discussion, which is instrumental in this endeavour, we give an argumentatively relevant characterization of each of these activity types in terms of the way of implementing the four stages of a critical discussion van Eemeren and Houtlosser (2005).

---

[19] In view of Walton's explicit choice for a plurality of normative frameworks consisting of the various activity types, it is hard to figure out how Walton and Krabbe's claim that the critical discussion is the most fundamental 'dialogue type' needed to evaluate argumentation (Walton and Krabbe 1995, p. 7) should be interpreted. The very recognition of a hierarchy among the various dialogue types carries an inconsistency regarding the idea of a plurality of normative frameworks.

[20] A general and coherent theoretical point of departure saves us not only some complications in the evaluation of argumentative discourse, but also enables us to identify the parts of an activity type that are indeed argumentative and to identify the peculiarities of that activity type relevant to the conduct of argumentative discourse.

[21] The claim to reasonableness that argumentation by definition involves can, of course, only be sustained if the argumentation is sound. A critical approach to argumentation is therefore required in all cases.

The function of argumentative activity types in the identification of fallacies

The general norms for sound argumentative discourse as expressed in the rules for critical discussion are – deliberately – formulated independently of the way in which argumentative discourse typically develops in the various argumentative activity types. This does not mean, however, that the context in which argumentative discourse occurs does not play a role in its analysis and evaluation. On the contrary, what Hymes (1972) calls 'goals' and 'outcomes,' as well as rules and regularities and other conventional constraints characterising a certain activity type, have an important function in explaining the strategic manoeuvring that takes place in argumentative discourse and the identification of fallacies that may be committed in situated argumentative discourse. In the pragma-dialectical method for identifying fallacies, the topic we are interested in here, argumentative activity types have a function on two different levels.

Firstly, contextual considerations such as those related to the execution of a certain activity type, have an important function in the reconstruction of implicit commitments undertaken in real-life utterances in terms of speech acts that play a part in a critical resolution process. As we have shown in earlier publications (e.g., van Eemeren et al. (1993b)), contextual criteria are indispensable for a pragma-dialectical reconstruction of unexpressed standpoints, unexpressed premises, and the structural organisation of argumentative discourse. It is, for instance, only due to the knowledge we have of the institutional goals of advertising that we can justify the attribution of a standpoint like 'You should buy my new product' to a commercial ad that introduces the product but leaves this standpoint unexpressed. Likewise, it is due to our knowledge of the role the prosecutor is supposed to play in a legal cross-examination that we can justify the reconstruction of a question the prosecutor asks at the opening stage as a strategic manoeuvre to elicit concessions or in the argumentation stage as an indirect argument in favour of a certain standpoint.

Secondly, argumentative activity types enter into the pragma-dialectical identification of fallacies because knowledge of the particular features of a certain activity type can have a function in the evaluation of strategic manoeuvring going on in argumentative discourse conducted in this activity type. In particular, the preconditions applying to the strategic manoeuvring that is allowed in a certain activity type can be of help because they establish the criteria for determining whether or not the soundness conditions for using a certain mode of strategic manoeuvring in that activity type have been met. The pragma-dialectical evaluation starts from the argumentative moves that are represented in the analytic overview based on the reconstruction process and makes use of soundness conditions incorporated in the normative framework of the rules for critical discussion. The application of these standards to the moves made in a certain mode of strategic manoeuvring, however, is dependent on the context of the activity type in which these moves are made in a particular case. With regard to argumentative moves involving an appeal to authority, for instance, the criteria that apply to judging the legitimacy of this mode of strategic manoeuvring, say by referring to a code of law, in a legal court case may be different from those applying, for instance, to an appeal to authority in a television debate by referring to opinion polls, or in an academic review by referring to the state of the art in the field. Although in all cases the ultimate standard is that argumentative discourse must comply with the rules for critical discussion, the criteria for making out whether this is in a certain kind of practice indeed the case may vary to some extent in accordance with the specific demands ensuing from the overall goals and conventions of the activity type concerned. If a divorced couple takes part in mediation talks with the aim to arrive at a mutually agreeable arrangement regarding custody of their daughter, using the pragmatic argument that the solution proposed is in the best interest of the child may be regarded a sound way to settle the matter because it agrees with what both parties consider desirable, whereas in the case of a parliamentary debate using a pragmatic argument will rather be accepted if the result achieved is desirable because it is in the best interest of the public at large.

In sum, the incorporation of contextual factors at the activity type level in the pragma-dialectical method for analyzing and evaluating argumentative discourse always involves the following ingredients: *empirical phenomena* consisting of real-life utterances or reconstructed speech acts; a *theoretical framework* to deal with these phenomena that includes a model for critical discussion, with pragmatic rules and principles, and a series of soundness conditions presented as rules for critical discussion; a set of *contextual criteria* to relate the phenomena that are examined with the theoretical framework. Fallacy judgments are in pragma-dialectics considered to be context-dependent in the sense that both the criteria for reconstructing real-life utterances as analytically relevant argumentative moves and the criteria for determining the evaluative relevance of these moves going by the soundness conditions inherent in the rules for critical discussion may vary according to activity type.

9.6 Conclusion

We agree with Walton that the assessment of arguments is to some extent context-dependent and that this context-dependency should be taken into account in building a theory of argumentation that is to serve as a basis for analysis and evaluation. In spite of these shared premises, however, our theoretical starting points are in various respects different from Walton's. As a consequence, the framework of analysis and evaluation we endorse is also different. As we have shown, these differences are reflected in how Walton and we deal with the problem of the contextuality of fallacies.

Walton's theoretical starting point in dealing with the contextuality of fallacies is that 'a good argument is one that contributes to the specific goal of a type of dialogue.' In the simplest translation, this means that an analyst must look at the actual contextualized language use, identify the type of dialogue that is being performed, and judge whether the argument under scrutiny furthers or inhibits the goals of the dialogue concerned. If the latter is the case, the argument is fallacious. The main problem we have with this approach is that Walton assigns a normative status to the dialogue types he appears to distinguish primarily on the basis of empirical observations of the kinds of argumentative discourse concerned. This means that Walton derives his models of what 'ought to be the case' in the various contexts of dialogue, without any further justification, from what 'is (apparently) the case.'

Pragma-dialecticians, like we are, take as their theoretical starting point in dealing with the contextuality of fallacies that 'a good argument is one that complies with the rules for critical discussion.' This approach is in agreement with the pragma-dialectical view of argumentation as aimed at resolving a difference of opinion in a reasonable way by testing critically the acceptability of the standpoint(s) at issue. The rules for critical discussion embody the norms for judging the reasonableness of argumentative discourse and – because reasonableness is intrinsically linked to argumentation – they apply to every context of argumentation. Whether or not in a particular case a certain rule for critical discussion has been violated, however, depends on the context of the argumentative activity type in which the argumentative discourse takes place. This is so because, simultaneously with being aimed at maintaining reasonableness, argumentative discourse is always aimed at being effective in the context in which it occurs. The criteria for determining whether the strategic manoeuvring involved in keeping a balance between these two aims derails into fallaciousness by going against the rules for critical discussion are context-dependent and may vary to some extent between the various activity types.

In our view, the identification of fallacies is not only context-dependent because the criteria for sound argumentative discourse are in this way connected with the argumentative activity type in which the discourse takes place, but also because the reconstruction of argumentative discourse made in the analysis preceding the evaluation is sensitive to the activity type. It is only on the basis of a careful analysis of the type of argumentative activity that is carried out that a rational judge can decide in a reasonable way how a given norm for critical discussion ought to be applied to the discourse so that it can be established whether or not a fallacy has been committed. This means that in pragma-dialectics argumentative activity types are *intermediates* in the process of analysis and evaluation rather than the *basis* of the evaluation of argumentation. This is indeed a very different approach from Walton's, which confronts actual argumentative discourse directly with divergent contextual norms.

CHAPTER 10

TRUST IN EXPERTS AS A PRINCIPAL-AGENT PROBLEM

Jean Goodwin

Abstract

Douglas Walton's work on the assessment of expert opinion has helped us distinguish the fallacy of using expertise to dominate others from the legitimate reliance on others' knowledge; it has further given us a much more elaborate account of critical questions through which the reliability of purportedly expert assertions can be tested. In this essay, I propose to extend Walton's line of work by grounding his critical questions within a more general theory of the communication of expert opinion. I argue that the layperson assessing expert opinion faces what in economics and related disciplines is called "the principal-agent problem." She must decide whether to trust what the expert says, although she is unable to assess fully either his expertise or his diligence in offering her his opinion. Participants in such transactions have a variety of practical strategies for managing principal-agent problems; I will show how in the case of appeals to expertise these strategies lead to the critical questions Walton has already enumerated, and also to additional considerations which deserve the attention of argumentation theorists.

Douglas Walton in his work on the *Appeal to Expert Opinion* (Walton (1997a), hereinafter cited by page number) acutely identifies the central puzzle of our reliance on experts. We need to consult experts because we don't know enough ourselves; but because we don't know, we can't really assess who has expertise. Walton's work also provides a rich set of resources layfolk can use when trying to reach practical resolutions to this puzzle. The goal of the present essay is to link more tightly the puzzle with the resources, by borrowing from neighboring disciplines a conceptualization of how expert/lay transactions can proceed even in the face of deep asymmetries of information.

As social epistemologists are fond of pointing out (e.g. Coady (1992)), many things we know, we know only by report: our birthdates, for example, or who our parents are; how much money we have in our bank accounts; what the capital of Kyrgyzstan is, and where. We are, in short, dependent on others for a wide range of ordinary knowledge.

We may be even more dependent on others whose knowledge is a result of experience, practice or study beyond the ordinary–i.e., on experts. Expertise is not, of course, a uniquely modern phenomenon; Plato was already fascinated by those who knew more. But while he recycled the same dozen or so examples–cobblers, helmsmen, generals, sculptors–contemporary society has increased knowledge production in the same way we have increased production of goods: by a division of labor. The study of biology at my university, for example, is divided between three departments, each of which is further divided into two to six main research clusters, not to mention interdisciplinary initiatives and independent research centers. My library reports receiving 66,195 issues of scholarly journals each year on paper, and an additional 53,018 in electronic form.

This contemporary "pin factory" of knowledge has had remarkable success in re-making the world. The objects which support everyday life are designed and maintained by experts with a vast array of specialties: consider the cell phone, for example. Indeed, our world has become so knowledge-soaked that many of the major challenges facing us today are the "unintended consequences" of knowledge-intensive activities. As Ulrich Beck famously put it, we live now in a "risk society," one in which "science is *one of the causes, the medium of definition and the source of solutions*" to our problems (Beck 1992, p. 155).

Because our world has been constituted through specialized knowledge, we cannot take care of business in matters small (e.g., personal financial decisions) or large (e.g., global climate change) without the help of experts. We find ourselves dependent on the expert opinion.

Epistemic dependency does not of course relieve us of epistemic obligations. Even if dependent, we still ought to try to make reasonable judgments about which experts to credit, and on what matters. But our assessment of expert opinion raises a puzzle, however–the central puzzle around which Walton frames his work. He opens the *Expert Opinion* by quoting an ancient dilemma:

> Who is to be the judge of skill?
> Presumably, either the expert or the nonexpert.
> But it cannot be the nonexpert, for he does not know what constitutes skill (otherwise he would be an expert).
> Nor can it be the expert, because that would make him a party to the dispute, and hence untrustworthy to be a judge in his own case.
> Therefore, nobody can be the judge of skills (p. xiii).

Walton comments:

> A very similar and closely related problem may be expressed . . . by asking the question: How can one rationally evaluate an appeal to expert opinion used to support an argument that has been put forward to support one side on a disputed issue? The problem, as the [ancient dilemma] indicates, is that the worth of such an opinion must either be evaluated by an expert in the field of the appeal or by a nonexpert (layperson). But the nonexpert is really in no position to evaluate the worth or acceptability of the opinion. He lacks the expertise (p. xiv).

Later in the work Walton develops this core idea further as the "inaccessibility thesis," that is, "the basic observation that the basis of the expert's conclusion is not accessible to the layperson who has sought her advice" (pp. 109-110). He explains:

> The most advanced expert, whose advice may be the most useful or most correct, may be the person least able to set out the line of reasoning–in a form accessible to the layperson, especially–that led her to use the knowledge in this field to arrive at this particular conclusion. This shielding of accessibility between the expert and the user is the basic problem of all appeals to expert opinion as a reasoned form of guidance, and is the root of numerous other regulated problems and pitfalls in the use of expert advice as a form of argumentation (p. 114).

Walton draws support for the inaccessibility thesis from contemporary work on psychology of expert judgment (see e.g. Ross (2006)). This research demonstrates that experts literally see the world differently than layfolk. Expert judgment thus proceeds along lines that only other experts can follow; to layfolk, it looks like the result of unaccountable "intuition." Another line of work–this time, in science and technology studies–also reinforces Walton's view. According to these scholars, scientific knowledge is not a token that one can hand to another; it is not severable from its conditions of production. Following the track originally laid down by Latour and Woolgar (1979), the only way to know (e.g.) biology is to be a biologist, long practiced at doing the things biologists do, and at talking with other biologists about them. Scientific knowledge is, as Collins and Evans (2002) have recently put it, primarily a form of expertise. Those outside the "core set" of scientists directly engaged with a particular problem cannot fully appreciate the nuances involved in the theories proposed to solve it–not even if they are scientists. Thus there can be no adequate assessment–often, no real understanding–of what a specialist says within her specialty, without becoming a specialist oneself.

This, then, is Walton's central puzzle. The layperson finds herself dependent on experts. But she lacks the knowledge and skill to determine who is an expert, and whether what they say is sound. It seems necessary, but not rational, to rely on expert opinion. And thus perhaps the philosophical tradition has been right in locating the appeal to expertise among the fallacies.

In some sense, however, the purported puzzle is not a puzzle at all. We are already relying extensively and at least somewhat successfully on expert advice, so apparently we already know how to assess it. As Walton comments, assessment is a task "done all the time in everyday reasoning" (p. xiv); a conclusion which he more than justifies by his detailed discussions of how experts are dealt with in a variety of settings, such as law courts. It simply cannot be fallacious to listen to experts; we must already possess methods for the reasonable assessment of expert opinion.

Much of Walton's book is aimed to give fuller articulation to these methods. After a comprehensive review of reasoning textbooks in philosophy, he lays out a series of "critical questions" which he recommends for use when determining whether to rely on an expert (E) who makes an assertion (A):

1. Expertise question: How credible is E as an expert source?

2. Field question: Is E an expert in the field that A is in?

3. Opinion question: What did E assert that implies A?[1]

4. Trustworthiness question: Is E personally reliable as a source?

5. Consistency question: Is A consistent with what other experts assert?

6. Backup evidence question: Is [E]'s assertion based on evidence? (p. 223)

For each of these major critical questions, Walton further lays out several additional subquestions which allow the layperson to check expertise in even greater detail.

---

[1] As Walton explains, we need to distinguish "two frameworks" involved in the use of expert opinion in argumentation (p. 17). The first is the transaction between expert and layperson, in which the expert opinion is conveyed; the second is the transaction between arguer and arguer, in which the expert opinion is deployed in argument. In this essay, I will follow Walton's lead in focusing on primary expert/lay transaction, leaving complexities of the arguer/arguer transaction for another time. Walton's critical question #3 is relevant primarily to this second transaction, and so will not be a focus of attention here.

The recommendations Walton makes are eminently practical. The layperson who uses these methods will make better judgments when figuring out whether to trust expert advice. Still, a significant question remains for argumentation theorists: the question, "why?" There appears to be a gap between the very general puzzle presented by expert opinion and the very specific methods developed to deal with it. It seems that we need a "theory of the middle range" to complement Walton's work. Such a theory would explain why the critical questions he recommends in fact help resolve the puzzle of expert opinion; it might also allow the questions he sketches to be articulated with even greater precision, ordered, and extended or critiqued. It is such a theory that I will sketch in the present essay.

I want to begin to fill in the gap between the puzzle and the practical recommendations by noting that argumentation theory is not the only field to grapple with the problems that arise when one person knows more than another. Economics, sociology, management theory, and political science have generated extensive literatures examining how transactions can proceed even when some participants are unable to adequately assess others' capacities and performances. These and other disciplines have developed a set of concepts "particularly valuable for understanding situations in which one party seeks . . . to achieve certain outcomes (such as profits) by relying on and structuring the behavior of various other actors" (Moe 1984, p. 755). The concepts provide useful models for understanding problems like that in the following case. Someone–the "principal"– needs to retain someone else–an "agent"–to do something she cannot or does not want to do for herself; for example, in a small-scale enterprise, to deliver goods that she has made. Can she rely on him to do so? Is he competent? Will he work hard? The principal needs to figure out whether she can trust the agent. But he, more than she, knows whether or not he is worthy of trust. How in the face of this asymmetry in information is it reasonable for her to hire him?

This is a simple case of what is called the "principal/agent problem," which, despite its name, is found in many, many relationships beyond that between employers and employees (Arrow 1985, Eisenhardt 1989, Mitnick 2008, Moe 1984, Shapiro 1987, 2005). Asymmetric information threatens to close down all sorts of transactions; for example:

- transactions between buyers who need to figure out whether to trust used car dealers' avowals, and dealers who know more about the cars they're selling than buyers do.

- transactions between stockholders who need to figure out whether to trust CEOs to conduct their business, and CEOs who know more about the operation of the business than stockholders do.

- transactions between health insurers who need to figure out whether to trust applicants by insuring them, and applicants who know more about their physical condition than insurers do.

- transactions between legislators who need to figure out whether to trust bureaucrats to carry out their policies, and bureaucrats who know better than legislators how they'll implement the laws on a day to day basis.

Expert advice provides yet another example of a transaction involving asymmetric information,[2] and it is no surprise to find doctors, lawyers, accountants

---

[2]Note that "transaction" here is meant generally, as an exchange in which both participants have an interest. We are most familiar with commercial and financial transactions, where at least part of the exchange involves something of value. But communicative exchanges in which no money is involved are also transactions: at a minimum, the speaker offers his utterance in exchange for his auditor's at-

and other knowledge professionals mentioned among the classic examples of situations raising the principal/agent problem. Note that the problem arises not simply because the expert–in these terms, the agent–knows more about the subject than the principal who consults him about it. Instead, as the ancient puzzle recognized, the principal/agent problem arises because the expert knows more about himself. In addition to the expert's first-order subject-matter knowledge, the expert also has unique access to second-order information about that knowledge: viz. how much he has, and how diligently he plans on applying it. This second-order information is what is necessary to accurately assess expertise, and it is this information which the layperson consulting the expert inevitably lacks.

Scholarship on the principal/agent problem commonly recognizes two broad categories of challenges participants face:

> [1] *Adverse selection* refers to the misrepresentation of ability by the agent. The argument here is that the agent may claim to have certain skills or abilities when he or she is hired. Adverse selection arises because the principal cannot completely verify these skills or abilities... For example, adverse selection occurs when a research scientist claims to have experience in a scientific specialty and the employer cannot judge whether this is the case... [2] *Moral hazard* refers to lack of effort on the part of the agent. The argument here is that the agent may simply not put forth the agreed-upon effort. That is, the agent is shirking. For example, moral hazard occurs when a research scientist works on a personal research project on company time, but the research is so complex that corporate management cannot detect what the scientist is actually doing (Eisenhardt 1989, p. 61).[3]

Again, both of these challenges are visible in expert/lay transactions, creating what have been called the "disabilities of expertise" (Shapiro 1987). The ancient puzzle focused on the layperson's inability to assess the expert's knowledge *prior* to entering into a transaction with him–that is, on adverse selection. But moral hazard presents just as serious a challenge, for the layperson is equally unable to assess expert's performance *during* the transaction, in applying his expertise to the situation the principal is interested in. Was a medical procedure badly done? Did the lawyers negotiating a complex deal drop the ball? In complex, highly contingent affairs, there are many possible accounts for poor outcomes; it is therefore difficult for the layperson to lay the failure at the expert's door. As one scholar has commented, in

> principal-professional exchanges . . . professionals have power over lay principals by virtue of their expertise, functional indispensability, and [the] intrinsic ambiguity associated with the services they provide. Such agency exchanges involve information asymmetry that is particularly severe, since principals do not possess the technical

---

tention. Transactions may occur face-to-face (as when I consult my doctor), or be highly mediated (as when I consult someone's article on WebMD). While undoubtedly important to a full account of the assessment of expert opinion, I aim here only to sketch the broad outlines of a theory, and will take up such distinctions only in passing.

[3] These terms are drawn from the principal/agent problem as it arises between insurers and insurees. The person most likely to seek insurance is the one who knows himself most likely to receive a pay-out from it; a sick person, for example, is more likely to apply for health insurance than the well. From the point of view of the insurer, this is an "adverse selection" of insurees. Once insured, the insuree has less incentive to take preventative actions, relying on the insurance to hold him harmless; thus fire insurance could lead to an increased risk of fire, or even to arson. Again from the point of view of the insurer, this is a "moral hazard."

knowledge to evaluate the effort invested or the outcome accomplished by professional agents (Sharma 1997, p. 768).

In addition to developing conceptions of the particular challenges facing those who want to proceed with a transaction in the face of asymmetrical knowledge, scholarship on the principal/agent problem has focused extensively on exploring solutions to them. The challenges are practical ones, and the solutions are therefore diverse, tailored to the immediate context. As it turns out, principals and agents have developed a very wide array of methods to address, work-around, hold harmless from, and re-distribute the risks of things going wrong. Consider:

The principal can offer *incentives*, sharing the risks or benefits of the transaction with the agent. Examples of this include low price/high deductible health insurance (sharing the cost of illness), and compensation schemes such as piecework rates and stock options (sharing the profits of the enterprise). Such incentives can align more tightly the principal's and the agent's goals, reducing the likelihood of shirking, risky behavior and the activities serving only the agent's individual interests (i.e., moral hazard).

In another attempt to deal with moral hazard, the principal can try to monitor the agent's performance in working on the principal' affairs using commonsense indicators, perhaps comparing some obvious features of the agent's activities or outputs with those of other, similarly-situated agents.

The principal can demand, or the agent offer, a *bond*, taking hostage something of value which the agent will lose if the transaction goes awry. The agent's willingness to offer something like a money-back guarantee or a 10 year warranty sends a strong signal of her confidence in her own information, thus addressing the challenge of adverse selection. Brand names can also function as a kind of reputational bond, since the agent's interest in preserving his good name for future transactions assures the principal that he is representing himself accurately now.

To conclude this much-abbreviated list, note finally that both participants can involve *third parties* in any of these solutions (see especially (Klein 2000, Shapiro 1987)). Principals and agents can engage further agents who specialize in alleviating information asymmetries by assessing competencies and performances. As one scholar has put it, "where principals must rely on strangers to act on their behalf and lack access, expertise, and clout to specify, evaluate, and constrain their performance, they often entrust a second tier of agents ... to be the gatekeepers to and watchmen over positions of trust" (Shapiro 1987, p. 639). In the U.S., for example, refrigerator buyers rely on Consumer's Union, used car buyers on Kelley Blue Book, and bond buyers on Moody's Investors Services. Of course, relying on these second order agents raises the principal/agent problem in a new guise; a difficulty to which I will return below.

It is here that reconceptualizing the appeal to expert opinion as an instance of the principal/agent problem begins to bear fruit. The critical questions Walton outlines for assessing expert opinion attempt to reveal and address the most important challenges to the appeal. The general conception of principal/agent transactions can help us to accomplish this task more systematically. We can reframe the critical questions as probes for the presence in a given transaction of adverse selection or moral hazard, or of some of the solutions just listed. This may enable us to re-organize some of the traditional wisdom, see why it is correct, and perhaps even propose new critical questions.

Consider first moral hazard–the danger that an expert will not in fact use his expertise for the layperson's benefit. The layperson can reasonably expect this to occur when the expert has interests which do not align with her own interest in seeking expert advice. Walton's list of critical questions captures this idea by recommending the person assessing expert opinion to ask whether the expert is biased (question 4, subquestion 2), since ideological or financial bias does provide an incentive for the expert to distort her opinion. In addition to outright bias, an expert always has an interest in expending as little effort as possible. We know all too well (to mention just one example) that physicians may avoid the labor of providing high quality medical advice by spending too little time with patients, by failing to listen to their experiences, by settling on the first diagnosis that comes to mind, or by prescribing the most common treatment. Indeed, in the U.S. the organization of health care systematically encourages many doctors to do what from the patients' perspective must be considered shirking. A prudent layperson assessing an expert opinion, in addition to checking for bias, should also in many cases question whether the expert seems to be telling her just what she wants to hear, or dismissing her concerns. Walton expresses the core of this idea by recommending as a critical question, "Is the expert conscientious?" (question 4, subquestion 3).

If there is reason to suspect the presence of moral hazard, the person consulting the expert will likely want to take some efforts to monitor the expert's performance. Since she is a layperson, she will have to rely on commonsense benchmarks that do not themselves require inaccessible expert knowledge. Walton captures one key form of monitoring in what he calls the "consistency question," the fifth on the list of critical questions. One accessible sign of the quality of an expert's opinion is that it is in accord with the announced views of other experts; thus we have the ordinary practice of seeking a second opinion. To detect and deter shirking, the layperson may want to monitor the effort the expert is putting in, either by direct observation (where possible), or by asking the expert to recount the process by which he made his judgment. Note that in doing so, the layperson is not seeking to second-guess the expert's judgment or review the evidence for herself. Instead, the expert's conspicuous ability to recite in detail the grounds of his judgment can give the layperson some confidence that he took pains to have grounds at all. Although Walton may not have intended it, something like this method of monitoring expert performance could be encapsulated in his "backup evidence" question (#6), which directs the layperson to ask "is the expert's assertion based on evidence?"

It appears that several of Walton's critical questions are aimed at dealing with the challenge of moral hazard, testing whether the expert is using his best efforts on the layperson's behalf. Consider now adverse selection–the basic challenge of determining whether to rely on someone as an expert at all. There are some hints in Walton's account of expert authority that despite his insistence that expertise is inaccessible, he expects laypeople to be able to assess whether or not the expert knows what he is talking about. For example, the first premise of his *ad verecundiam* argument scheme is that "E is an expert" in the relevant field (p. 210). Elsewhere Walton recommends the layperson to ask, "is he a genuine expert," which requires "giving evidence that the individual cited has attained a level of skill in this field" (p. 226). Now, it may be true that in some cases, the layperson consulting the expert has enough knowledge of the subject matter to directly assess the expert's knowledge, as this version of the scheme seems to require.[4] As I have been arguing throughout this essay, however–based in large part on Walton's own views–it is more likely that the layperson does not know enough to directly assess expertise. Instead, the more the layperson needs the expert's

---

[4] Alternately, the person consulting the expert may possess "referred expertise" (Collins and Evans 2002)–an expertise specifically in understanding and managing others' expertise. For example, a scholar in one area may develop an understanding of some very general categories by which to assess good scholarship generally. Although this possibility is certainly worth following up, such referred expertise cannot be a very common possession, and so cannot give us a general solution to the principal/agent problem as it arises in expert/lay transactions. This is another thread I therefore leave for future work.

advice, the less able she is to assess what he knows; that was the core insight of the classical puzzle of expertise, and that also is the crux of the principal/agent problem.

Walton moves towards a more promising approach, however, when he frames the first critical question for the appeal to expert opinion as an inquiry into the purported expert's *credibility as an expert*. (p. 223). This formulation directs the layperson not to assess the extent of knowledge the purported expert possesses, but instead assess his worthiness for credit, or reliance, or trust in the immediate situation.[5] And although some people may be better at making such assessments than others, accomplishing this does not require any specialized knowledge or expertise. Figuring out whether those we deal with are trustworthy is an everyday task, one that we have many strategies to accomplish.

For one thing, the layperson may have information allowing her to assess the purported expert's trustworthiness. If, for example, she has dealt with the expert in the past, she may have personal experience that following his advice led to success. Further, if the layperson expects her relationship with the purported expert to extend beyond the immediate transaction, she may infer that preserving that relationship gives him an interest in representing himself accurately. In this case, the purported expert's reputation with the layperson functions as a bond. In putting himself forward as an expert, the expert stakes his "brand name" on what he says; he can expect to lose his good reputation if his advice turns out to be bad. In some circumstances, the expert's offering such a reputational bond can justify the layperson's trust in his advice.[6]

The critical subquestions Walton proposes for determining a person's "credibility as an expert" point to a further range of potential considerations. Walton properly advises the layperson assessing the expert to examine such things as his employment, his degrees, the testimony of his peers, and his publications (p. 223). Note that in each of these, the layperson is relying on third parties, those who are likely to share the relevant expertise themselves and thus are capable of assessing the purported expert's real knowledge. These third parties act as gatekeepers, presumably only hiring, publishing, granting degrees or good reputations to those who are truly qualified. As one scholar noted:

> scrutiny by professional peers neutralizes the agent's advantage of unique access to an esoteric body of knowledge and exposes the behavior of agents for comparison against the work and ethical standards of their respective community of professionals. Absent any asymmetry of knowledge, professional control of agents is accomplished by such means as determination and enforcement of a code of ethics, formal organization to periodically audit individual members, informal relations among colleagues, and licensure (Sharma 1997, p. 780).

---

[5]To clarify: the difference proposed here is between "directly" assessing another's expertise and "indirectly" coming to rely on another's expertise, based on assessing her as being trustworthy. When I review an essay in argumentation theory, I feel myself capable of assessing the author's expertise "directly"; when I read an article in a medical journal, I feel myself incapable of "direct" assessment, and instead proceed "indirectly," deciding whether or not to rely on it based on my assessment of whether the source is trustworthy. Elsewhere, I have attempted somewhat unsuccessfully to capture this distinction as one between "epistemic" and "pragmatic" justifications (Goodwin Forthcoming).

[6]See further (Hardin 2001) on trust as "encapsulating" another's interest in maintaining a relationship. Of course, in other circumstances reputational bonding will provide insufficient grounds for trust. If (for example) the layperson suspects that the purported expert doesn't care about his reputation, or if she realizes that will be unable to hold him responsible if things go badly, then she would be imprudent to trust him.

Or as Walton put it, "the question of who the real experts are may be best answered by the other experts" (p. 112). Institutions promoting lay access to expert-on-expert assessments are a major feature of contemporary life. In the U.S., for example, anyone who is a physician has been accepted to and has completed medical school, has been accepted to and has completed a residency at a teaching hospital, has passed examinations by the Board of Medical Examiners, and likely has been hired to work in a medical organization of some sort. At each step, his capacities for giving good medical advice have been scrutinized by other physicians, often on the basis of quite extended evidence-gathering. In many circumstances, it is entirely reasonable for the layperson to trust someone who has been thus multiply certified by those in a position to know.

Of course, in proceeding this way the layperson confronts another set of principal/agent problems: Why should she trust medical schools? What's the difference between the Board of Medical Examiners and other industry associations– the National Cattlemen's Beef Association, for instance? She can no more assess the knowledge and skill of these groups of experts than she can the knowledge and skill of the individual expert she's examining. Although full consideration of this question is beyond the scope of this essay, let me suggest that the layperson can rely on the same repertoire of strategies for addressing principal/agent problems that I have already discussed. Where she may not have experience with this physician, for example, she does have experience with the medical system as a whole; and if her experience is good, she has some reason for confidence in the judgments of the Board of Medical Examiners and other professional organizations (for a similar argument, see (Turner 2001)). In addition, the Board (and others) are offering her a reputational bond: she can presume that they are highly motivated to act well in order to retain her trust–and more generally, the public respect for their profession or "brand name"–in the future. The Board (and others) are also themselves subject to the scrutiny of yet further agents: principally, state and federal legislatures, who would likely pounce on any systematic malfeasance or harm to the public.

I have been contending here that argumentation theorists would do well to consider the puzzle of lay reliance on expert opinion as one instance of a more general problem in transactions between principals and agents. Borrowing from the existing scholarly literatures on the topic, we have seen how a layperson has the practical means to manage the challenges of adverse selection (determining who is an expert) and moral hazard (shirking and other bad behavior). The account I have given has preserved very much of what Walton has laid out in *Expert Opinion*: his understanding of the core puzzle of inaccessibility, as well as many of the critical questions he has developed for the practical task of assessment. At the same time, the account has emphasized, perhaps more than Walton did, that in judging whether to rely on what someone says, the layperson is assessing not the knowledge, but the trustworthiness of the purported expert. It is this shift of attention from assessing knowledge to assessing trustworthiness which allows non-experts to escape the two horns of the ancient dilemma. Although undoubtedly some people are better trust-ers than others, the skill of judging trustworthiness is widely distributed; indeed, it's available to anyone who is willing to devote some time to practicing it in their everyday life.

Reframing expert opinion in terms of principal/agent theory has other consequences as well. For one thing, it suggests an overall rough organization of the many possible critical questions into two loose clusters, one associated with adverse selection and the other with moral hazard. It also suggests a new perspective on what is occurring in the expert/lay transaction. While there is very little that the expert and layperson can do within the scope of a single transaction to redress the asymmetry of their knowledges, there are many methods for enhancing trust. What I have called here reputational bonding is an important example. By openly committing himself–by staking his reputation as an expert on the opinion he is giving–the expert can actually create for the layperson a new reason for trust. This strategy of reputational bonding shows us that the problem of assessing expert opinion is not only the *layperson's* problem. The expert also can have an interest in making an expert/lay transaction succeed: in order to flourish in his profession, to meet his professional responsibilities, or out of concern for the individuals or groups who might benefit from his advice. Experts thus have incentives to discover or create practical communicative means to signal their expertise in such a way that layfolk will be able to assess it with confidence.

This creativity of the expert/lay transaction is going to generate new problems and puzzles for argumentation theory. Because experts and layfolk can in practice invent new reasons for trust, theory will never be able to keep up–if by "keeping up" we mean compiling a list of the necessary and sufficient critical questions which establish the presumptive soundness of any appeal to expert opinion.[7] Of course there will always be good pedagogical reasons to simplify matters. Our textbooks will continue to provide undergraduates with brief, and therefore usable, lists of important considerations that will help them deal with the sorts of situations which are the focus of that particular argumentation course. But we cannot expect those lists to have more than heuristic value. Let me close by sketching briefly one example of the limitations of lists, and the creativity of communicative transactions.

As noted above, Walton recommends that the person assessing expert opinion should inquire after the purported expert's employment, his degrees, the testimony of his peers, and his publications. These are indeed good questions–ones I endorse myself–for traditional publications. They seem radically restrictive, however, for evaluating some of the sources of information available online. On popular informational websites such as eHow and Yahoo! Answers, it is generally impossible to verify even the identity of those contributing their opinion, much less their credentials. Should we therefore spurn these sites entirely? If we did so, we would be overlooking the quite intriguing strategies for securing trust that these sites have created, and continue to refine. The sites involve third party volunteers–other internet users–in commenting on and rating the worth of the answers. They also provide mechanisms which allow answerers to accumulate a reputation for answering, commonly by a system which awards points for answers, with more points for higher-rated answers and penalties for conduct that violates local norms. Yahoo! Answers, for example, associates each login name with a "level" from 1 to 7, based in part on how many user-selected "best answers" he or she has provided. EHow has a similar point system, which allows answerers to be ranked as Novices, Enthusiasts, Authorities, or even Masters. It seems safe to presume that answerers, some of whom have responded to thousands of questions, care about the reputations they have so laboriously accumulated. The person consulting the site can therefore reason that a highly-rated answerer would not put his reputation at risk by answering on a topic he knew nothing of, or by failing to put good effort into answering. This may not be much, but it is some reason for trust–and a reason created essentially from nothing. These websites enable something like reputational bonding to occur between experts and layfolk who did not know each other and will likely never communicate again.

---

[7]There nevertheless may be specific practical strategies for creating trust for which necessary and sufficient conditions can be established. I believe, for example, that *expert* (or *epistemic*) *authority*, a subtype of expert opinion more generally, is one such (Goodwin 1998), (Goodwin 2001), and expert *testimony* is another (Kauffeld and Fields 2005).

The systems developed by these websites may or may not yet be successful; each of us can make our own decisions about whether to consult them, and for what. But they should be of interest to anyone attempting to theorize the assessment of expert opinion. Conceptualizing the transaction between expert and layperson as one between agent and principal allows us to draw on a rich literature exploring the strategies for proceeding in the face of information asymmetries. It also permits us to appreciate the ancient dilemma as a practical problem, one for which inventive minds are even now creating new, practical solutions.[8]

[8]I want to thank Fred Kauffeld, John Fields and an anonymous reviewer for their useful commentaries and provocations. Research and writing of this essay was made possible by generous assistance from Iowa State University's Center for Excellence in the Arts and Humanities.

CHAPTER 11

AN OVERVIEW OF THE CARNEADES ARGUMENTATION SUPPORT SYSTEM

Tom Gordon

11.1 Introduction

Carneades is a set of Open Source software tools for supporting a range of argu-mentation tasks, based on a mathematical model of Doug Walton's philosophy of argumentation and developed in collaboration with him over the course of several years, beginning in 2006.[1] Work on Carneades is a research vehicle for studying argumentation from a more formal, computational perspective than is typical in the field of informal logic, and for developing prototypes of tools de-signed to be useful for supporting real-world argumentation in practice. Thus there have been several versions of Carneades, as we experiment with different formal models of various argumentation tasks and with different ideas for tools which might be useful for helping people to argue more effectively. Carneades is work in progress. There is still much to do and feedback from an empirical evaluation of Carneades tools may lead to further changes in the system.

We began this project by doing a use-case analysis of common argumentation tasks, as illustrated in Figure 11.1.[2] This is another three-layered model of argu-mentation tasks (Brewka and Gordon 1994, Prakken 1995), where the three layers are inspired by Aristotle's distinctions between logic, dialectic and rhetoric. We do not claim, however, that our conceptualization of these three layers is per-fectly faithful to Aristotle's views on these topics. Indeed, we anticipate that there may be some disagreement about how the tasks have been distributed among the layers in this diagram. For us, the important thing is not so much whether a particular task is considered logical, dialectical or rhetorical, as the simple recognition of the task as being important for the performance of some role in argumentation dialogues, and to try to capture which tasks depend on the results of which other tasks.

The logical layer, at the bottom of the diagram, covers the constructon of argu-ments from data, information, models and knowledge. We intend the sources of arguments to be very broad, ranging from sensory data, witness testimony and others kinds of evidence, across arguments from the interpretation of natural language texts, up to purely formal derivations of arguments from propositions expressed in some formal language, such as predicate calculus. We view argu-mentation schemes (Walton et al. 2008) not only as a useful tool for reconstructing and evaluating past arguments in natural language texts, but also as templates

---

[1] http://carneades.berlios.de

[2] See also (Macintosh et al. 2009), which applies this use-case diagram in a survey of argumentation software, to classify available tools.

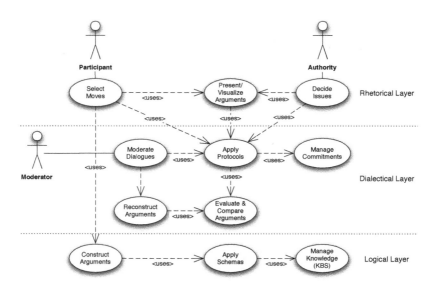

Figure 11.1: Argumentation Uses Cases

helping to guide users as they construct, 'invent' or generate their own arguments to put forward in ongoing dialogues (Gordon and Walton 2009a).

The dialectical layer, in the middle of the diagram, covers tasks relevant for comparing and aggregating potentially conflicting or competing arguments, put forward by opposing parties in argumentation dialogues, such as legal procedures before courts. Procedural rules, often called 'protocols', regulate the allocation the burden of proof among the parties, the assignment of proof standards to issues, resource limits, such as due dates for replying or limiting the number of turns which may be taken, and criteria for terminating the process, among other matters.

Finally, the rhetorical level, at the top of the diagram, consists of tasks for participating effectively in argumentation dialogues, taking into consideration the knowledge, experience, temperament, values, preferences and other characteristics of audiences, in particular one's opponent in a dispute. However, rhetoric is not only concerned with methods for taking advantage of an opponent to win a dispute. It is also about expressing arguments in clear ways which promote understanding, given the needs of the audience. We include at this level techniques for visualizing sets of interrelated arguments, i.e. *argument graphs*, as a particular class of methods for presenting arguments in ways which promote understanding.

Notice that the application scenarios which interest us, and which we want to support with software tools, are centered around dialogues, typically with two or more parties, in which claims are made and competing arguments are put forward to support or attack these claims. Following Walton, we recognize that there are many kinds of dialogues, with different purposes and subject to different protocols (Walton 1998b). This focus needs to be contrasted with the mainstream, relational conception of argument in the field of computational models of argument, typified by Argumentation Frameworks (Dung 1995), which views argumentation not as a dialogical process for making justified decisions which resolve disputed claims in the face of resource limitations, but as a method for inferring consequences from an inconsistent set of propositions, by maximizing in some way, depending on the semantics of the particular approach, the number of propositions in the set which can be assumed to be true simultaneously. To see the difference between these conceptions of argument, notice that a proposition which has not been attacked is acceptable in this relational model of argument, in all common semantics, whereas in most dialogues, in particular persuasion dialogues, a proposition which has not been supported by some argument is typically not acceptable, since most protocols place the burden of proof on the party which made the claim.

In the rest of this paper we describe the work we have done so far to support argument evaluation, construction and visualization before concluding and discussing our plans for future work.

11.2   Argument Evaluation

We begin in the middle, dialectical layer of Figure 11.1, because it is central to our work, and not just in the diagram. Since the main task of the bottom, logical, layer, is to construct arguments, and the main task of the top, rhetorical, layer is to present arguments, we first need to define what we mean by arguments and how they are evaluated.

Informally, an argument links a set of statements, the premises, to another statement, the conclusion. The premises may be labelled with additional information, about their role in the argument. Aristotle's theory of syllogism, for example, distinguished major premises from minor premises. The basic idea is that the premises provide some kind of support for the conclusion. If the premises are accepted, then the argument, if it is a good one, lends some weight to the conclusion. Unlike instances of valid inference rules of classical logic, the conclusion of an argument need not be necessarily true if the premises are true. Moreover, some of the premises of an argument may be implicit. An argument with implicit premises is called an *enthymeme* (Walton 2006c, p. 178).

We developed the mathematical model of argument which serves as the foundation for the Carneades software tools at the dialectical level in a series of papers (Gordon 2005, Gordon and Walton 2006, Gordon et al. 2007, Gordon and Walton 2009b). Let us focus here on the later, more mature papers. In (Gordon et al. 2007) we presented a formal, mathematical model of argument structure and evaluation which applied proof standards to determine the acceptability of statements on an issue-by-issue basis. The model uses different types of premises (ordinary premises, assumptions and exceptions) and information about the dialectical status of statements (stated, questioned, accepted or rejected) to allow the burden of proof to be allocated to the proponent or the respondent, as appropriate, for each premise separately. Our approach allows the burden of proof for a premise to be assigned to a different party than the one who has the burden of proving the conclusion of the argument, and also to change the burden of proof or applicable proof standard as the dialogue progresses from stage to stage. Useful for modeling legal dialogues, the *burden of production* and *burden of persuasion* can be handled separately, with a different responsible party and applicable proof standard for each. Finally, following Verheij (2003a), we showed another way to formally model critical questions of argumentation schemes as additional premises, using premise types to capture the varying effect on the burden of proof of different kinds of questions.

In Gordon and Walton (2009b), we developed this model further, with the aim of integrating the features of prior computational models of proof burdens and standards, in particular the model of Prakken and Sartor (2009) into Carneades.

The notions of *proof standards* and *burden of proof* are relevant only when argumentation is viewed as a dialogical process for making justified decisions. During such dialogues, a theory of the domain and proofs showing how propositions are supported by the theory are collaboratively constructed. The concept of proof in this context is weaker than it is in mathematics. A proof is a structure which enables an audience to decide whether a proposition statisfies some proof standard, where a proof standard is a method for aggregating or accruing arguments. There are a range of proof standards, from *scintilla of evidence* to *beyond reasonable doubt* in the law, ordered by their strictness. The applicable standards depend on the issue and the type of dialogue, taking into consideration the risks of making an error. Whereas finding or constructing a proof can be a hard problem, checking the proof should be an easy (tractable) problem, since putting the proof into a comprehensible form is part of the burden and not the responsibility of the audience. Argumentation dialogues progress through three phases and different proof burdens apply at each phase: The burdens of claiming and questioning apply in the opening phase; the burden of production and the tactical burden of proof apply in the argumentation phase; and the burden of persuasion applies in the closing phase.

The Carneades software, which is implemented in a functional style using the Scheme programming langauge (Dybvig 2003), enables arguments and argument graphs to be represented and proof standards to be assigned to statements in a graph. Argument graphs are immutable and all operations on argument graphs are non-destructive, as dictated by the functional programming paradigm. Every modification to an argument graph, such as asserting or deleting an argument, or changing the proof standard assigned to a statement, returns a new argument graph, leaving the original unchanged. The acceptability of statements in a graph is computed and, if necessary, updated at the time the graph is modified. Dependency management techniques, known from reason maintenance systems (Doyle 1979, De Kleer 1988), are used to minimize the amount of computation needed to update the labels of statements in the graph, as changes are made. Quering an argument graph, to determine the acceptability of some statement in the graph, just performs a lookup of the pre-computed label of the statement, and can be performed in constant time. An XML syntax for encoding and interchanging Carneades arguments, inspired by Araucaria's Argument Markup Language (Reed and Rowe 2004), has been developed, as part of the Legal Knowledge Interchange Format (ESTRELLA Project 2008). The Carneades software is able to import and export argument graphs in this LKIF format.

11.3   Argument Construction

Argumentation schemes are useful for reconstructing, classifying and evaluating arguments, after they have been put forward in dialogues, to check whether a scheme has been applied correctly, identify missing premises and ask appropriate critical questions. Argumentation schemes are also useful for constructing new arguments to put forward, by using them as templates, forms or, more generally, procedures for generating arguments which instantiate the pattern of the scheme. We elaborated the role of argumentation schemes for generating arguments in a series of papers (Gordon 2007a, 2008, Gordon and Walton 2009a), focusing on computational models of argumentation schemes studied in the field of Artificial Intelligence and Law for legal reasoning, including Argument from Defeasible Rules, Argument from Ontologies, and Argument from Cases.

Argument from Rules

The term "rule" has different meanings in different fields, such as law and computer science. The common sense, dictionary meaning of rule (Abate and Jewell 2001) is "One of a set of explicit or understood regulations or principles governing conduct within a particular sphere of activity." It is this kind of rule that we are interested in modeling for the purpose of constructing arguments. In the field of artificial intelligence and law, there is now much agreement about the structure and properties of rules of this type (Gordon 1995, Prakken and Sartor 1996, Hage 1997, Verheij 1996):

1. Rules have properties, such as their date of enactment, jurisdiction and authority.

2. When the antecedent of a rule is satisfied by the facts of a case, the conclusion of the rule is only presumably true, not necessarily true.

3. Rules are subject to exceptions.

4. Rules can conflict.

5. Some rule conflicts can be resolved using rules about rule priorities, e.g. *lex superior*, which gives priority to the rule from the higher authority.

6. Exclusionary rules provide one way to undercut other rules.

7. Rules can be invalid or become invalid. Deleting invalid rules is not an option when it is necessary to reason retroactively with rules which were valid at various times over a course of events.

8. Rules do not counterpose. If some conclusion of a rule is not true, the rule does not sanction any inferences about the truth of its premises.

One consequence of these properites is that rules cannot be modeled adequately as material implications in predicate logic. Rules need to be *reified* as terms, not formulas, so as to allow their properties, e.g. date of enactment, to be expressed and reasoned about for determing their validity and priority.

In the Carneades software, methods from logic programming have been adapted and extended to model legal rules and build an inference engine which can construct arguments from rules. Rules in logic programming are Horn clauses, i.e. formulas of first-order logic in disjunctive normal form, consisting of exactly one positive literal and zero or more negative literals. The positive literal is called the 'head' of the rule. The negative literals make up the 'body' of the rule. A rule with an empty body is called a 'fact'. In logic programming these rules are interpreted as material conditionals in first-order logic and a single inference rule, resolution, is used to derive inferences. Since there is no way to represent negative facts using Horn clauses, rules do not counterpose in logic programming, even though they are interpreted as material conditionals and the resolution inference rule is strong enough to simulate modus tollens. In Carneades, we do not interpret rules as material conditionals, but as domain-dependent inference rules. Both the head and body of rules are more general than they are in Horn clauses. The head of a Carneades rule consists of a set of literals, i.e. both positive and negative literals. The body of a Carneades rule consists an arbitrary first-order logic formula, except that quantifiers and biconditionals are not supported. Variables in the body and head of a rule are interpreted as schema variables. Using de Morgan's laws, Carneades compiles rules into clauses in disjunctive normal form. Given an atomic proposition P, a rule can be used to construct an argument pro or con P if P or $\neg P$, respectively, can be unified with a literal in the head of the rule.

The burden of proof for an atomic proposition in the body of a rule can be allocated to the opponent of the argument constructed using the rule, by declaring the proposition to be an exception. The syntax of rules has been extended to allow such declarations. Similarly, a proposition in the body of a rule can be made assumable, without proof, until it has been questioned by the opponent of the argument. These features make it possible to use Carnedes rules to model a broad range of argumentation schemes, where exceptions and assumptions are used to model the critical questions of the scheme. Whether a critical question should be modeled as an exception or an assumption depends on whether the "shifting burden" or the "backup evidence" theory of critical questions is more appropriate (Gordon et al. 2007).

The Carneades inference engine uses rules to construct and search a space of argument *states*, where each state consists of:

topic. The statement, i.e. proposition, which is the main issue of the dialogue, as claimed by its proponent.

viewpoint. Either 'pro' or 'con'. When the viewpoint is pro, the state is a goal state if and only if the topic of the state satisfies its proof standard. If the viewpoint is con, the state is a goal state only if the topic does not satisfy its proof standard. Notice the asymmetry between pro and con. The con viewpoint need not prove the complement of the topic, but need only prevent the pro viewpoint from achieving its goal of proving the topic.

pro-goals. A list of clauses, in disjunctive normal form, where each clause represents a set of statements which might be useful for helping the proponent to prove the topic.

con-goals. A list of clauses, in disjunctive normal form, where each clause represents a set of statements which might be useful for helping the opponent to prevent the proponent from proving the topic.

arguments. A graph of the arguments, representing all the arguments which have been put forward, hypothetically, by both the pro and con roles during the search for arguments.

substitution. A substitution environment mapping schema variables to terms. The scope of variables is the whole argument graph. Variables in rules are renamed to prevent name conflicts when they are applied to construct arguments.

candidates. A list of candidate arguments, which have been previously constructed. A candidate argument is added to the argument graph, and removed from this list, only after all of its schema variables are instantiated in the substitution environment. This assures that all statements in the argument graph are ground atomic formulas.

The space of states induced by a set of argument generators, such as the generator for the scheme for argument from rules, may be infinite. Carneades is implemented in a modular way which allows the space to be searched using any heuristic search strategy. Common strategies have been implemented, including depth-first search, breadth-first and iterative-deepening. For all of these strategies a resource limit, restricting the number of states which may be visited in the search space, may be specified, to assure termination of the search procedure. The system is extensible. Further heursitic-search strategies, including domain-dependent strategies, can be implemented and plugged into the search procedure. By default, Carneades uses a resource-limited version of depth-first search.

The heuristic search strategies are implemented in a purely functional way. Given a state and set of argument generators for computing successor states, the successor states are modeled as a stream of states, where each successor state is generated, lazily, just before it is visited by the search strategy to check whether or not it is a goal state.

The Legal Knowledge Interchange Format (LKIF) also includes an XML language for rules, as well as arguments (ESTRELLA Project 2008). The Carneades software is able to import and export both arguments and rules in LKIF format.

Argument from Ontologies

In computer science, an *ontology* is a representation of concepts and relations among concepts, typically expressed in some decidable subset of first-order logic, such as description logic (Baader et al. 2003). Such ontologies play an important role in integrating systems, by providing a formal mechanism for sharing terminology, and also in the context of the so-called *Semantic Web* (Berners-Lee et al. 2001) for providing machine-processable meta-data about web resources and services. There is a World Wide Web standard for modeling ontologies in XML, called the Web Ontology Langauge (OWL) (McGuinness and van Harmelen 2004). The Carneades software includes a compiler from OWL ontologies, encoded using the syntax of the Knowledge Representation System Specification (KRSS) (Patel-Schneider and Swartout 1994), into Carneades rules, based on the Description Logic Programming (DLP) mapping of Description Logic axioms into Horn clause rules (Grosof et al. 2003). The latest version of OWL includes a rule language profile, called "OWL 2 RL", which is also based on DLP. We may in the future make an effort to assure that Carneades is compliant with the RL profile of the OWL 2 standard. Work is in progress on a compiler from the standard RDF/XML format for OWL directly into Carneades rules, to avoid the currently necessary intermediate step of translating OWL ontologies in RDF/XML first into KRSS format.

Ontologies and rules may be used together to construct arguments with Carneades. LKIF uses OWL to define the language of individual, predicate and function symbols, represented as Uniform Resource Identifers (URIs), which may be used in rules. URIs provide a world-wide way to manage symbols, avoiding ambiguity and name clashes. This enables very large knowledge bases to be constructed, in a distributed and modular way. LKIF files can import OWL ontologies and, recursively, other LKIF files.

Reasoning with OWL ontologies is defeasible in Carneades. Any argument derived from axioms of an ontology, using the translation of these axioms into Carneades rules, may be undercut or rebut by other arguments. Thus, unlike some nonmonotonic logics, such as Defeasible Logic (Nute 1994), Carneades does not distinguish between 'strict' and defeasible rules. All rules are defeasible, also those derived from the axioms of an ontology. We believe this simplification is justified for the kinds of domains we are interested in supporting, where practical reasoning about the "real world" is more important than reasoning about abstract mathematical concepts. Outside of mathematics, generalizations are rarely universal.

Argument from Cases

In the Artificial Intelligence and Law community a great deal of research has been conducted on case-based legal reasoning, particularly in common law jurisdiction such as the United States, where arguing with precedent cases plays a very central role in both legal education and practice. When trying to apply legislation to a particular case, at least two kinds of related interpretation problems must be faced. The first interpretation problem is to construct general legal rules from legislation expressed in natural language. The goal is to determine the operative legal facts which must be established when applying a legal rule in order to construct an argument for the conclusion of the rule. In the law of contracts, for example, there is a rule stating, essentially, that a contract requires an offer, acceptance and an exchange of promises ('consideration'). Here, 'offer', 'acceptance' and 'consideration' are the operative legal facts. They are abstract technical terms of law. The second interpretation problem arises when trying to apply these operational legal facts to the concrete facts of a particular case. Has an offer been made? Has consideration been exchanged? This problem, called the subsumption problem, also requires the interpretation of the legislation, in the light of past precedent cases. Operative legal facts typically denote 'open-textured' concepts (Hart 1961).

These interpretation problems are at the root of the continual debate about the proper role of judges and the relationship between the legislative and judicial branches of government, as we witnessed again recently during the confirmation hearings of Judge Sotomayor to the US Supreme Court. Since judges are not typically elected by the people, and may be thought to lack the same level of democratic legitimacy, many people insist that judges should interpret legislation narrowly. Judges should merely apply the law and not usurp the exclusive right of the legislature to make law. But an overly simplistic view of judging fails to appreciate the distinctions between legal norms and their necessarily imperfect and ambigious representation in natural language in legislation and the hermeneutic difficulties of interpreting natural language, even given the best of intentions and conscientious effort.

In the field of Artifical Intelligence and Law, computational models of case-based reasoning are arguably still at an early developmental stage, reflecting presumably the lack of a deep and widely accepted theory of legal reasoning with cases in legal philosophy. The most influential work on case-based reasoning in AI and Law is the HYPO system (Rissland 1989, Ashley 1990) and its successors, CABARET (Skalak and Rissland 1992) and CATO (Aleven and Ashley 1997).

In Carneades, we have implemented a reconstruction of CATO by Wyner and Bench-Capon (2007). For some legal issue, such as whether or not information was a trade secret, the legal domain modeled in HYPO, the precedent cases are analysed to collect the set of *factors* which were found relevant for deciding the issue. A factor is a proposition which tends to favor either one side or the other regarding the legal issue. For example, efforts to keep the information secret tend to support a finding that the information was a trade secret, whereas disclosure of the information to people outside the company tends to support a finding that the information was not a trade secret. Each precedent case is modeled as a set of such factors together with the decision of the court regarding the issue of interest, such as, in our example, whether the information was found to be a trade secret or not. Given the set of factors known or assumed to be true in the current case, Carneades constructs, for each precedent case, a set of six partitions of the union of the factors of the current case and the precedent case. For example, partition P1 is the intersetion of the pro-plaintiff factors in the precedent case and the pro-plaintiff factors in the current case. And partition P5 is the set of pro-defendant factors in the current case which are not in the precedent case. These partitions are used in computational models of six argumentation schemes to analogize cases, distinguish cases argued to be analogous as well as downplay these distinctions.

There is a problem integrating arguments from ontologies or rules with case-based arguments using this reconstruction of CATO, since ontologies and rules are modeled at a finer level of granularity, using predicate logic, than the factors in this model of case-based reasoning, which are at a more abstract, propositional level of granularity. This problem is overcome in Carneades using bridging rules, similar to the output/input transformers of Prakken (2008a), to map predicate logic formulas to factors. Since different instantiations of schema variables in the predicate logic formulas can get mapped to the same factor, losing any distinctions between these various formulas in the resulting argument graphs, this solution only works when one can safely assume that at most one predicate logic formula will get mapped to each factor during the analysis of a particular case. For example, this solution may not be adequate if two different pieces of information must be evaluated to be determined whether they are trade secrets in a single case, unless the problem can be reduced to two problems that can be analysed separately.

11.4   Argument Visualization

We have been experimenting with methods for visualizing Carneades argument graphs and designing graphical user interfaces for working with argument graphs. An important difference between our work and most prior work on argument visualization, with the exception of Verheij (2005), is that our diagrams are views onto a mathematical model of argument graphs and the user interfaces provide ways to modify, control and view the underlying model. Argument diagramming software for Wigmore (1908), Beardsley (1950) or Toulmin (1958) diagrams, such as Araucaria (Reed and Rowe 2004) or Rationale[3], lack this mathematical foundation. Essentially, the diagrams are the models in these other systems, rather than views onto a model.

Our approach gives us much freedom to experiment with different diagramming methods and user interfaces for manipulating Carneades argument graphs, without changing the underlying model of argument. In Gordon (2007b), we described a couple of different approaches, including one which is very close to Wigmore's style of argument diagramming.

One diagramming issue we have been discussing again recently, is how best to represent undercutters, i.e arguments which directly attack the inferential link between the premises of an argument and its conclusion. An undercutter claims that the premises of the argument being undercut do not give us reason to derive the conclusion of the argument, not even presumptively. In the Carneades mathematical model of argument graphs, undercutters are modeled as attacks on the major premise of an argument. If the major premise has been left implicit, in an enthymeme, then it is first made explicit by adding it to the argument, before adding the undercutting argument. Some diagramming tools, such as ArguMed (Verheij 2005), use a technique called 'entanglement' to visualize undercutters, in which the undercutting argument points to the arrow between the premises and conclusion of the argument being undercut. It would be possible to use this diagramming method in views of Carneades argument graphs as well, by visualizing the major premise of an argument as the arrow or link from the minor premises to the conclusion of the argument.

---

[3]http://rationale.austhink.com/

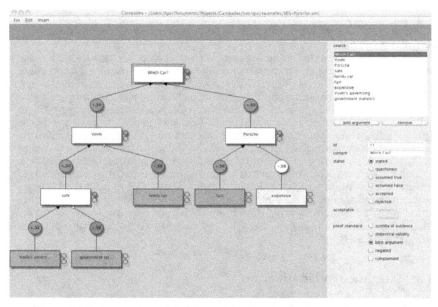

Figure 11.2: Screenshot of the Carneades Argument Diagramming Tool

The Carneades software includes a library for generating diagrams of argument graphs using Graphviz (Ellson et al. 2001), which can produce graphs in various file formats, including PDF and SVG[4]. These diagrams, however, are not interactive and thus do not provide a user interface for creating or editing argument graphs. Work is underway on an new argument diagramming tool for Carneades which will be much more interactive, provide better support for argumentation schemes and be integrated with other Carneades tools, such as the inference engine for generating arguments from ontologies, rules and cases. Figure 11.2 shows a screen shot of a prototype of this new tool. The Carneades argument diagramming tool will provide users with a variety of views onto Carneades argument graphs, including high-level, abstract views which show only attack and support relations among arguments, similar to diagrams of the kind typically used to visualize Dung Argumentation Frameworks (Dung 1995).

11.5 Conclusions and Future Work

The Carneades software is a toolbox for supporting various argumentation tasks, designed and built in collaboration with Doug Walton on the foundation of his philosophy of argumentation. The Carneades argumentation toolbox is work in progress. Here we summarized the tools currently available in Carneades for reconstructing, representing and evaluating arguments, for using a knowledge-base of ontologies, rules and cases to search for and construct arguments for both sides of an issue and for visualizing complex networks of arguments in any effort to make relationships among arguments clearer and more understandable, especially to lay audiences.

---

[4]http://www.w3.org/Graphics/SVG/

Much work remains. Although the Carneades inference engine for the LKIF was validated in pilot applications during the ESTRELLA project, further pilot applications and case studies are needed, especially with regard to other parts of the system, such as the argument visualization tools. Currently the system is too difficult to install and requires too much specialist knowledge to use. Our work on the new argument diagramming tool is a first step in the direction of developing a better integrated, easy-to-install and easy-to-use version of Carneades.

In addition to consolidating existing features, improving their usability with a rich graphical user-interface, we have plans for additional features, to support further argumentation tasks:

- The next version of the Carneades argument diagramming tool will include support for using argumentation schemes, both to classify and evaluate existing arguments and to construct new arguments, similar to the support for argumentation schemes provided currently by Araucaria.

- As discussed in (Gordon and Walton 2009a), Carneades currently also has a module for interacting with users to generate arguments from witness testimony. This serves as a dialogue component for Carneades which enables it to be used as an expert-system shell. As Carneades applies ontologies, rules and cases while searching for arguments to answer a query posed by the user, in a backward-chaining way, this component enables Carneades to ask the user questions to obtain information about the facts of the case. This component currently only has a command-line user interface and is thus little more than a proof-of-concept. A graphical user-interface for this feature is planned, for use in both web and desktop applications.

- Similarly, we would like to extend Carneades with a web-based user interface comparable to Parmenides (Atkinson et al. 2006a), to enable Carneades to be used as a collaboration tool on the web for collecting arguments, by generating surveys which help lay users to apply argumentation schemes and ask critical questions.

- The argument diagramming tool and the inference engine tool for generating arguments from knowledge-bases should be more tightly integrated, to implement a kind of graphical debugger and tracer. As the inference engine searches for arguments, the argument graph of each state in the search space could be visualized, along with the other information of the state, such as the list of open goals. We would like to provide users with a way to guide the search for arguments, for example by manually ordering the goals or backtracking to some prior state. This system would take us a step further towards our goal of making Carneades a tool for assisting users with argumentation tasks. Currently the argument construction module is too much like a fully automatic theorem prover.

- We would like to continue our collaboration with Lauritsen on document assembly to extend Carneades with a tool which uses the information in an argument graph to generate outlines of explanatory or justificatory documents, such as court opinions (Lauritsen and Gordon 2009).

- At the rhetorical level, we have begun work on modeling assumptions about the beliefs, values and preferences of audiences, along with methods, using a kind of abduction, which make use of this information to select goal statements to try to prove or disprove in dialogues.

- One of the anonymous reviewers suggested an interesting topic for future research: modeling knowledge about whether it is worthwhile to engage in argumentation in the first place. As he pointed out, arguing is not always the best way to resolve conflicts or coordinate actions.

- We are working with Brewka to apply his work on Abstract Dialectical Frameworks (Brewka and Woltran 2010) to clarify the formal relationship between Carneades and Dung Argumentation Frameworks (Dung 1995).

- Finally, we plan to revisit the research topic we addressed in the Pleadings Game (Gordon 1995), by extending Carneades with tools supporting the procedural aspects of argumentation dialogues. But unlike the Pleadings Game, which modeled a particular type of dialogue, our aim here is to provide Carneades with a way to define and use protocols for a variety of dialogue types. We would like to investigate the suitability of prior work on business process modeling tools and workflow engines for this purpose.

11.6   Acknowledgements

The work reported here was supported in part by a grant from the European Commission, in the ESTRELLA project (IST-4-027655). I would like to acknowledge my colleague, Stefan Ballnat, for his substantial contributions to the implementation of the Careades inference engine and Matthias Grabmair for his work on the implementation of the new argument diagramming tool for Carneades. Finally, I would like to thank Doug Walton. It has been an honor and a pleasure to be able to collaborate with him closely the last few years.

CHAPTER 12

THE GENERATION OF ARGUMENTATION SCHEMES

David Hitchcock

12.1 Introduction

Doug Walton's work on argumentation schemes is one of his central contributions to the theory of argumentation. His co-authored *Argumentation Schemes* (Walton et al. 2008) includes (pp. 308-346) a "user's compendium" of 60 argumentation schemes, some with alternative versions or sub-schemes. Most come with associated "critical questions" and references to publications, most by Walton himself, where the scheme is discussed in detail. Notably, Walton's *Argumentation Schemes for Presumptive Reasoning* (Walton 1996b) describes and analyzes (pp. 46-110) 25 argumentation schemes that Walton takes to involve what he calls "presumptive reasoning", defined as a tentative sort of reasoning that in a dialogue shifts a "weight of presumption" from the proponent of a thesis. The opponent can shift the weight of presumption back by asking a "critical question" associated with the scheme implicit in the proponent's argument. The proponent can in turn shift the weight of presumption back again by giving a satisfactory answer to the question. And so on (Walton 1996b, p. 46).

An argumentation scheme is a pattern of argument, a sequence of sentential forms with variables, with the last sentential form introduced by a conclusion indicator like 'so' or 'therefore'. The scheme becomes an argument when each variable is replaced uniformly at all its occurrences with a constant of the sort over which the variable ranges. A simple example is the pattern, 'x is human, so x is mortal', which becomes the argument 'Socrates is human, so Socrates is mortal' when the variable 'x' ranging over names and definite descriptions is replaced with the name 'Socrates'. As far as we know, this simple argumentation scheme is valid, in the sense that it is impossible for the conclusion of an argument fitting the scheme to be untrue while its premiss is true; the impossibility, we might suppose, is physiological rather than semantic. The argumentation schemes identified by Walton and others are at a higher level of abstraction, but not so abstract that they become purely formal schemes like the valid scheme for modus ponens arguments: 'If p then q; and p; therefore q'. Some writers, such as (Verheij 2003a, p. 171), distinguish intermediate-level argumentation schemes from highly specific domain rules, but admit that the boundary between them is fluid. At the intermediate level of abstraction typical of Walton's argumentation schemes, arguments fitting a given scheme sometimes have a good inference from premiss(es) to conclusion and sometimes do not, depending on whether certain conditions are met. For example, arguments fitting the generic composition scheme, 'All the parts of x are F; therefore, x is F' (Walton et al. 2008, p. 316), are valid if and only if the variable M of which F is a value is compositionally hereditary with respect to some kind K of aggregate to which x belongs, in the sense that, when all the parts of an aggregate of kind K have the same value of the variable M, then so does the whole aggregate. For previous formulations of this validity condition, see (Walton et al. 2008, p. 316) and (Walton 1989, Woods and Walton 1989).

As an aid to evaluation of the inference in an argument fitting a certain scheme, the theorist will provide a list of so-called "critical questions" to be asked, corresponding to the conditions under which arguments of the scheme in question have a good inference. In the case of generic composition arguments, there is a need for just one such critical question: Is the variable of which F is a value compositionally hereditary with respect to some kind to which the aggregate x belongs? In the case of the argument, 'All the parts of this chair are brown, so the chair is brown', the answer is positive, since colour is compositionally hereditary with respect to articles of furniture. But the answer is negative for the argument, 'All the parts of this chair are small, so the chair is small', since size relative to some fixed benchmark is not compositionally hereditary with respect to physical objects or to any other kind to which the chair belongs.

Where do these argumentation schemes come from? Arguments and reasoning do not come with labels indicating the argumentation scheme or schemes to which they belong. So how do theorists and textbook writers dream them up? How should they generate them?

First, we should note that writers on argumentation schemes often get their schemes from other writers on argumentation schemes. As Blair (2001, p. 375) points out, Perelman and Olbrechts-Tyteca (1958), Walton (1996b) and Kienpointner (1992a) all cite preceding writers as having articulated the schemes that they include in their lists. But this observation simply pushes the question of generation back in time, for somebody must have been the first person to have formulated a particular scheme. It is possible, of course, that we cannot take ourselves back to such an ur-moment, that we have lost our primeval innocence about patterns of argument through the incorporation of standard classifications into our everyday language, where we talk readily of signs, similarities, causes, testimony, experts, generalization and other concepts central to commonly recognized argumentation schemes. In that case, as a referee pointed out, tradition is a more basic source of argumentation schemes than either of the two approaches that I am about to distinguish.

12.2   Bottom-up generation

One can get an argumentation scheme from any argument, simply by replacing each of one or more "content expressions" (Hitchcock 1985) by a variable with a specified range that includes the replaced expression. Thus the argument 'Socrates is human, so Socrates is mortal' can be viewed as an instance of the scheme 'the individual object x is human, so x is mortal', but also of such schemes as 'Socrates is of kind K, so Socrates is mortal' or 'the individual object x is of kind K, so x has attribute F'. Working from the ground up in this way, from actual arguments selected in some manner, one has choices of how general to make one's abstraction. At the highest level of abstraction, the argument just mentioned is of the scheme 'p, so q', where the variables p and q range over sentences. At an only slightly lower level of abstraction, the argument has the scheme 'x is F, so x is G', where the variable x might range not only over names of individual objects but also over names of classes of individual objects, properties of individual objects and so on; and the variables F and G might range over predicates of all sorts, including classificatory, descriptive, evaluative and prescriptive predicates. This scheme has the advantage that it encompasses a very wide range of arguments, for each of which the rule that would license the inference has a single form: From the information that something is F, you are entitled to conclude that it is

G. However, rules of this form can be of many types, not only those licensing an inference from a classification to a descriptive attribute ('Socrates is human, so Socrates is mortal'), but also those licensing inferences from an infallible sign to a descriptive attribute ('she is giving milk, so she has given birth'), from a fallible sign to a descriptive attribute ('he is sneezing, so he has an allergy'), from a present causal factor to a predicted future effect ('he is arrogant, so he will lose friends'), from a descriptive attribute to an evaluation ('the novel has an implausible plot, so it is not very good'), from an effect to a prescription ('donating to Mdecins sans Frontires would help relieve a lot of suffering, so it's something you should consider seriously'), and so forth. To establish the legitimacy of a rule of inference for any of these types, one needs to establish that members of some specified class of objects that are F must also be G, either universally or for the most part or in the absence of overriding or undermining circumstances. So the goal of evaluation is the same for all of them. But the way in which one establishes the generalization varies from one type of inference to another. To establish that every member of a kind must have a certain attribute, one can appeal either to a definition of what it is to be of that kind or to empirically based well-established non-accidental generalizations about members of the kind. To establish that an attribute F is a sign of another attribute G, one needs empirically based good information that having attribute G typically causes members of some specified kind to which the individual x mentioned in the argument belongs to have attribute F. And so on.

There are choices not only of how general to make one's abstraction but also of how to group particular arguments into classes for the purpose of generalization. Sign arguments of the form 'x is F, so x is G' can be grouped with other sign arguments that are not obviously of that form, such as the argument 'There are dark clouds in the western sky, so it will soon rain'. One can reformulate this argument to fit the form 'x is F, so x is G', by treating the time and place of the utterance as the referent of x: 'There are now dark clouds in the sky to the west of here, so it will rain here soon after now', or, more formally, 'This place now has dark clouds to the west, so soon after now this place will experience rain.' Such recastings require some ingenuity on the part of a theorist trying to generate argumentation schemes from the bottom up.

12.3   A bottom-up approach: the "argumentative schemes" of Pereleman and Olbrechts-Tyteca

The variability inherent in a bottom-up approach to generating argumentation schemes is well expressed by Chaim Perelman and Lucie Olbrechts-Tyteca, in their ground-breaking work *La nouvelle rhétorique* (Perelman and Olbrechts-Tyteca 1958):

> What we wish to analyze in the following chapters are argumentative schemes of which the particular cases examined serve only as examples ... these same argumentative statements could be analyzed differently, in accordance with other planes of cleavage. And this is because there is no reason why a single statement cannot be regarded as capable of expressing several schemes which would act at the same time on the minds of different personseven on a single hearer. (Perelman and Olbrechts-Tyteca 1969, p. 188)

The descriptive approach of Perelman and Olbrechts-Tyteca illustrates a risk of the bottom-up approach. They pay no attention to how an arguer might establish or an audience evaluate the inferential component of a scheme. There is no mention of "critical questions". They are describing how people actually argue outside demonstrative contexts, and how rhetorical handbooks recommend that they argue. It is of no concern to them whether a form of argument actually establishes the truth of a factual statement, the wisdom of a recommendation, or the merits of an evaluation. All that counts is whether the scheme is effective in securing or intensifying the adherence of the intended audience.

As a result, they abstract forms of argument that some theorists think never have a good inference, such as the argument from waste:

> The argument from waste consists in saying that, since one has already begun a task and made sacrifices which would be lost in case of renunciation of the enterprise, it is necessay to carry on in the same direction. This is the justification furnished by the banker who continues to lend to his insolvent debtor in the hope of ultimately getting him on his feet again. (Perelman and Olbrechts-Tyteca 1969, p. 279)

Arguments from waste tend to be persuasive; people do not like to see their past efforts go to "waste". But previously expended resources are, in general, totally irrelevant to whether it is wise to continue a certain course of action. Economists refer to the argument from waste as the fallacy of "sunk costs" (see Mankiw et al. (2006, pp. 303-304)). According to standard economic analysis, it does not matter how much time or how many resources one has spent on a task. All that counts is whether continuing the task will bring its intended beneficiaries more benefits than costs. The example of the banker which Perelman and Olbrechts-Tyteca use illustrates the point perfectly. As a prudent lender, the banker must base his decision whether to lend more money to his insolvent debtor on the expected future gain or loss to the bank from continuing to lend on the one hand and ceasing to lend on the other. The money already loaned does not enter into the calculation.

Walton has argued Walton (2002b) that the argument from waste is not always fallacious, that it can be construed as an argument from pre-commitment to a course of action, which in some cases is a legitimate form of practical reasoning. His treatment is more nuanced than that of Perelman and Olbrechts-Tyteca, and is supported by careful consideration of the scholarly literature on the rationality of taking sunk costs into account, as well as by a typically nuanced discussion of a varied group of examples where people appeal to sunk costs in their reasoning about what to do. Without such theoretical investigation, bottom-up generation of argumentation schemes can baptize worthless patterns of argument. As Blair puts it, if argumentation schemes are in the first instance descriptions of patterns to be found in actual argumentation, "their normative force requires an explanation, for from the fact that people's arguments happen to exhibit a particular pattern, it does not follow that the conclusions of such arguments are warranted by their premises." (Blair 2001, p. 368)

12.4   Other bottom-up approaches

One can get argumentation schemes from traditionally recognized informal fallacies, which historically have been identified by abstraction from actual arguments. This approach was the one first taken by Douglas Walton. In their early collaborative work, John Woods and Douglas Walton brought the resources of non-classical formal logics to bear on the analysis of such argumentative moves as circular reasoning (*petitio principii*), reasoning from a sequence to a causal connection (*post hoc propter hoc*), projecting an attribute of a whole to its parts and *vice versa* (division and composition), and appeals to deference to authority, to force, to the person, to ignorance and to popular approval (*ad verecundiam, ad baculum, ad hominem, ad ignorantiam, ad populum*). Their co-authored papers, collected in (Woods and Walton 1989), commonly note that the argumentative move stigmatized as fallacious is in fact quite respectable under certain conditions. The move is only a fallacy when its associated critical questions have unsatisfactory answers. This general pattern of analysis continued in Walton's subsequent solo work on fallacies, which often took the form of an entire book on an individual fallacy or related group of fallacies. In his 1996 book an argumentation schemes for presumptive reasoning, Walton makes the link to his work on fallacies clear: "Many of the fallacies are misuses of presumptive inference." (Walton 1996b, p. ix)

Another indirectly bottom-up approach to generating argumentation schemes is implicit in Bart Verheij's "method for the investigation of argumentation schemes" (Verheij 2003a, pp. 174-181), which involves determination of the types of sentences in a scheme, of the scheme itself, of the exceptions blocking its use and of the conditions for its use. Verheij's method does not generate argumentation schemes, but rather presupposes an initial sense of them as objects for formal description. That initial sense comes from examples of their use in a given field, such as law.

12.5   Top-down generation

Bottom-up approaches to generating argumentation schemes presuppose the existence of arguments from which an argumentation scheme can be generated, and thus go naturally with the use of argumentation schemes for purposes of evaluation and analysis of pre-existing arguments. Top-down approaches do not require pre-existing arguments, and go naturally with the use of argumentation schemes to guide a search for evidence relevant to a question to be investigated (inquiry) or a proposition to be proved (invention). These different uses of argumentation schemes have been nicely distinguished by (Garssen 2001, p. 81).

Inquiry begins with a question, which may need to be refined as investigation proceeds. Invention is directed at support of a thesis, which can be regarded as an answer to a question. Argumentation schemes for inquiry and invention can thus be generated from a taxonomy of questions, which ideally would be based on a single principle of division that creates a set of jointly exhaustive and mutually exclusive classes of questions, where each class has in common a single quasi-algorithmic pathway to determining the correct answer to the question. Working out such a pathway would require answering Blair's trenchant questions about argumentation schemes conceived as in the first instance a priori prescriptions for cogent or presumptive argumentation: " ... on what principles are they formed? Where do they get their probative force?" (Blair 2001, p. 368)

The ideal of a neat Porphyrean tree of species of questions, each with its associated quasi-algorithmic pathway to its corresponding answer, is however practically unachievable. For example, the forum in which the results of inquiry and invention are to be articulated can influence the nature of an argumentation scheme. In scientific inquiry, the goal is to eliminate decisively all but one of the plausible explanations of a phenomenon under investigation. This goal may take decades to achieve, as in biologists' investigation of the way in which energy is released in the cells of living organisms, an investigation ably described in Chapter 4 of (Weber 2005). In a legal proceeding, on the other hand, the goal varies according to whether the proceeding operates on an adversarial or an inquisitorial model and according to whether the inquirer or inventor is an advocate for one side or a judge. In a criminal case in a common law jurisdiction, the goal of the prosecution is to show beyond a reasonable doubt that the accused is guilty as charged, the goal of the defence is to show that there is at least a reasonable doubt about the same proposition, and the goal of the judge is to determine whether the evidence, testimony and arguments presented by the two sides collectively establish beyond a reasonable doubt that the accused is guilty as charged. In a civil case, on the other hand, the judge is to determine the issue (e.g. whether there has been a breach of contract) on a balance of probabilities.

Furthermore, in specialized fora the search for evidence that would support a favoured explanatory hypothesis may be only part of the task of invention. The *stasis* theory of the ancient rhetorician Hermagoras of Temnos distinguishes four types of issues (*staseis*) that the defence can raise in response to an accusation: conjecture (whether the accused performed the deed), definition (whether the deed meets the definition of the alleged crime), quality (whether the deed was justified), objection (whether the procedure for bringing an accusation has been followed); see for example the description of this system in Cicero's *De Inventione* I.10-16 (Cicero 1949). This system could be used to generate an argumentation scheme for establishing the guilt of an accused, but the scheme would quickly ramify into a large number of sub-schemes, depending on the elements of the crime alleged to have been committed and the procedural requirements for bringing a charge.

Further, some argumentation schemes straddle different types of questions. Reasoning by analogy, for example, is appropriate in any inquiry where there is an inadequate theoretical basis on which to work out an answer to the question. Such situations arise with a variety of types of questions, including questions about possible explanations of a puzzling phenomenon, questions about what is morally or legally required in a novel or perplexing type of situation, and questions about the current market value of a piece of real estate. The ideal of a systematic division of argumentation schemes for invention and inquiry on the basis of a principled division of questions into mutually exclusive and jointly exhaustive classes thus seems at most partially attainable.

12.6 A top-down approach: Grennan's combinatorial generation

Perhaps the most systematic purely top-down approach to generating argumentation schemes is that of Wayne Grennan in his *Informal Logic: Issues and Approaches* (Grennan 1997). Without any reference to a corpus of arguments from which schemes are abstracted, Grennan tries to generate a complete list of all forms of single-premiss arguments that have the potential to give inductive support to their conclusion, in the sense that under certain circumstances the conditional probability that the conclusion of an argument of the given form is true given that the premises are true is greater than 0.5 but less than 1. In applying this approach, Grennan ignores other evidence for or against the conclusion, in effect assuming for the evaluation task that the conclusion has a prior probability of 0.5, in other words that it is no more likely to be true than to be false. Grennan calls such potentially probability-increasing forms "presumptively valid"; one could also call them "potentially inductively strong".

Grennan starts by distinguishing eight sorts of claims that may occur as a premiss or as a conclusion: obligation claims, supererogatory recommendations, prudential recommendations, evaluative claims (which may be either gradings or rankings or comparisons), physical-world claims (which may be either brute facts or institutional facts), mental-world claims, constitutive-rule claims (based on necessary truths and falsehoods, e.g. definitions), and regulative-rule claims (expressing obligations or permissions in a system) (p. 162). Each of the 64 re-

sulting possible combinations could potentially fit into any of nine types of argument patterns: cause to effect, effect to cause, sign, sample to population, parallel case, analogy, population to sample, authority, and ends-means. Thus ab-
stractly there are 576 (9 times 64) combinations of premiss-type, conclusion-type
and argument-pattern-type to be examined to see if the argumentation scheme
so constituted could be presumptively valid. However, many combinations are
impossible; by ruling out such impossible combinations as an argument from authority whose premiss is a supererogatory recommendation, Grennan narrows
the field to 227 combinations, some of which however are never presumptively
valid. For the rest, Grennan provides a "sketch" that includes the premiss form,
the conclusion form, an example, the warrant backing and rebuttal factors. His
treatment of the first of four valid patterns with obligation premisses will indicate
the nature of his treatment:

> *Arguments with Obligation Premisses ... Sample-to-Population Version*
> Premiss Form: $N\%$ of x's must do A.
> Conclusion Form: $N\%$ of X's must do A.
> Example: "Seventy percent of the 100 15-year-olds polled in Hali-
> fax must be home by 10:00 P.M. on weekday nights. Therefore, 70%
> of Canadian 15-year-olds must be home by 10:00 P.M. on weekday
> nights."
> Warrant Backing: x is representative of X.
> Rebuttal Factors: (1) The sample is too small; (2) there is systematic
> bias in the sample selection. In the example it is plausible to think
> that a systematic bias results from conducting the poll in a small ge-
> ographic area. (p. 166)

It may be doubted whether there is much benefit to distinguishing as separate argument patterns eight different ways of arguing from sample to population, according to which type of claim occurs as premiss and conclusion. In fact,
Grennan takes sample-to-population arguments with a supererogatory (p. 170)
or prudential (p. 174) premiss and conclusion to be useless for proof, because
anyone doubting the conclusion would be just as likely to doubt the premiss,
and he notes that there cannot sensibly be sample-to-population arguments with
a constitutive-rule (p. 195) or regulative-rule (p. 197) premiss and conclusion.
The valid patterns of sample-to-population reasoning have a premiss and conclusion that are both either obligation (p. 166) or grading (p. 177) or ranking
(p. 178) or comparison (p. 179) or physical-world (p. 186) or mental-world (p.
190) claims. For each of these six valid patterns, Grennan proposes as the warrant
backing that the sample is representative of the population and as rebuttal factors
that there is a sample-selection bias and that the sample is too small. Apart from
the apparently accidental omission of the rebuttal factor of small sample when
the premiss is a ranking claim, and the duplication of the warrant and rebuttal factors for the two things being compared when the premiss is a comparison
claim, there is no difference in the evaluative questions proposed for the six argumentation patterns. So what is the point of distinguishing them? It would be
more useful, both theoretically and practically, to treat sample-to-population reasoning as a single argumentation scheme, to note that such reasoning can make
sense and be probative only with six types of claims as premiss and conclusion, to
elaborate in more depth on the warrant backing and rebuttal factors, and to note
the duplication of the evaluation questions when an argument projects a comparison from a sample to a population. In fact, Grennan's proposals for the warrant
backing and rebuttal factors leave much to be desired. The proposition that the
sample is representative of the population is not the backing for the warrant, but
is the warrant itself: If $N\%$ of a sample x has property F, then approximately $N\%$

of the population X has property F. The backing for such a warrant is complex, and can vary from one type of sample-to-population projection to another. If all members of the sample have property F, then the projection of the property to the population can be justified by theoretical reasons for taking the population to be uniform with respect to the variable of which F is a value; for example, all samples of a chemical compound can be expected to have the same solubility in pure water, so that testing one sample for solubility is enough, or perhaps two to check for contamination of the compound or the water or deficiencies in lab technique. Or one may establish representativeness by pointing out that the sample was selected by a genuinely random method from the population to which the property is being projected, where a genuinely random method is one that gives every member of the population an equal chance of being selected for the sample. Or, in cases where the method of selection was not random, one can weight the contribution of members of the sample to determining the percentage with property F so as to make the distribution in the sample of various properties thought to be associated with having property F correspond to the known distribution in the population of these properties. In fact, it is entirely artificial to put biased selection and small sample as rebuttal factors and representativeness as backing for the warrant. The size and manner of selection of the sample are required to establish representativeness; they are not just rebuttal factors.

Rather than 227 candidates for presumptively valid argument patterns, then, there are nine such patterns. In discussing each of them, one can recognize different sorts of premiss-conclusion combinations that fit the pattern. But, unless the evaluation questions differ substantially from one group of such combinations to another, there is no point in separate treatment of the combinations.

12.7   A mixed approach: Hastings' "modes of reasoning"

Several theorists, including Hastings (1963), Schellens (1985) and Kienpointner (1992a), take a mixed approach that combines a theoretically based taxonomy of claims or rules of inference with reference to a corpus of arguments. Hastings, for example, intended his "modes of reasoning" as guides to debaters, whose task is to frame and deliver arguments for or against a given proposition. Thus he identified six of his nine modes of reasoning by the type of conclusion to be argued for, in each case with a corresponding type of premiss suited to establish the conclusion, and in this respect his generation of argumentation schemes was top down. However, it was in another respect bottom up, in that he abstracted his nine modes from a corpus of 250 arguments selected from a variety of rhetorical sources: "persuasive speeches, informative speeches, real-life debates, discussions, legal discourse, and academic debates" (Hastings 1963, p. 12). The first six modes were as follows (pp. 25-93), with the frequency of their occurrence in Hastings' sample (p. 175) indicated in parentheses:

1. from example to a descriptive generalization (26%)

2. from criteria to a verbal classification (20%)

3. from definition to characteristics (7%)

4. from sign to an unobserved event (5%)

5. from cause to effect: prediction (10%)

6. from circumstantial evidence to hypothesis (6%)

The debater can identify the proposition being debated as one of the six types of conclusion distinguished in these modes, and can use the schemes to determine what type of evidence would support that type of conclusion and what type of evidence could be used to undermine or override evidence for that type of conclusion.

Hastings characterized the remaining three processes only in terms of their starting point, on the ground that they are usable in proving conclusions of various types (Hastings 1963, pp. 93-139):

7. from comparison (3%)

8. from analogy (2%)

9. from authority (testimony) (18%)

The remaining 3% of the arguments in Hastings' sample were unclassified.

12.8   Choices in the generation of argumentation schemes

As previously mentioned, bottom-up generation can produce schemes at various levels of abstraction. Greater generality makes a scheme more widely applicable and leads to a more manageable typology, but at the price of a certain crudity in the recipes the schemes provide. For example, the scheme for reasoning by analogy, if it is to be used for invention, directs someone who wishes to argue that a case of interest has a queried property (e.g. that a piece of information passed on to a company's competitor was a trade secret) to look for similar cases that are known to have the queried property, as well as for partly similar cases known to lack the queried property that can be distinguished from the case of interest. One may even include in the scheme the suggestion to look for similarities that are relevant, in the sense that they are values of variables that stand in more or less tight determination relations to the variable of which the queried property is a value (Ashley 1988). But such advice is too general, for example, for a real estate appraiser who is searching for recent sales of comparable properties in order to determine by analogy the market value of a property being appraised. The real estate appraiser needs to know *what* variables are relevant, such as the location, lot size, floor area of the building on the lot, and so forth. A general scheme for reasoning by analogy may be accurate as far as it goes, but its application as a tool for invention requires supplementation by knowledge of the particular field in which one is reasoning analogically. The same point holds if the scheme is to be used for analysis or evaluation of an already existing argument by analogy.

As a tool for analysis, a system of argumentation schemes should be based upon an empirical study of the arguments that people actually produce and the reasoning that they actually engage in. An important constraint on the development of argumentation schemes as tools for analysis is that they should not distort the form of the arguments from which they are abstracted. But the empirical adequacy of a system of schemes is no guarantee that scholars will or should accept it as a fruitful analytical tool. Perelman and Olbrechts-Tyteca, for example, developed a system of schemes on the basis of a 10-year empirical study of texts from the European rhetorical, literary and philosophical tradition. But their system has not won general acceptance, with the exception of its identification of dissociation as a distinctive form of argument. A further difficulty with the generation of argument schemes as tools for analysis is that the same argument may be in one and the same respect an instance of two different argumentation schemes, related to one another as genus and species. Walton, for example, raises this difficulty about abductive reasoning and argument from appearance. Argument from appearance, whose scheme is 'this object looks like an x, so it is an x' (Walton 2004, p. 244), is a species of abductive reasoning, whose scheme is either 'A is the most successful of the successful accounts that would explain data D, so A is the most plausible hypothesis' or 'an argument with conclusion C is the most plausible of the plausible arguments from data D, so C is the most plausible conclusion.' (Walton 2004, pp. 217-218) A possible solution to this difficulty is to mark explicitly in one's taxonomy the fact that one scheme is a species of another, leaving the genus in the taxonomy to cover arguments that do not belong to any of its identified species.

As tools for evaluation, argumentation schemes should reflect a well-grounded theory of good inference. Whatever one's preferred taxonomy of general ways of legitimately inferring conclusions from reasons, however, there is a difficulty with using such a taxonomy as a basis for generating argumentation schemes: some argumentation schemes straddle different types of legitimate inference. Arguments by analogy, for example, share the common form of projecting a queried property from one or more analogues to a case of interest on the basis of assumed similarities. Their inference is good if the assumed similarities are relevant to the possession of the queried property and are not "outweighed" by unmentioned relevant dissimilarities. But the determination relations that make a similarity or dissimilarity relevant vary in how tight or loose they are. Some are exceptionless, as in the determination of the province or territory in which an address in Canada is located by the first letter of its postal code, which licenses nondefeasible analogical inferences. Others reflect loose causal relationships where many factors affect a result of interest, like the market value of a piece of real estate, and the corresponding analogical inference is probabilistic and defeasible. Still others express criteria for attributing supervenient properties like classifications, evaluations or prescriptions, where inductive approaches do not naturally apply and corresponding analogical inferences are best understood as cases of Walton's plausible reasoning.

12.9   Summary and conclusion

One can generate argumentation schemes purely from the bottom up, first collecting a heterogeneous corpus of arguments that might be expected to exemplify the ways in which people actually argue and then grouping the arguments by perceived similarities of form unguided by any theoretical insight into criteria for a good form of inference. Such theoretically naïve groupings are somewhat arbitrary and are likely to prove unsatisfactory as a guide to understanding and evaluating arguments. At the opposite extreme, one can generate argumentation schemes from the top down, starting from taxonomies of statements and of rules of inference, in each case generated by epistemological considerations. Such empirically unrooted templates risk a combinatorial explosion of unrealized abstract logical possibilities. A more fruitful approach is to combine a framework of types of statements and of reasonable inference with an empirical base of actual arguments, with the goal of constructing a usable instrument for inquiry, invention, analysis or evaluation. The works of Hastings (1963), Schellens (1985) and Kienpointner (1992a) described in section 7 above provide good models of how such a combined approach can be carried out. The resulting system of schemes need not be complete. But it should be comprehensive, and the schemes should be distinguished in a natural way according to the set of critical questions belonging to each scheme.

CHAPTER 13

ON THE NATURE OF ARGUMENT SCHEMES

Henry Prakken

13.1 Introduction

Since the 1980s, computer science, especially artificial intelligence (AI) has developed formal models of many aspects of argumentation that since the work of Toulmin and Perelman were thought of as belonging to informal logic. Doug Walton is one of the argumentation theorists who has recognised the relevance of this body of work for argumentation theory. One of the concepts on which recent work in AI has shed more light is that of argument schemes (sometimes also called 'argumentation schemes'), which features prominently in Walton's work. A study of argument schemes from the perspective of AI is therefore very appropriate for this volume in honour of his work. More precisely, the aim of this paper is to use insights from AI to propose an understanding of the nature of argument schemes as a means to evaluate arguments, and to compare this understanding with Walton's own account of argument schemes.

Walton regards argument schemes as essentially dialogical devices, determining dialectical obligations and burdens of proof. Using this account, a procedure for evaluating arguments should take the form of a set of dialogue rules. I shall instead argue that argument schemes are essentially logical constructs, so that a procedure for evaluating arguments primarily takes the form of a logic. More specifically, I shall argue that most argument schemes are defeasible inference rules and that their critical questions are pointers to counterarguments, so that the logic governing the use of argument schemes should be a logic for nonmonotonic, or defeasible reasoning. The dialogical role of argument schemes can then be modelled by embedding such a logic in a system for dialogue, so that in the end argument evaluation with argument schemes is a combination of logical and dialogical aspects.

However, after having developed this account, I shall also argue that not all argument schemes naturally fit the format of defeasible inference rules and that there is an often overlooked distinction between two types of argument schemes. One type fits the model of defeasible inference rules, i.e., elements of a reasoning method, but another type can better be seen as a reasoning method in itself. To model reasoning with such argument schemes, an account is needed of how reasoning methods can be combined, and I shall argue that this goes beyond the usual nonmonotonic logics and dialogue systems.

13.2 Walton on the nature of argument schemes

A common description of argument schemes is that they are stereotypical non-deductive patterns of reasoning, consisting of a set of premises and a conclusion that is presumed to follow from the premises. Uses of argument schemes are evaluated in terms of a specific set of critical questions matching each scheme. More precisely, on a "standard account" (Godden and Walton 2007), uttering an instance of an argument scheme in a dialogue creates a presumption in favour of its conclusion and a corresponding burden of proof for the other side in the dialogue to defeat this presumption by asking critical questions. Asking a critical question defeats the presumption and shifts the burden to answer the question to the proponent of the argument. If he gives a satisfactory answer to the question the presumption is reinstated and the burden shifts back to the opponent of the argument to ask another critical question, and so on.

Recently, Walton has refined this "standard account" of the effect of critical questions (Walton and Reed 2003). Some critical questions ask whether a premise of the argument is true (and if this premise was not stated by the other side, such a question in fact identifies a hidden premise of the argument). Asking such a question creates as described above a burden on the other side to back the premise with further grounds. However, other critical questions point at possible exceptional circumstances which, if true, would defeat the use of the argument scheme. Asking such questions does not shift the burden back to the other side: instead, the one who asks such a question must back it up with some evidence as to why the exception would be true. Only if such evidence is provided, the burden of proof shifts back to the proponent of the original argument.

Let me illustrate this with the *argument scheme from the position to know*, as given by Walton (1996b, pp. 61–3):

a is in the position to know whether A is true
a asserts that A is true

A is true

which has the following three critical questions:

(1)   Is a in the position to know whether A is true?
(3)   Is a an honest (trustworthy, reliable) source?
(2)   Did a assert that A is true?

In Walton's standard account, asking any of these three questions puts a burden on the proponent of the argument to back the challenged premise with further grounds. However, in his refined account this only holds when the first or third question is asked: asking the second question instead puts a burden on the opponent of the argument to provide grounds why a would not be honest (trustworthy, reliable). Of course, sensible dialogue rules can permit the opponent to also combine the first and third question with grounds why the answer would be negative, but in Walton's account they should not require this.

As said above, on this account (in both the standard and more refined form), the role of argument schemes in evaluating arguments is essentially dialogical: an argument is a move in a dialogue and the scheme that it instantiates determines the allowed and required responses to that move by the other side. In other words, for an argument to be justified it is not sufficient for it to fit a recognised argument scheme: it should also survive a dialogical process of critical examination.

13.3   Logics for defeasible argumentation

For some thirty years now, logic-based AI has been concerned with the logical formalisation of 'nonmonotonic' or 'defeasible' reasoning. The first nonmonotonic logics were meant to formalise 'quick-and-dirty' reasoning with empirical 'default rules' (such as 'Italians are usually Catholic' or 'witnesses usually speak the truth'), where one applies a default rule if nothing is known about exceptions, but one is prepared to retract a conclusion if further knowledge tells us that there is an exception. Later the focus was broadened to other forms of defeasible reasoning, such as various forms of causal or temporal reasoning, induction and abduction. One might even say that AI has replaced the old philosophical distinction between deductive, inductive and abductive reasoning with a new distinction between deductive and defeasible reasoning, where induction and abduction are just some of the species of defeasible reasoning.

Technically, nonmonotonic logics have been defined in several forms. Originally, fixpoint, consistency-based and preferred-model approaches were very popular. Later argument-based approaches were introduced, according to which defeasible reasoning takes the form of constructing arguments, attacking these arguments with counterarguments, and adjudicating between conflicting arguments on grounds that are appropriate to the conflict at hand. For an overview see Prakken and Vreeswijk (2002).

Two kinds of such logics exist. Some, e.g. Bondarenko et al. (1997) and Besnard and Hunter (2008), locate the defeasibility of arguments in the uncertainty of their premises, so that arguments can only be attacked on their premises. Others, e.g. Pollock (1995) and Vreeswijk (1997), instead locate the defeasibility of arguments in the riskiness of their inference rules: in these logics inference rules are of two kinds, being either deductive or defeasible, and arguments can only be attacked on their applications of defeasible inference rules. Moreover, such attacks can have a weak or a strong form: weak attacks (which Pollock calls *undercutters*) only say that there is some exceptional situation in which the inference rule cannot be applied, without drawing the opposite conclusion, while strong attacks (which Pollock calls *rebuttals*) do draw the opposite conclusion.

For present purposes a combination of both approaches is needed, since as we saw above with the scheme from the position to know, presumptive arguments can be attacked both on their premises and (in a weak and strong form) on their presumptive inferences[1]. So from now on I will assume that arguments can be attacked in three ways: on their premises, on their inference and on their conclusion. If arguments consist of several inference steps, then the last two attacks may also be launched against an intermediate step in the argument. (Attacking an intermediate premise is the same as attacking an intermediate conclusion.)

What is the effect of these attacks on the justification of an argument? When one argument undercuts another then (other things being equal) we are clearly justified in accepting the undercutting and rejecting the undercut argument. However, with premise and rebutting attack this depends on an additional assessment of the relative strength of the two conflicting arguments. In many cases reasonable criteria are available for making such assessments. For example, sometimes a conflict between two contradictory applications of the position-to-know scheme can be adjudicated in terms of knowledge about the relative reliability of the witnesses or experts who are in the position to know. Similarly, two instances of the practical syllogism on how to achieve a certain goal might be compared on the positive and negative side effects of the two ways to achieve the same goal. Some argument-based logics even allow arguments on how other arguments should be compared.

This leads to an important notion of defeasible argumentation, namely the relation of *defeat* between conflicting arguments. Whatever conflict resolution criteria are appropriate, logically speaking we always end up in one of two situ-

---

[1] Recently I formalised such a combination in Prakken (2010).

ations: either the conflict cannot or it can be resolved. In the first case we say that both arguments defeat each other and in the latter case we say that the preferred argument defeats the other and not vice versa (or that the first argument strictly defeats the other). So 'X strictly defeats Y' means 'X and Y are in conflict and we have sufficient reason to prefer X over Y' while 'X and Y defeat each other' means 'X and Y are in conflict and we have no sufficient reason to prefer one over the other'. It should be noted that this binary nature of the outcome of the comparison does not preclude the use of comparative standards which are a matter of degree: even with such standards it must in the end still be decided whether a certain difference in degree is sufficient to accept one argument and reject the other.

However, the binary defeat relation between arguments is not enough to determine which arguments we can accept and which ones we must reject. Above I used the phrase "other things being equal" for a reason. Consider two rebutting arguments A and B based on two conflicting witnesses Alice and Bob. Even if, other things being equal, we would prefer Bob's testimony given that, say, he is an adult and Alice a child, the argument using Bob's testimony may be undercut by a third argument C expressing Carl's testimony that Bob's testimony is unreliable since he has a strong reason to hate the suspect. Then we have that B strictly defeats A but C strictly defeats B (since C undercuts B). Clearly in this case we are justified in accepting A and rejecting B even though B strictly defeats A, since A is 'reinstated' by argument C. This case is still intuitive but defeat relations within sets of arguments can be arbitrarily complex, so one cannot resort to intuitions and a calculus is needed.

What does this calculus look like? Currently there is no single universally accepted one and there is an ongoing debate in AI on what is a good calculus. However, for present purposes their differences do not matter: I therefore briefly sketch one simple and intuitive calculus that suffices for many applications, and which has the form of an *argument game* between a proponent and an opponent of an argument. (It is sound and complete for Dung's (1995) so-called grounded semantics.) Proponent starts the game with the argument to be tested and then the players take turns: at each turn opponent must defeat or strictly defeat proponent's last argument while proponent must strictly defeat opponent's last argument. Moreover, proponent is not allowed to repeat his arguments. A player has won if the other player has run out of moves. Now an argument A is *justified* if the proponent has a winning strategy (in the game-theoretic sense) in a game beginning with A, i.e., if he can make the opponent run out of moves no matter how she plays. In fact, the evaluation of arguments is three-valued: arguments that are attacked by a justified argument are *overruled* but sometimes two conflicting arguments are neither justified nor overruled, since they are equally strong and all other things are equal: then they are both *defensible*. Of course, all this is relative to a given information state. New information may give rise to new arguments that change the result. For example, new, previously unknown information may become available on the reliability of experts or witnesses, as in our above example of Alice, Bob and Carl.

The ideas can also be defined in terms of the so-called dialectical tree of all possible ways to play an argument game. The initial argument is *justified* if the proponent can choose his arguments in the dialectical tree in such a way that he always ends in a leaf with one of his own arguments. Figure 13.3 (taken from Prakken and Sartor, 2009) illustrates this calculus with an example dialectical tree. Each node in this tree is an argument. (The figure abstracts from their internal structure: in the simplest case an argument just has a set of premises and a conclusion, but when the argument combines several inferences, it has the structure of an inference tree as is familiar from standard logic.) Each link between two arguments is a defeat relation (mutual or strict depending on the arrows). Each branch of the tree is one way in which the game on proponent's top argument can be played. The colours have the following meaning. The idea (due

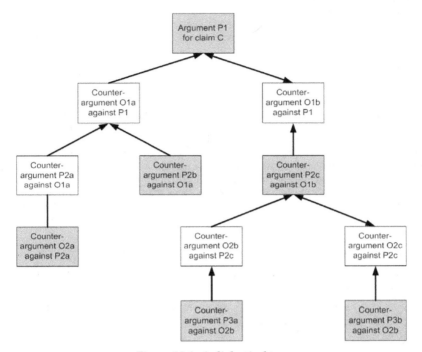

Figure 13.1: A dialectical tree

to Simari et al., 1994) is to label all arguments in the tree as *in* or *out* according to the following definition:

1. An argument is *in* if all its counterarguments are *out*

2. An argument is *out* if it has a counterargument that is *in*

In the figures *in* is coloured as grey and *out* as white. It is easy to see that because of (1) all leaves of the tree are trivially *in*, since they have no counterarguments. Then we can work our way upwards to determine the colour of all other arguments, ultimately arriving at a colour of the initial argument. If it is grey, i.e., *in*, then we know that proponent has a winning strategy for it, namely by choosing a grey argument at each point where he has to choose. If, on the other hand, the initial argument is white, i.e., *out*, then it is opponent who has a winning strategy, which can be found in the same way. So in Figure 13.3 proponent has a winning strategy, since he can always choose a grey reply and end up in a grey leaf node. To illustrate that all this is relative to a given information state, imagine that new information gives rise to a new defeater of P2b: then instead opponent could win.

13.4   The logic of argument schemes

In this section I give a logical account of argument schemes. I shall propose that argument schemes can be seen as defeasible inference rules and their critical questions as pointers to counterarguments. I shall then discuss two ways to formalise this proposal, and discuss an alternative logical account, in which argument schemes are types of generalised conditional premises of an argument.

Argument schemes as defeasible inference rules

As is hopefully clear by now, the concept of an argument scheme and a defeasible inference rule are very similar: both have premises and a conclusion, both are schematic instead of specific, and critical questions seem intimately related to the possibility of counterargument. The relation between the two concepts is particularly apparent in the work of John Pollock, e.g. (1987, 1995), who based his set of defeasible inference rules on general cognitive and epistemological principles, such as principles of perception, memory, induction and temporal persistence. Moreover, for each defeasible inference rule Pollock defined typical ways to undercut it, in the form of inference rules which specify circumstances under which the original inference rule does not apply. These undercutters are clearly close in spirit to some of the critical questions of argument schemes. Here are two examples. The first is Pollock's *perception principle*:

R_1: Having a percept with content p is a reason to believe p.

Pollock (1987) formulates a general undercutter for perception, which I paraphrase as follows:

U_{R1}: 'The present circumstances are such that having a percept with content p is not a reliable indicator of p' is a reason for not applying R_1.

Clearly, this undercutter is just the tip of the iceberg of theories on the reliability of perception. Pollock's favourite example of perceptual reasoning is: 'if an object looks red, it is red', which is undercut by 'the object is illuminated by a red light'.
    Another example is Pollock's *memory* scheme:

R_2: Recalling p is a reason for believing p.

for which Pollock (1987) defines the following two undercutters:

U_{R2}: 'p was originally based on beliefs of which one is false' is a reason for not applying R_2.
U'_{R2}: 'p was not originally believed for reasons other than R_2' is a reason for not applying R_2

(An aside: Bex et al., 2003 model reasoning with witness testimonies as a combination of Pollock's perception and memory schemes with Walton's position-to-know scheme. The reason is that the statement uttered by a witness is usually of the form 'I remember that I saw that q', so that to infer q from the witness statement, these three schemes must be successively applied.)
    Now at this point my proposal for defining the nature of argument schemes should come as no surprise: argument schemes are defeasible inference rules and critical questions are pointers to counterarguments (I earlier defended this interpretation with others in Bex et al., 2003 and a similar account was given by Verheij, 2003a, 2003b). More precisely, the three kinds of attack on arguments correspond to three kinds of critical questions of argument schemes. Some critical

questions challenge an argument's premise, others point to undercutting counterarguments, while again other questions point to rebutting counterarguments.
The above position-to-know scheme only has questions of the first (1,3) and second (2) type. An example of a scheme that has questions of all three types is
the *argument scheme from expert opinion* (cf. e.g. Walton, 1996b; Godden and Walton, 2007):

> E is an expert in domain D
> E asserts that A is true
> A is within D
> _____
> A is true

This scheme has six critical questions:

1. How credible is E as an expert source?
2. Is E an expert in domain D?
3. What did E assert that implies A?
4. Is E personally reliable as a source?
5. Is A consistent with what other experts assert?
6. Is E's assertion of A based on evidence?

Questions (1), (2) and (3) challenge a premise (respectively, the first and second
premise), questions (4) and (6) point to undercutters (the exceptions that the
expert is biased and that he makes scientifically unfounded statements) while
question (5) points to rebutting applications of the expert opinion scheme. The
Pollock-style undercutters corresponding to questions (4) and (6) are:

> U_E: 'E is not personally reliable as a source' is a reason for not ap-
> plying the expert testimony scheme.
> U'_E: 'E's assertion of A is not based on evidence' is a reason for not
> applying the expert testimony scheme.

Actually, Pollock also regards each valid inference scheme from propositional
or first-order logic as a reason, (which he calls *conclusive reasons*) and, to reflect
the deductive nature of these schemes, he disallows any undercutting or rebutting attack on them. Accordingly, such reasons can (as also suggested by Verheij, 2003b) be regarded as argument schemes of which the only critical questions
are whether their premises are true.

This logical account of argument schemes enables a simplification of the sets
of critical questions of argument schemes (likewise Verheij, 2003a, 2003b). Since
the kind of logic assumed in this paper allows for any premise and conclusion
attack, critical questions that point at such attacks need not be explicitly listed but
are automatically generated by the logic. Therefore, the only critical questions
that need to be listed are those that point to specific undercutters. So the above
version of the expert testimony scheme only needs questions (4) and (6).

An alternative logical account: argument schemes as generalised conditional premises

Before I further pursue the inference-rule interpretation of argument schemes, I
must first discuss proposals to regard them instead as generalised conditional
premises. Walton and Reed (2003) and Katzav and Reed (2004) add to each argument scheme a conditional premise *If other premises then conclusion* and then
apply modus ponens to derive the conclusion. Katzav and Reed (2004) defend

this treatment of argument schemes by saying that arguments express "relations of conveyance" between states of affairs instead of inferential relations between statements, so that argument schemes define types of conveyance relations.

To illustrate the difference with my inference-rule account, while I would formulate the scheme from the position to know as follows:

a is in the position to know whether A is true
a asserts that A is true

A is true (false)

the alternative proposal is to formulate it as:

a is in the position to know whether A is true
a asserts that A is true
If a is in the position to know whether A is true, and a asserts
that A is true, then A is true

A is true

Although at first sight, this alternative account would seem to preclude the need for defeasible inference rules, this is actually not true. Clearly, the 'if-then' in the latter version of the scheme cannot be a material implication, since the connection between the premises and the conclusion of the scheme is meant to be presumptive. So the conditional must be defeasible, which means that in this alternative account the logical nature of argument schemes boils down to the logical nature of defeasible conditionals. In particular, critical questions that in my inference-rule account point to undercutting, respectively, rebutting attacks must now be formulated in terms of a nonmonotonic-logic mechanism for handling exceptions to, respectively, conflicts between defeasible conditionals.

Now in the field of nonmonotonic logic many logics for defeasible conditionals have been developed, but a consensus is still far away. Moreover, unlike statements like 'birds fly' or 'the poor must be helped', argument schemes express general cognitive and epistemological principles, which makes it more natural to regard them as inference rules. Finally, in natural language argument schemes are almost never uttered as part of an argument, so that on the alternative account almost all natural-language arguments must be regarded as incomplete (likewise Govier, 1987). For all these reasons it seems more natural to regard argument schemes as (metalinguistic) inference rules.

Against this it might be argued that if argument schemes are formalised as inference rules, they can never be put into question as such (note that this is different from rebutting or undercutting them in specific situations, since such attacks still acknowledge the general plausibility of the attacked scheme). For example, it has been argued that witness statements are so often unreliable, that they should never be used to make inferences (Kaptein 2009). If instead the scheme is regarded as a premise, it can always be challenged or attacked as such, so the argument goes. My response to this is that perhaps this can be used as a criterion for deciding what are argument schemes and what are (contingent) conditional statements: if of a candidate scheme it is rationally conceivable that it is challenged as such, regardless of specific facts, then it should be formalised as a conditional premise, while if this is not conceivable, it should be formalised as an inference rule.

Summarising my account so far, there are two ways to regard argument schemes as logical constructs: as defeasible inference rules in a logic of defeasible argumentation, or as premise types in a nonmonotonic logic for defeasible conditionals. In both accounts critical questions reduce to three types of attack: in the first account on arguments, in the second on conditional premises. The state-of the art is that the inference-rule account is better formalised than the premise-type account, but this may, of course, change over time. I have given other reasons to prefer the inference-rule account, but they are more a matter of pragmatic convenience than of logical correctness.

Two ways to formalise the inference-rule account

In the remainder of this paper I will pursue the inference-rule approach to argument schemes. However, even within this approach there are two options for representing argument schemes, depending on whether the absence of exceptions is added as additional premises. Recall that there are three types of critical questions: those that ask whether a premise is true, those that ask whether there is an exception (undercutter) to the scheme and those that ask whether there is an argument with a contradictory conclusion (rebuttal). Sometimes the issue is raised whether critical questions of the second type point at implicit premises of the argument scheme.

Within the inference-rule approach, both answers to this question can be given a plausible formalisation (although again a negative answer seems closer to natural language). For instance, here is how the position-to-know scheme is formalised if the absence of exceptions is added to the premises:

> a is in the position to know whether A is true
> a asserts that A is true
> a is an honest (trustworthy, reliable) source
> _____
> A is true

The idea of this method is that undercutting attacks are reduced to premise attacks: for example, an argument that person a is dishonest now attacks the third premise of a position-to-know argument. This is a sensible idea, but it requires that two types of premises are distinguished. Since in the new method a positive answer to a type-two critical question is merely presumed, a counterargument that attacks such a presumption should always strictly defeat the argument using the scheme: no relative assessment should be needed to reach this conclusion. In the method that I have used so far, where positive answers to type-two questions are not added as premises, this result is automatically obtained, since undercutting counterarguments always strictly defeat their target. However, if undercutters are reduced to premise attacks then the problem arises that in the logics assumed in this paper the defeating nature of premise-attackers depends on a relative assessment of the premise and its attacker. So to retain in the new method the strictly defeating nature of negative answers to type-two questions, a second type of premise, which in Prakken (2010) I called *assumptions*, must be introduced in addition to ordinary premises. For example, in the new version of the position-to-know scheme the first two premises are ordinary ones, while the third one is an . More importantly, the effect of a premise attack should differ as to whether an ordinary premise or an is attacked: in case of attack on an ordinary premise a genuine choice must be made between the premise and the attacker, while attack on an always succeeds, i.e. always leads to strict defeat, by the very nature of an assumption. (In the next section we will see that this difference is also required to give a proper dialogical treatment of critical questions.) The method with two types of premises is used in the Carneades system of Gordon et al. (2007). (It also has a third type of premise; see further Section 13.5.)

Again the choice between the two methods is a matter of pragmatic convenience instead of logical correctness. In fact, it is easy to show that the two methods are logically equivalent. The main pragmatic difference is that the original method seems closer to natural-language, since natural-language arguments seldom make the presumed positive answers explicit. I will therefore below use the first method, but because of their equivalence everything will also apply in obvious ways to the second.

To end my logical account of argument schemes, I return to the issue discussed in Section 13.3 of how attack determines defeat, since the analysis of the present section allows me to make things more precise.

If two arguments rebut each other, then two things of the same kind oppose: two arguments that apply (possibly different) argument schemes to different premises, supporting contradictory conclusions. This attack is symmetric and should therefore be adjudicated on substantial grounds. So the defeat relations can go either way.

If one argument undercuts another one, then the one attacks a presumption of the other, namely a presumption that some exception does not arise. This is asymmetric attack: a presumption conflicts with evidence to the contrary, and the nature of presumptive reasoning then requires that the evidence to the contrary wins. So here the defeat relation always goes one way, from the undercutter to the undercut argument (except if two arguments undercut each other; for an example see Section 13.6).

Finally, what if one argument attacks a premise of another? Like undercutting attack, this attack is not symmetric, but unlike undercutting attack it is not automatically successful: whether this is so, depends on whether the attack is against an ordinary premise or an assumption. So to know which argument defeats the other we must examine the nature of the premise: where does it come from, is it assumed or is it a hard fact? If it is assumed, then the attacker strictly defeats it by the nature of presumptive reasoning. However, if it is a hard fact, the answer depends on the reasons why we made it so. In a dialogical setting often further reasons for the attacked premise will be given: then the counterargument takes the form of a rebuttal of a subargument, and we are back to the first case. So in reality it will not often happen that premise attacks have to be adjudicated on substantial grounds: either the attack is on an assumption so automatically succeeds, or the attacked premise is supported by a further argument so that the conflict transforms into a rebutting attack.

To give an example, suppose the prosecution in a murder case wants to prove that the victim had a fractured skull since the suspect had hit him with an angular object, and uses the following argument (by Walton, 1996b called an argument from evidence to hypothesis):

> A fractured skull of this kind can be caused by being hit by an
>   angular object
> A fractured skull of this kind was observed with the victim
> _____
> The victim was hit by an angular object

Suppose next that the defence challenges the first premises of this argument. Then the prosecution might answer with the following argument from expert testimony:

> The pathologist who investigated the victim said that a fractured skull of
>   this kind can be caused by being hit by an angular object
> The pathologist is an expert on this matter
> _____
> A fractured skull of this kind can be caused by being hit by an
>   angular object

Now a premise attack on the first argument is in fact a rebuttal of the second argument, which is a subargument of the first.

With these last observations we have reached the point that the use of argument schemes in dialogue must be discussed.

13.5   The role of argument schemes in dialogue

Above I agreed with Walton's view that fitting a recognised scheme is a necessary but not a sufficient condition for correctness of an argument. However, at first sight, Walton and I would seem to disagree on what else is needed. While I have argued above that an argument has to survive the *logical* competition with its counterarguments, Walton argues that an argument must survive a *dialogical* testing process with critical questions. So at first sight, Walton and I would seem to have incompatible views on the nature of argument schemes.

Yet there is a way to reconcile these two views. When above I said that an argument has to survive the competition with its counterarguments, I abstracted from the origin of the relevant counterarguments. Now one way in which these counterarguments can be identified is through a process of dialogue. This is one way in which dialogue is relevant for the justification of arguments. Another way in which dialogue is relevant for this purpose is that it provides additional ways to attack arguments beyond those provided by nonmonotonic logic. In particular, in dialogue an opponent of an argument could ask the proponent to provide further justification for an ordinary premise, and if proponent fails to provide such justification, opponent has defeated the argument without providing a counterargument. (Cf. the law, where in civil procedures plaintiff must usually provide evidence for any claim disputed by defendant on the penalty of losing on the issue.) To model such phenomena, we must move beyond logic and study dialogue systems for argumentation.

Embedding logic in dialogue systems

The formal study of dialogue systems for argumentation was initiated by Charles Hamblin (1970), and Doug Walton played an important role in its further development: see e.g. Woods and Walton (1978), Walton and Krabbe (1995). From the early 1990s several areas of computer science also became interested in the dialogical side of argumentation, such as artificial intelligence & law, multi-agent systems and intelligent tutoring. Of particular interest for present purposes are so-called persuasion dialogues, where two parties try to resolve a conflict of opinion. Dialogue systems for persuasion aim to promote fair and effective resolution of such conflicts. They have a *communication language*, which defines the well-formed utterances or speech acts, and which is wrapped around a *topic language*, in which the topics of dispute can be described. Walton and Krabbe (1995) call the combination of these two languages the 'locution rules'. The topic language is governed by a *logic*, which in argumentation theory is usually assumed to be standard, deductive logic, but which in AI & Law and MAS usually is a nonmonotonic one. The communication language usually at least contains speech acts for claiming, challenging, conceding and retracting propositions and for moving arguments and (if the logic of the topic language is nonmonotonic) counterarguments. It is governed by a *protocol*, i.e., a set of rules for when a speech act may be uttered and by whom (Walton and Krabbe, 1995 call them 'structural rules'). It also has a set of *effect rules*, which define the effect of an utterance on the state of a dialogue (usually on the dialogue participants' commitments, which is why Walton and Krabbe, 1995 call them 'commitment rules'). Finally, a dialogue system defines *termination* and *outcome* of a dispute. In argumentation theory the usual definition is that a dialogue terminates with a win for the proponent of the initial claim if the opponent concedes that claim, while it terminates with a win for opponent if proponent retracts his initial claim (see e.g. Walton and Krabbe, 1995). However, other definitions are possible. In my own work, e.g. Prakken (2005), I assume termination by agreement or external intervention, and I then define the outcome by applying an argument-based logic to the arguments exchanged in the dialogue. The benefits of such a definition will become apparent below.

Now what is important for us is that systems for persuasion dialogues allow for changing an information state by making new claims or stating new arguments (and sometimes by retracting earlier claims or arguments). This is an important respect in which dialogue systems go beyond the above logical account, which assumed a given and fixed information state. It is relevant for the evaluation of arguments in the following way. Since defeasible arguments can be defeated by counterarguments, the process of searching for information that gives rise to counterarguments is an essential part of testing an argument's viability: the harder and more thorough this search has been, the more confident we can be that an argument is justified if we cannot find defeaters.

Accordingly, a dialogue system for persuasion should be designed such that it furthers this goal: its speech acts should support changes of information states, and its protocol rules should promote that this happens in a fair and effective way. Dialectics is a driving principle here (Loui 1998): when one side is winning, the resources should be given to the other side, to give it the chance to overturn this intermediate result. This is fair since both sides can have their say, and it is effective since it avoids one-sidedness and tunnel vision. It also allows us to give a combined logical and dialogical account of the burden of proof (Prakken and Sartor 2009): if one side in a dialogue is winning in a certain information state (as determined by applying our argument-based logic to that state), the burden is on the other side to create a new information state where she is winning.

How do, on this account, logic and dialogue systems interact in the evaluation of arguments? In Prakken (2005) I proposed the following account (partly inspired by Gordon, 1995). The dialogical procedure provides the relevant arguments and counterarguments at each stage of the dialogue. Of those only the arguments that have no retracted premises and no challenged premises that are not supported by a further argument, are made input to the underlying logic. (An additional subtlety is that concessions raise the status of the conceded claim or premise to that of a hard fact). This logic then determines the justified arguments at that stage, after which the dialogical principle of dialectics shifts the burden of proof to the other side to make herself win at a new dialogue stage, and so on. Termination of a dialogue is largely governed by pragmatic considerations (how much time and other resources are left, how likely is it that new relevant evidence can be found?) and the ultimate justification of an argument is determined by applying the logic to the final information state. Thus the ultimate justification of an argument depends on both logic and dialogue, or more generally on both logic and investigation. This seems close to Walton's view, who in Godden and Walton (2007) says that evaluation of an argument is never complete until the dialogue itself is closed.

Argument schemes in dialogue systems

Let us against this background zoom in on the dialogical role of critical questions of argument schemes. Which of these questions should a dialogue system for persuasion allow? If we take the idea of argument schemes as presumptive forms of inference seriously, then it does not make much sense to allow critical questions that point to undercutters or rebuttals. Since argument schemes are presumptive inference rules, a negative answer to these questions is presumed and the opponent should present an undercutting or rebutting counterargument. So, for example, the opponent of an expert testimony argument should not just ask 'is the expert reliable?' or 'is the expert contradicted by other experts?' but present an undercutting argument for why the expert is not reliable or a rebutting argument that another expert said otherwise.

So in persuasion dialogue only critical questions that challenge a premise make sense (if the method with multiple types of premises is chosen then this must be further restricted to ordinary premises). Now this type of question is, of course, already built into dialogue systems for persuasion as the possibility to challenge any premise of any argument. Therefore, such dialogue systems need no special provision for critical questions of argument schemes: questions of the first type are part of their general structure, and questions of the second and third type are catered for by the possibility of moving undercutting or rebutting counterarguments.

This analysis fits well with Walton's recent refined account of the dialogical effect of critical questions, according to which only critical questions that ask whether a premise of the argument is true create a burden on the other side to back the premise with further grounds, while critical questions that point at possible exceptional circumstances must be backed up with some evidence as to why the exception would be true; only if such evidence is provided, the burden of proof shifts back to the proponent of the original argument. This is perfectly captured in the present account. Firstly, a critical question that challenges an argument's (ordinary) premise removes the argument from the input of the logical testing procedure and so makes its proponent currently lose, so that the burden of proof shifts to him to provide a further argument for the challenged premise, after which it re-enters the logical testing procedure. Secondly, moving an undercutter or rebuttal (or an assumption attack) adds the moved argument to the input of the logical testing procedure and so makes its mover currently win, so that the burden shifts back to the other side (which can then either challenge a premise of the counterargument or move a counter-counterargument).

I end this section with a few asides on complicating factors. Firstly, sometimes the type of question or premise is not clear beforehand but can itself become the subject of a (meta-) dialogue (like the question of who has the burden of proof can be at issue in legal procedure). Prakken et al. (2005) present a dialogue system that allows for such metadialogues. It is more complicated than the usual ones, but the big picture still applies: first the dialogue outputs a set of arguments and counterarguments, then the logic is applied to draw conclusions from it, then the burden is shifted to the party who is currently losing.

Secondly, Gordon et al. (2007) refine the distinction between ordinary premises and assumptions with a third kind of premise called *issue*, which must always be backed with a further argument, even if it is not challenged. Before such further backing has been given, an argument with an issue premise will not enter the logical testing procedure at all. Issue premises occur, for instance, in legal procedures, where legal claims that are not backed by evidence are usually ignored. While this complicates the analysis, the big picture as sketched above still applies.

Finally, above I assumed that information on the relative strength of arguments (which as we saw is sometimes needed to determine defeat relations) is always available at the time the conflict arises. However, this is not always the case. For example, in legal procedure conflicts are usually not decided until after the adversaries have exchanged all their arguments and counterarguments; only then the trier of fact (judge or jury) adjudicates the conflicts. In Prakken (2008b) I studied so-called adjudication dialogues, a variant of persuasion dialogues where the adversaries try to persuade not each other but a neutral third party. Again the resulting dialogue system is more complicated, but again it still fits the big picture sketched above.

13.6   Argument schemes that compress reasoning methods

At this point the reader might think that I propose to regard all argument schemes as defeasible inference rules, to be treated as sketched above. However, this is not the case. Some argument schemes turn out to be abstractions of more complex lines of reasoning, which may not always be naturally reduced to reasoning with defeasible inference rules. I shall illustrate this with some fragments from a recent Dutch murder trial (profiting from a case study by Bex, 2009).

In this case the proof that the accused was guilty of murder involved various intermediate conclusions. One of them was that the victim's fractured skull and brain damage were both caused by being hit by an angular object. The court proved this from the statements of a pathologist, who had declared that a fractured skull like the one of the victim can be caused by being hit by an angular object, and that brain damage like that of the victim can be caused by being hit by an angular object or by falling. Clearly, the court's proof combines several arguments from expert testimony with abductive causal explanations of the observed injuries. The expert testimony arguments are straightforward (one of them was listed above at the end of Section 13.4) but it is not immediately obvious how the causal reasoning can best be reconstructed. In fact, Walton has proposed various versions of his scheme for this kind of reasoning. In Walton (1996b) he gives the following *argument scheme from evidence to hypothesis*:

> If P is the case, then Q will be observed
> Q has been observed
> _____
> P is the case

which has the following critical question (in addition to those that ask whether the premises are true):

> Could there be another reason why Q has been observed?

The court's explanation of the fractured skull easily fits this format (if we are willing to read 'P can cause Q' as equivalent to the first premise). It was listed above at the end of Section 13.4. However, the court's explanation of the brain damage contains a subtlety, since it states two possible explanations of the brain damage (being hit and falling) and then chooses the first without explaining why. Let us first reconstruct both possible explanations as an instance of the scheme from evidence to hypothesis:

> *Explanation 1:*
> Brain damage of this kind can be caused by being hit by an angular object
> Brain damage of this kind was observed with the victim
> _____
> The victim was hit by an angular object

> *Explanation 2:*
> Brain damage of this kind can be caused by falling
> Brain damage of this kind was observed with the victim
> _____
> The victim had fallen

Note that because of the critical question of the scheme, both arguments undercut each other, so that a choice has to be made. The reasons for the court's preference will be clear: if we accept explanation 2, then we must still also accept the conclusion of explanation 1, since the latter explains the other observed injury, namely, the fractured skull, while the 'falling' hypothesis does not explain the fractured skull. By contrast, explanation 1 explains both injuries. So by the principle of parsimony (explanations should be as simple as possible), which is generally accepted in any theoretical account of abductive reasoning, explanation 1 is preferred.

Both in AI and philosophy detailed formal and computational models of this kind of reasoning have been developed. All have the following general form of a two-steps procedure (though they may differ in detail): given a set C of causal generalisations and a set F of observed findings, in the first step all possible abductive explanations of F are constructed by identifying all H such that:

H combined with C implies F; and
H combined with C is consistent.

Then in the second step suitable preference criteria are applied to select the best explanation of F.

Now the point is that in our example this kind of reasoning is not most naturally modelled in the account that I gave in Sections 13.3 and 13.4: in an argument-based logic the two explanations of the brain damage undercut each other so it must be determined which one defeats the other. However, in doing so we above referred to a third argument: the 'hit' explanation of the brain damage was preferred since it also explains the fractured skull. Now in an argument-based logic this third argument, which explains the fractured skull, is fully unrelated to the two arguments explaining the brain damage. Only in abductive models as just sketched does this relation become apparent, since these require that *all* findings in F are explained by H. For this reason I will from now on refer to these models as 'global' models of abduction. The word 'global' is meant to contrast with the 'local' application of defeat criteria to binary conflicts between arguments in the argument-based model described in Sections 13.3 and 13.4.

The reader may think that an argument-based account can still be natural if the Q in the scheme from evidence to hypothesis is instantiated with

A fractured skull *and* brain damage of this kind was observed with the victim.

However, this ignores that the court considered two possible explanations of the brain damage but only one of the fractured skull: so we cannot simply instantiate the second premise of the scheme with

A fractured skull and brain damage of this kind can be caused by being hit by an angular object.

nor with

A fractured skull and brain damage of this kind can be caused by being hit by an angular object or by falling.

So upon closer examination it turns out that the court did not really apply the scheme from evidence to hypothesis at all: instead, it seems that the conclusion that both injuries were caused by being hit was drawn on the basis of a global abductive reasoning process as just explained.

Interestingly, in later publications Walton proposed versions of an abductive reasoning scheme that better reflect the global nature of abduction. For example, Walton (2001a) proposes the following *abductive argumentation scheme*:

F is a finding or given set of facts
E is a satisfactory explanation of F
No alternative explanation E' given so far is as satisfactory as E
_____
E is plausible, as a hypothesis

with the following critical questions:

(1)   How satisfactory is E itself as an explanation of F, apart from the alternative explanations available so far in the dialogue?
(2)   How much better an explanation is E than the alternative explanations available so far in the dialogue?
(3)   How far has the dialogue progressed? If the dialogue is an inquiry, how thorough has the search been in the investigation of the case?
(4)   Would it be better to continue the dialogue further, instead of drawing a conclusion at this point?

To start with, it is important to see that the critical questions of this scheme differ in nature from those of all other schemes listed in this paper. Leaving the first two questions for a moment and focusing on the the third and fourth, we see that they are not typical for this particular scheme but are in fact a special case of the general embedding of nonmonotonic logic in dialogue systems that I proposed above in Section 13.5 (in fact, the same holds for the phrase "available so far" in the first two questions). So they should not be part of particular argument schemes, since they are automatically generated by a proper embedding of argument-based logic in dialogue systems. The same holds for Walton's tentative phrasing of the conclusion: by saying that E is plausible as a hypothesis, he wants to leave room for rejection of E after further investigation, but as explained above, this too is captured by the embedding of nonmonotonic logic in a dialogical context.

Moreover, the scheme itself is also different in nature from the schemes discussed above. To see this, let us reconstruct our example as an instance of this scheme. For the brain damage we obtain:

{ fractured-skull, brain-damage } is a given set of facts
{ hit-by-angular-object } is a satisfactory explanation
        of { fractured-skull, brain-damage }
{ hit-by-angular-object } is more satisfactory
        than all other explanations given so far
_____
{ hit-by-angular-object } is plausible, as a hypothesis.

Now note that the second premise in fact compresses an application of Walton's original scheme from evidence to hypothesis (note that the causal generalisation has disappeared), while moreover, the third premise at once compresses all other constructed applications of that scheme (in our example just one but it could be any number) plus an application of preference criteria to make a choice between the alternative explanations. Note also that the critical question of the original scheme is now hidden in the third premise. So there seems no meaningful way in which this argument can be undercut or rebutted, since all conflict resolution is done inside the premises. In other words, this is not just a single argument but a compression of a more involved reasoning process.

In fact, we may say that the truth of the second and third premise of the argument is verified by applying definitions of a global model of abduction. To make this point more clear, let us see how a persuasion dialogue could evolve if the premises of this argument are challenged. Consider first a challenge of the second premise. Then a reply could be

We know that `hit-by-angular-object` can cause `fractured-skull`
We know that `hit-by-angular-object` can cause `brain-damage`
`hit-by-angular-object` combined with our causal knowledge
   implies both `fractured-skull` and `brain-damage`
`hit-by-angular-object` is consistent with our causal knowledge
––
{ `hit-by-angular-object` } is a satisfactory explanation of
   { `fractured-skull, brain-damage` }

This makes sense, but only since this argument is nothing but a restatement of the first step of the two-steps global model of abduction sketched above. Essentially, the argument restates the global model's definition of an explanation, so its justification derives from that definition.

Likewise for the argument that can be given to support the third premise. In fact, this argument is even more complex:

We know that `hit-by-angular-object` can cause `fractured-skull`
We know that `falling` can cause `brain-damage`
{ `hit-by-angular-object, falling` } combined with our causal
   knowledge implies both `fractured-skull` and
   `brain-damage`
{ `hit-by-angular-object, falling` } is consistent
   with our causal knowledge
––
{ `hit-by-angular-object, falling` } is a satisfactory explanation
   { of `fractured-skull, brain-damage` }

Moreover,

{ `hit-by-angular-object` } is more parsimonious than
   { `hit-by-angular-object, falling` }
More parsimonious explanations are more satisfactory
––
{ `hit-by-angular-object` } is more satisfactory than
   { `hit-by-angular-object, falling` }

Moreover,

No further explanations of { `fractured-skull, brain-damage` } have
   been given
––
{ `hit-by-angular-object` } is more satisfactory than
   all other explanations so far

This argument also derives its justification from being part of the global model of abduction: the first part restates the global model's definition of an explanation, while the second part applies the global model's preference criteria.

In sum, the arguments that can be given in support of the premises of Walton's new abductive argumentation scheme do not instantiate separate argument schemes but are part of a single global model of abduction. We can therefore conclude that the new scheme abstracts from, or compresses a complex reasoning form, hiding its internal structure.

But then the question arises why we would take this detour via constructing arguments at all. Why don't we directly apply the global model of abduction instead of decomposing it into separate arguments? Perhaps a proponent of argument schemes would say that such a decomposition allows us to apply the mechanism of critical questions to critically evaluate the arguments. However, if we look at these questions, we see that their nature has changed. We have already seen that the third and fourth question of the new abductive scheme are in fact special cases of a general way of embedding logic in a dialogical context. Moreover, its first two questions do not really point at exceptions to a defeasible rule: it is not generally presumed that if one explanation is better than another, it is better to a certain degree. Likewise for the degree of satisfactoriness of an explanation in itself. These things can just as well be dealt with within a global model of abduction.

This leads to a generalised view on the relation between the logical and dialogical levels of persuasion dialogue. In Section 13.5 I embedded an argument-based logic in a dialogue system for persuasion, so that during a dialogue the participants in fact build a dialectical tree of arguments and counterarguments (which is then used to compute the intermediate or final outcome of the dialogue, according to the logic). However, we now see that sometimes the logic embedded in a dialogue system is of a different type: it can also be, for instance, a global logical model of abduction, so that during a dialogue the dialogue participants build a global causal model (which again can be used to compute the outcome of the dialogue, according to the abductive model).

This can be generalised even further. For example, in the Dutch murder case another intermediate conclusion was that a sledgehammer found at the crime scene had the victim's blood on it. This was proved on the basis of two estimates of conditional probabilities provided by DNA experts. Lack of space prevents me from analysing the court's statistical reasoning in detail, but here too it can be said that an argument-based reconstruction of the court's reasoning would conceal that the court actually applied a Bayesian model of this subproblem. So the 'logical' model that is built during a persuasion dialogue could be of any kind, as long as some notion of inference applies to it: it could be a logical, abductive, statistical, or even connectionist model (see Nielsen and Parsons, 2007 for a dialogue system for building a Bayesian network).

The Dutch murder case also motivates a final generalisation: since the court used different reasoning models on different subissues, the inferential model built during a dialogue in general consists of connected submodels of different kinds. Therefore, to compute the (intermediate or final) outcome on the final problem, we need a logical theory of combining different modes of reasoning. Recently, I proposed such a model in Prakken (2008a). Space does not permit me to give details here, but I hope that the above has convinced the reader that such a theory is needed.

Does all this mean that there is no place for argument schemes at all? On the contrary: in our murder example the input of both the abductive and Bayesian models was partly based on expert testimonies and witness testimonies, and this may well be modelled as argumentation with argument schemes. (Incidentally, this also shows that challenges of input elements of different reasoning models should be allowed.) Moreover, the output of an abductive or Bayesian analysis may be input to a new reasoning process with argument schemes, for example with a rule application scheme as proposed by Gordon and Walton (2009a). Indeed, in our murder case the court further used its factual conclusions to draw normative conclusions on the suspect's punishability and punishment. So in fact in our example the court combined three different kinds of reasoning: global abduction, Bayesian statistics, and reasoning with argument schemes.

Finally, I just said that we need a logical theory of combining different modes of reasoning. However, we also need a dialogical theory of how logical models of this kind can be built during dialogues, and how these models can be critically examined. A general theory is not yet available, but in Bex and Prakken (2008) a special case is developed for a combination of abductive reasoning and defeasible argumentation in the context of crime investigation: first a logical theory is defined of combined abduction and argumentation and then a dialogue system is defined for building a combined abductive and argument-based model of a criminal case.

13.7 Conclusion

In this article I have investigated the nature of argument schemes as they appear in the work of Doug Walton. Walton and I disagree in that while he regards argument schemes as dialogical devices, I have argued that they primarily are logical constructs and that their dialogical aspects are a special case of embedding an argument-based logic in a dialogical context. However, Walton and I agree in that we both hold that for an argument to be justified, it is not enough to fit a recognised argument scheme: it must also survive a critical testing process. Moreover, we agree that evaluation of an argument is not complete until this process is closed.

I have also argued that not all schemes appearing in the work of Walton are of the same logical nature: many can be regarded as defeasible inference rules in an argument-based nonmonotonic logic, but some appear to be compressions, or abstractions, of other kinds of reasoning. This calls for the need to study how different kinds of reasoning can be logically related and embedded in dialogue systems.

Finally, I do not claim that all argument schemes in the literature are of one of these two types: further investigation may reveal even more types of argument schemes. For example, maybe some can best be seen as templates for conditional premises (as in Katzav and Reed, 2004). And there may be still other kinds of argument schemes. At least I hope that the present investigation has convinced the reader that it is important to be aware that constructs that are presented in the literature as argument schemes may not always be of the same nature.

Part III

Theory of Reasoning

CHAPTER 14

REFLECTIONS ON REASONING, ARGUMENT AND LOGIC

J. Anthony Blair

14.1   Introduction

At first it might seem that this should be a one-short-paragraph paper, for reasoning, argument and logic and their connections appear to be straightforward. Reasoning is inferring, or the drawing of inferences or implications. Argument is the expression of inferences, that is, of reasoning. Logic is the study of the norms of good reasoning as it is expressed in arguments, hence, equivalently, the study of the norms of good arguments. Nothing could be simpler.

On this understanding, *'reasoning'* denotes a kind of mental act, namely one whereby *inferences* are made or *implications* are drawn. The propositional content of the act of inferring or of drawing an implication can be described or expressed verbally. The verbal expression of an act of inferring is (termed) an *argument*. A verbal expression of an inference has the general form, "from *this* may be inferred *that*" or "*this*, therefore *that*" (where "this" refers to propositions or sentences that are the basis or grounds of the inference, often termed the *premise(s)* of the inference or argument and "that" refers the proposition or sentence inferred from the grounds, often termed its *conclusion*). The verbal expression of the act of noticing an implication, which is also termed a *conditional* statement, has the general form, "if this, then that" or "this implies that." Inferences can vary in quality, depending on whether and to what extent the premises license the drawing of the conclusion. *Logic*, it is said, is the discipline that provides standards in terms of which inferences or implication claims may be assessed.

Well, not so fast, says Harman (for one, (Harman 1986)). Suppose you accept some proposition, *p*, and you see that *p* implies another proposition, *q*, so you reason as follows: "If *p* (which is plausible), then, given that *p* implies *q*, it follows that *q*." That is, you reason that *q* may be inferred from *p*, given that *p* implies *q*. The expression of your reasoning – the corresponding argument – is: "*p*, and if *p*, then *q*: therefore *q*." According to logic, this is an instance of a valid argument form, *modus ponens*, and so is a valid argument. So do you not then have a good argument for accepting *q*? And should you not infer *q*, given that *p* implies q? Not necessarily. If *q* is obviously false, then what you have is a good reason (and a good argument) – assuming that *p* does indeed imply *q*–not for inferring *q*, but for rejecting *p* in spite of its plausibility. So a logically good (valid) argument (i.e., expression of an inference) is not necessarily a logically good argument for (i.e., good reason to infer, or accept) its conclusion.

Harman's example might be taken to show that a logically valid argument is not necessarily a logically good argument. Yet in some sense of 'logic' a valid argument is a logically good argument, i.e., is an argument with a logically good implication from premises to conclusion. Is there more than one sense of 'logic'?

Or is there more than one sense of 'argument' – an implication relationship between propositions and a reason for accepting a proposition? Also, Harman's example shows that a valid logical implication does not necessarily map a logically good inference, so if the former counts as an argument and the latter as reasoning, is there a distinction to be drawn between argument and reasoning?

These considerations raise the questions whether the words 'reason' and 'reasoning,' 'inference,' 'argument' and 'good argument' (and perhaps also 'logic') can have different meanings or uses, or at least differing references. I shall try to sort them out.

14.2   Reasoning

"Reasoning," writes Adler (2008, p. 1), "is a transition in thought, where some beliefs (or thoughts) provide the ground or reason for coming to another." It would be natural to call such a transition in thought an inference. This is a useful starting point, assuming that Adler's description may be taken to cover hypothetical or suppositional transitions and beliefs.

'Reasoning' can mean the activity of reasoning, an event that occurs over time; it can mean a record of that activity, an expression of it in some linguistic or other communicable form; or it can mean the abstract entity that is the propositional content of the linguistic expression.[1] A hiker in Colorado who recognizes he has lost his way might engage in reasoning recorded as follows: "I'm facing the sun, it's noon, so I'm facing south. I think there's a road to the west and if I can find a road I can find my way, and I want to find my way, so, since the west is 90 degrees to the right when facing south, I should travel in that direction. But since it's been a mostly cloudy day, the sun might not reappear, so, because it is so easy to get disoriented when walking in the mountains, I need to have a landmark in the west that I can head towards. That distinctive mountain peak roughly due west of me is a landmark I can readily identify, so I should start to travel in the direction of that mountain peak." This reasoning process might have taken the lost hiker five seconds at most. I have *expressed* the hiker's reasoning in three sentences and about 100 words that can be read aloud in about 30 seconds.

In his episode of reasoning the lost hiker made, by my count (and on this description of it) seven inferences. (1) He inferred that he is facing south from his observation of the sun and the time of day and his knowledge that the sun lies to the south at noon (in Colorado). (2) He inferred that he should try to reach a road from his desire to find his way and his belief that if he finds a road he will be able to find his way. (3) He inferred that he should travel at right angles to the right from the direction of the sun from his knowledge that the sun is in the south and that west is 90 degrees to the right when facing south, from his belief that a road is to the west and from his desire to find a road. (4) He inferred that the sun might not reappear from his observation that it has been a generally cloudy day and his knowledge that on a generally cloudy day the sun appears rarely, if at all. (5) He inferred that he should identify a landmark in the west to travel towards from his belief that the sun might not reappear to provide directional orientation and his belief that it is easy to get disoriented when hiking in mountains, and his desire to proceed west. (6) He inferred that the mountain to the west would make a good landmark from his belief that he can readily identify its distinctive shape. (7) He inferred that he should start to walk in the direction of the distinctive mountain peak from his attitudes that doing so is a good way to proceed to the west, which is a good way to find a road, which is a good way to find his way, which is what he ultimately wants to do.

---

[1]There may be other meanings; these three are prominent.

By the way, in describing the lost hiker's reasoning, if we count each separate item in the considerations from which he made his inferences as a distinct reason, then I have mentioned four different kinds of *reasons* the hiker had for the inferences he made. Some are knowledge he possesses (e.g., that the direction of the sun is south at noon [in that part of the world]; that when facing south, west is 90 degrees to the right; that it is easy to get disoriented when walking in the mountains; that the sun appears rarely, if ever, on a cloudy day). Some are beliefs he holds (e.g., that there is a road to the west, that if he finds a road he can find his way, that he will be able readily to identify the distinctive mountain peak to the west while he travels). Some are observations he makes (e.g., there is the sun, it's a mostly cloudy day, that mountain has a distinctive shape). Some are desires he has (e.g., the desire to find his way, the desire to find a road, the desire to travel west, the desire to avoid becoming disoriented while trying to travel west). Alternatively, we could count each combination of considerations from which the hiker drew each inference as a single (compound) reason. In that case we might describe each reason as consisting of one or some combination of these four kinds of consideration: knowledge, beliefs, observations and desires. Whether reasoning is purely cognitive so that only knowledge and beliefs, and this only *knowledge of* or *beliefs about* observations and desires figure in it (as an emphasis on "beliefs" in Adler's account implies), and sensory or affective factors such as observations and desires do not in a direct way, is a question I will not take up.

14.3 Argument (in one sense)

If, following standard philosophical practice, we call each separate kind of consideration from which the hiker drew each of his inferences a *premise* (P1, P2, P3...) and we call what he inferred from each group of considerations a *conclusion* (C), and we call each such premise-conclusion set an *argument*, then we can transcribe the hiker's inferences as arguments.[2] I won't write them all out, but here are the first three:

(1) P1   The sun is in that direction.
     P2   It is noon.
     P3   The sun lies to the south (in Colorado) at noon.
     P4   I am in Colorado now.
     _____
     C   That direction is to the south.

(2) P1   I want to find my way.
     P2   If I find a road I will be able to find my way.
     _____
     C   I should try to find a road.

(3) P1   The direction of the sun is south.
     P2   West is 90 degrees to the right when facing south.
     P3   I believe there is a road to the west.
     P4   I want to find a road.
     _____
     C   I should travel at right angles to the right from the
          direction of the sun.

Notice that these "arguments" have not been thought, expressed or uttered by the lost hiker. They were not rehearsed in his mind in the process of reasoning

---

[2] I suspect it was this sort of usage that Copi had in mind when he wrote, in *Symbolic Logic*, that "Corresponding to every possible inference is an argument,..." (Copi 1954, p. 3).

about what to do, upon discovering himself lost, in order to find his way. The 107-word description of his reasoning, quoted above, is a much closer record of what went on in his mind as he reasoned than is any of these arguments. These "arguments" are expressions of the propositional contents of the inferences he made in his reasoning. That is one sense of 'argument' or kind of argument and it is in common use in philosophy. However, it is a technical sense and quite different from other senses of the term or kinds of argument found in ordinary language and non-technical understandings of the concept of argument. Since there are those other senses of 'argument' in common use, I will call arguments in the sense that (1), (2) and (3) are arguments, propositional content arguments ("PC-arguments," for short).

14.4   Logic

If logic is the study of the norms of good reasoning as expressed in arguments,[3] we should be able to apply logical norms to the lost hiker's reasoning as expressed in the above PC-arguments. The first argument's premises deductively imply its conclusion, so the argument is logically valid. That makes (1) a logically good argument according to the norms of deductive logic.

The second argument, however, is not deductively valid, for if we add such further information consistent with the given premises as that it looks like a thunderstorm is approaching and that the hiker is exposed to lightening on the ridge where he is standing, or that the hiker has a badly sprained ankle, then the conclusion that he should head west to try to find a road does not follow. He should get off that ridge and seek shelter out of danger from lightning, or he should stay put so as not to further injure his ankle and wait for help to come. Similarly, the third argument is not deductively valid, for if the new information that there is an impassible river between the hiker's position and the road, or that there is a high cliff to the hiker's west, is added, then the conclusion does not follow. But their deductive invalidity does not make (2) and (3) consist of bad reasoning or bad reasons for their conclusions. They are in fact both plausible arguments. Their support just happens to be *pro tanto* support: their conclusions follow from their premises in the sense that they are well supported by their premises or that they are consequences of their premises – provided that other things are equal.

If logic is to constitute the source of the norms by which to judge such arguments, then we need to have a logic for such *pro tanto* arguments. Suppose such a logic existed. Two questions immediately occur. One is, What would its features be? The other is, Would it be sufficient to assess the merits of arguments – would an argument qualify as good if it met the criteria of such a logic?

14.5   Argument (in a second sense)

Let me take up the second question first. The lost hiker's arguments may be understood as arguments in the sense of being based on expressions of the propositional contents of reasoning (i.e., as PC-arguments). But that is not the only sense of 'argument.' Suppose the lost hiker, S, has a companion, H, who does not agree that they should head west because she does not believe there is a road to the west. The two hikers agree they should stay together, but now S might want to

---

[3]"Logic is the study of the methods and principles used to distinguish good (correct) from bad (incorrect) reasoning. ... corresponding to every possible inference is an argument, and it is with these arguments that logic is chiefly concerned" (Copi 1982, pp. 3,6).

try to convince H that they should head west or to persuade H to head west. [4]
He cannot use the arguments that were based on his own reasoning, because H
rejects a key premise in those arguments. S must either persuade H that there is
probably a road to the west or else that they should head west for other reasons
that H is willing, or can be persuaded, to accept.

We can infer from such situations two conditions that an argument intended
to persuade or convince another person must meet in order to be a good argument. First, either all its premises must be immediately acceptable to the other
person, or else it must include acceptable rebuttals of objections that the other
person has to some of its premises. To work, such an argument needs to start
from the beliefs and other attitudes (i.e., the commitments) of the party or parties
to whom it is addressed. This is not a psychological requirement; it's a conceptual one. Being convinced by an argument consists of, in part, finding its
premises congenial. Second, and for the same reason, the inferences that the argument relies on have to be immediately acceptable to the other party(ies), or
else any objections to them must be rebuttable using arguments whose premises
and inferences are immediately acceptable to the other party(ies). When the aim
or purpose of expressing an argument is to persuade or convince another party
of something, then an argument that fails to accomplish that aim or purpose, for
the reason that some or all of its premises or inferences are unacceptable to that
party, is not a good argument for that purpose. These are necessary conditions
of a good "persuasive" or "convincing" argument, or "persuasive use" or "convincing use" of argument.

These conditions seem to be of a different order from those that apply to the
arguments expressing the lost hiker's reasoning. They are functions of putting
arguments to use to attain an objective – in this case, of the speaker, S, expressing
them to a hearer, H, in order to attain the objective of convincing H to travel
west. They might be termed pragmatic conditions, because they have to do with
the use of language or with communication, although since 'pragmatic' is used
in a great many ways, that label risks being confusing or uninformative. They are
also often termed dialectical conditions because they are functions of discourse
or discussion involving two or more parties, which is what the term 'dialectic'
referred to in its original sense in ancient Greek.[5]

Pragmatic or dialectical criteria do not enter into the assessment of PC-arguments, which are by definition abstracted from any context of use. However,
these criteria apply to the assessment of the *ceteris paribus* assumptions of *pro
tanto* arguments that have been used to try to convince or persuade. That is, these
assumptions have to be acceptable to anyone well informed of the context in
which the arguments are used. For instance, if a sow grizzly bear with two small
cubs was recently observed by S grazing in a meadow in the westerly direction,
the inference that S should travel west is undermined, for the inference follows
only if (among other things) it is safe to travel to the west, and travelling into
the vicinity of a sow grizzly with cubs is not safe, a fact that should be known to
anyone venturing into the Colorado mountains.

But that is not all. Suppose S is a park ranger, and now (safely back from his
misadventure) he is at the park's education center, addressing a group of bored
teenagers from the city chafing to get going on a hike in the mountains, and he is
trying to persuade them to take certain safety precautions, some basic do's and

---

[4]In this paper I use "S convinces H that p" to mean something like "S, by using reasons, brings H
to believe that p (is true)" and I use "S persuades H to a" to mean something like, "S, by using reasons,
gets to perform action a." One way to persuade someone to do something is to convince him or her that
it is the best thing to do, provided that he or she wants to do what is best.

[5]'Dialectic' comes from the Greek διαλεκτικός – *dialektikos* "of or pertaining to discourse or discussion" (Burchfield 1971).

don'ts. Assume that S is conscientious about his job and earnestly wants to get through to these young people and persuade them to follow his advice. We may say that his argument then needs not only to be logically sound and dialectically satisfactory, but also to grab and keep the teenagers' attention, be forceful and vivid enough to make an impression on them, speak to their felt interests and concerns so that they will remember its details and heed its admonitions once they are off hiking. S will have to present himself to the teenagers as authoritative, and S's argument will have to employ vivid terminology and dramatic metaphors and illustrative anecdotes that will resonate with his hearers. His argument and his presentation of it must, in short, be *rhetorically* effective as well in order to be a good argument (understanding rhetoric to include the art of effective communication, even though it denotes much more than that[6]).

We may conclude that whatever the logical criteria of good persuasive or convincing arguments turn out to be, however necessary they are for determining the goodness of such arguments, they will not be sufficient. Pragmatic or dialectical, and rhetorical criteria will need to be satisfied as well. That fact implies that whatever the correct understanding of the relations among reasoning, argument and logic turns out to be, there is more (and indeed, much more) to be said about argument in use than can be provided just by an understanding of reasoning and of logic as it has traditionally been conceived by logicians, namely excluding pragmatic and rhetorical norms from its domain. Thus it is not true that, *without qualification*, logic is the study of the norms of good argument.[7]

14.6   Further complications

Based on the discussion so far, we can distinguish the following. The process of the mental activity called "reasoning" can consist of someone drawing an inference from some assumed information. This inference can be expressed in language and that expression is also called the person's "reasoning"; moreover, it may also be termed the person's "argument." So there is reasoning (the mental act) and reasoning (the symbolic expression of the mental act). The latter is not synonymous with PC-argument (recall the difference between the 107-word "transcription" of the lost hiker's reasoning and the PC-arguments contained in it), but a person can formulate the propositional contents of an expression of reasoning as a PC-argument and use it to try to convince or persuade another person of its conclusion, in other words, to invite the other person to draw the same inference. The expression of reasoning to try to persuade or convince the other is also called an "argument"; so there is a second sense of 'argument'–the argument that is uttered or otherwise expressed (I will called it a U-argument). A U-argument is not the same as a PC-argument, since one and the same PC-argument can be expressed differently (in word order, sentence order, or word choice).

Moreover, the *communicative activity* of uttering arguments and responding to them is also called "argument" (a.k.a. "argumentation"), so there is a third sense of 'argument'–call it *argument-interchange*. An argument-interchange is an exchange of U-arguments. By an argumentative exchange I mean a conversation between two (or among three or more) parties in which one party (or each in turn) offers arguments (utters or expresses arguments) intended to convince or persuade the other(s) and offers U-arguments in response to the other's (or others') U-arguments. In an argumentative exchange, by definition the U-arguments

---

[6]For an account of rhetoric's complexity, see (Hauser 2002).
[7]Cf. Carney and Scheer (1980, p. 3): "Logic is primarily the study of arguments and of methods to determine whether arguments are correct or incorrect."

are expressed by particular parties on particular occasions for particular purposes. Here is an invented, but not unnatural, exchange between the imaginary hikers, H and S. The numbers refer to the turns in the exchange.

(4) H1: We're truly lost. What shall we do?
     S1: I think there's a road to the west, and if we find a
          road I know we can find our way, so let's head west.
     H2: I'm pretty sure the only road (around here) is north
          of where we are now, not west.
     S2: You might be right, though that's not my sense. But
          since the terrain to the north, east and south of us
          looks very rough and it looks fairly easy-going to the
          west, I suggest we head west to start with, just in
          case there is a road there. We can reassess our decision
          after a few kilometers if we haven't found a road. What
          d'you think?

This conversation can also be portrayed in a dialogue box, as follows:

Dialogue Parties		
Turns	Opponent (H)	Proponent (S)
1	We're truly lost. What shall we do?	I think there's a road to the west, and if we find a road I know we can find our way, so let's head west.
2	I'm pretty sure the only road (around here) is north of where we are, not west.	You might be right, though that's not my sense. But since the terrain to the north, east and south of us looks very rough and it looks fairly easy-going to the west, I suggest we head west to start with, just in case there is a road there. We can reassess our decision after a few kilometers if we haven't found a road. What d'you think?

Notice that this exchange is dialectically well regulated. By that I mean its participants conform to the norms that seem likely to be instrumental for a productive interactive exchange of used-arguments. Each party responds relevantly to the speech act of the other at the previous turn [a norm of orderly turn-taking and a norm of conversational relevance]. A question at H1 (about what to do) is responded to at S1 with a relevant answer (a suggestion about what to do, accompanied by a reason supporting that suggestion) [another norm of conversational relevance]. At the next turn (H2), a doubt about a premise of the reasoning at the previous turn (S1) is registered, along with a reason supporting that doubt [a norm of appropriately discharging the burden of proof]. Also at H2, H stops short of arguing for heading north instead of west, thereby acknowledging the doxastic standoff between her and S's senses of where the road is [a norm of reflecting the epistemic status of one's beliefs in claims appealing to them]. At S2 the doubt registered at H2 is acknowledged (the possibility that the premise is mistaken is conceded) [another epistemic status norm] and responded to with reasons [a norm of appropriate burden of proof discharge again].[8]

---

[8]To be sure, any defense of such norms requires a systematic analysis. One such analysis is offered in van Eemeren and Grootendorst (1984b, 2004).

The exchange is also rhetorically well regulated. By this I mean that the character of the contributions to the exchange is what might be expected to lead to the productive maintenance of that exchange and a productive outcome. H invites a suggestion about what to do from S, thus exhibiting respect for S and so the prospect of a cooperative discussion. S cooperates by reciprocating that respect not only by giving H a reason to support his suggestion (and thereby opening himself to a rebuttal by H at two places: the suggestion made and the reason offered in support of it), but also by appropriately qualifying his doxastic attitude towards that reason by saying, "I think," rather than asserting his opinion categorically and so (by implication) dogmatically. H responds in kind at H2 by not simply denying S's premise but by giving a reason for her denial, and by qualifying her reason ("I'm pretty sure") in a way that leaves it open to rejoinder, and. Also at H2, H leaves S to draw the appropriate conclusion from H's doubt, thereby inviting S to share her reasoning. At S2, S shows respect for H by conceding the possibility that H is right and he is wrong, and introduces a new reason that takes that concession into account. S also shows respect for H by ending S2 not only by inviting H's response, but also by doing it in an open-ended way – that is, in a way that's open to H's agreement or disagreement equally. The tendering of respect for the other by each party invites reciprocal open-minded consideration of points of view and arguments. This is an example of effective cooperative communication.

It also seems clear that in this exchange each party invites the other to make an inference. (Assuming S and H to be honest and candid, these are also inferences to which each of them, respectively, is committed.) Following Walton (1996b), Walton et al. (2008) we can discern in each case the general pattern or "scheme" of reasoning employed or exhibited in each argument. Viewed one way, we can say that the schemes are generalizations abstracted from the PC-arguments. S's and H's arguments can be generalized; that is, each can be seen to exhibit, and to be expressed, in an abstract form that makes no reference to particular people, places or times. Viewed another way, we can say that the PC-arguments instantiate the schemes. These schemes could also be instantiated on other occasions by other arguments made by other people.

Below, for illustrative purposes, I write out S's U-argument, the corresponding PC-argument, and one way of expressing its corresponding argument scheme.

(5) S's U-argument:
     I think there's a road to the west, and if we find a road I know
     we can find our way, so let's head west.

(6) S's PC-argument:
     P1 There is a road to the west
     P2 If we find a road, we can find our way.
     (P3. We want to find our way.) [Unexpressed premise.]
     So, C. We should travel west.

(7) One formulation of the scheme of S's PC-argument:
     1. Objective O is desirable.
     2. X is a means to O.
     3. Doing A is a means to X.
     Therefore, 4. Doing A is desirable.

Reasoning or arguing according to this scheme employs or relies on an inference license or warrant that might be expressed as the following:

W: If A is a means to X and X is a means to a desirable objective, O,
then doing A is desirable, other things being equal.

The reasoning relies on this inference license because one could not reasonably argue or reason from 1, 2 and 3 and conclude 4 unless W were true. W is
thus an unexpressed assumption of S's argument. However, it is not an unexpressed *premise* of the argument, for its function is not to serve as another item
of data from which the conclusion is inferred, but rather to license the inference
from the given (plus clearly assumed) data to the conclusion. (Counting W as an
unexpressed premise would require finding some further inference license, W',
that would warrant drawing the conclusion from the premises 1, 2, 3 and W. But
if W must count as an unexpressed premise of the argument, then by the same
reasoning so must W'. In that case, it would be necessary to find some other inference license, W'', that would warrant drawing the conclusion from the premises
1, 2, 3, W and W'. A vicious infinite regress would be the result – vicious, because
it would in principle never be possible to produce the warrant that licenses the
inference that S relies on in his argument; yet S clearly makes a reasonable inference in formulating his argument and invites H to make the same inference by
accepting his argument.[9])

The *ceteris paribus* proviso must be part of the warrant, for unless so qualified
the warrant would clearly be subject to fatal counter-examples. (E.g., "Shooting
one's husband is a means of killing him, killing one's husband is a means of es-
caping an unhappy marriage, and it is desirable to escape an unhappy marriage,
therefore if in an unhappy marriage it is desirable to shoot one's husband.")

An argument scheme's warrant provides a means of assessing arguments that
instantiate it. It supplies the information needed to formulate questions investigating whether indeed other things are equal in any particular instance of an
argument relying on that warrant. For instance, about an instance of W, above,
the following questions need to be asked to test its merits:

Q1: In the circumstances, will A in fact bring about X?

Q2: In the circumstances, will X in fact bring about O?

Q3: In the circumstances, is there a competing objective that is preferable to O?

Q4: In the circumstances, is there a better (preferable) means to X than A?

Q5: In the circumstances, is there a better (preferable) means to O than X?

Q6: Regardless of its being a means to X, are there in the circumstances any
     objections to trying to do A (e.g., unwanted unintended consequences, vi-
     olations of moral norms)?

Q7: Are there, in the circumstances, any independent objections to bringing
     about X?

If, when they are asked of S's argument in the circumstances in which he
makes it, the answer to any of these questions is affirmative, then S's argument
is problematic: the conclusion's support by the premises is thrown into doubt.
Perhaps S can refute or overcome the objections behind affirmative answers to
Q6 and Q7, but it seems clear that he bears the burden of proof to do so.

There is a difference between the first three and the last four questions on
the above list. Whether any U-argument is a good one depends on whether its
premises should be accepted and on whether its inference is justified. The first

---

[9]Some might dispute this analysis. There is a debate about how best to understand the role of
warrants that I cannot go into in this paper (see (Hitchcock 1998, 2006, 2007b, Bermejo-Luque 2006,
Pinto 2006, Verheij 2006)).

three questions address the issue or premise acceptability; the last four address
the issue of inference adequacy.

The test of such an argument's inference adequacy can be thought of as a test
of its logic, provided that the domain of logic is not restricted to the conditions
of necessary consequence. If the latter usage is adopted, then there is no logic
for *pro tanto* arguments, since they cannot be deductively valid. A terminological
fiat against calling the cogency of the inferences of such arguments their "logic"
does not remove the need to assess such inferences, nor does it render worthless
such assessments, which are in fact constantly and usefully made. If we do bow
to a narrow or strict construction of logic, then we need to distinguish between
the (strictly-speaking) *logic* of such arguments and what might be termed their
inference-cogency.

Whether any U-argument is a good argument in the sense that it is likely to
be (or was) understood, its merits appreciated and its conclusion accepted on the
basis of its premises by its intended audience is an empirical question. Recent
research suggests that logically good or inferentially cogent arguments tend to
be more persuasive than logically or inferentially weak ones O'Keefe (2002). It
seems plausible to expect that dialectically well-ordered arguments – those that
speak to the beliefs and attitudes of the target audience, respond to doubts, discharge the burden of proof, and so on – will tend to be more persuasive as well.
Finally, it also seem reasonable to expect that rhetorically well-designed and well-
presented arguments will tend to be more effective than the alternative, since the
canons of the art of rhetoric are based on the study of effective arguments.

14.7   Concluding remarks

We have seen in the foregoing discussion that each of the terms 'reasoning,' 'argument,' 'good argument' and 'logic' can be used to refer to different things.
Reasoning can refer to a mental activity, a report of such a mental activity, or
the propositional components of such a mental activity and their ("premise" –
"conclusion") relationship to each other. *Argument* can refer to reasoning in that
third sense (PC-arguments); it can refer to the use of such reasoning to try to
persuade or convince another (U-arguments); and it can be used to refer to argumentation, a form of communication in which two or more parties exchange
arguments for some purpose (exchange-argument). *Good argument* can be meant
to refer to a rhetorically good, dialectically good, logically good, or effective argument. *Logic* can refer narrowly to the study of the norms of necessary or deductive relations between propositions, or more widely to the norms of justified
inferences or premise-conclusion relationships in reasoning or arguments generally. I make no claim that this list of the ways these terms are used is exhaustive.
In my view none of these uses of these terms can reasonably be specified as *the
correct one*; they are just different. They all have the authority of wide use. Nothing prevents anyone from stipulating some different use for one of these terms
for some particular purpose, and that stipulation would be justified if there were
a benefit to it that could not be more conveniently realized. However, to assert
that "Reasoning is ... ," "Argument is ... ," "A good argument is ... ," or "Logic
is ... ," completing the sentence by one of the above or by a completely different
conception from any of the above, and contending that this stipulation is "the
correct conception," seems to me at the very least to require defense in the face
of the variety of uses described above.

Thus, when Walton (1997c, p. 605) writes that "Reasoning is a sequence of
inferences", and "An inference is a set of propositions and a Toulmin warrant
joining the premise propositions to the conclusion", I suggest he is most char-
itably understood to be suggesting that it is helpful to analyze reasoning as a

sequence of inferences, when an inference is analyzed as a set of propositions such that accepting one of them on the basis of the others is justified by a warrant, understanding 'warrant' in the way Toulmin proposed. Even so, we are owed reasons for the helpfulness of understanding or analyzing 'reasoning' and 'inference' in these ways.

Similarly, when Walton (1990b, p. 411) writes, "Argument is a social and verbal means of trying to resolve, or at least to contend with, a conflict or difference that has arisen or exists between two (or more) parties." and adds that "An argument necessarily involves a claim that is advanced by at least one of the parties." It is best to take these as recommendations about how best to understand argument, or about one useful way to use 'argument'. By the way, notice that Walton's definition here is about argument interchanges, and not about PC-arguments or U-arguments. Still, Walton owes a justification for understanding the activity of argument as a means of trying to resolve a conflict between (or among) parties. For instance, why couldn't its objective be to demonstrate verbal acuity, or to embarrass an opponent, or to perpetuate or accentuate a disagreement? And how is this sense of argument different from negotiation, which seems also to be a social and verbal means of trying to resolve a conflict or difference that exists between two (or more) parties? In other words we need some reason to accept and think about these concepts in terms of one of these stipulative definitions.

One problem with working from stipulative definitions in trying to understand concepts (such as reasoning, argument and logic) is that the definition frames the inquiry with its own perspective when that perspective is what needs to be defended. The reliance on stipulation can be a way of avoiding the needed task of justifying the perspective, and if it is, then the definition is question-begging. In such a case it is problematic whether the theoretical claims made from that perspective fully illuminate the phenomena. For instance, Walton *et al.* define *argument* in another place as "a sequence of reasoning made up of a chaining together of inferences that is used to contribute to the settling of an unsettled issue in a dialogue" (Walton et al. 2008, p. 268). What about arguments that don't occur in dialogues – the arguments in a book or in a political speech before a television audience, for instance? And what about such arguments that don't contribute to settling an unsettled issue, but instead push on an open door or flog a dead horse – when the writer or speaker is just piling on, or flailing away at a lost cause? The reply will be either that such contexts are *really* dialogues (they just don't appear to be what they really are) and there *must* be some dispute (for otherwise why would two parties argue?); or else it will be that it is fruitful to analyze such arguments *as if* they were occurring in dialogues between the author and (certain of) his or her readers or the speaker and (certain members of) his or her audience as if there were some dispute. Since many of these contexts are really *not* dialogues and there is *no* unsettled issue involved, the second reply is preferable. But given the definition of argument stipulated, the properties that will be focused on in the theorizing will be dialogue-relevant and dispute-resolution-relevant properties. The risk is that there might be other salient properties of, for instance, arguments of books or of television speeches that will be overlooked as a result of the focus entailed by the definition of argument modeled as occurring in a dialogue that frames it. For instance, a politician in a two- or three-party state, in framing a speech that will be televised or reported nationally, must address an audience of potential supporters who among them occupy diverse and conflicting interests, beliefs and attitudes. There is no single dialogue interlocutor or interlocutor-type. In evaluating such a speech, it would not be legitimate to hold the speaker responsible for addressing all the questions or doubts of each type of potential supporter. Of course the analyst must understand the speech as directed to an audience, but thinking of this audience as a dialogue partner will blind the analyst from the truly significant properties of the politician's argumentation, such as his or her attempt to address widely-shared underlying hopes and fears. The point is not at all to deny the usefulness of modeling some – even much – argumentation as if it were dialogue. It is, rather, to resist *defining* the concepts of argument and argumentation so as to limit their application or perspective to dialogues.

In general, this paper is a plea for understanding reasoning, argumentation and logic in the full complexity of their various uses[10].

---

[10]I am grateful to Geoff Goddu for pointing out numerous serious blunders in an earlier version of this paper, which I have tried to remove in revising it.

CHAPTER 15

CORROBORATIVE EVIDENCE

David Godden

Abstract

Corroborative evidence can have a dual function in argument whereby not only does it have a primary function of providing direct evidence supporting the main conclusion, but it also has a secondary, bolstering function which increases the probative value of some other piece of evidence in the argument. It has been argued (Redmayne, 2000) that this double function gives rise to the fallacy of double counting whereby the probative weight of evidence is overvalued by counting it twice. Walton has proposed several models of corroborative evidence, each of which seems to accept the fallaciousness of double-counting thereby seeming to deny the dual function of corroborative evidence. Against this view, I argue that the bolstering effect is legitimate, and can be explained by recourse to inference to the best explanation.

15.1 Introduction

Deductively valid arguments are monotonic. That is, supplementing the initial premise set with additional premises cannot weaken the inferential strength of the argument. Indeed, the opposite is also true. Just as the premise-conclusion link in a monotonic argument cannot be weakened with additional information, it cannot be strengthened either.[1]

Non-monotonic arguments lack both of these properties. First, the inferential link in non-deductive arguments is defeasible: the truth of their premises does not guarantee the truth of their conclusions. Thus, non-deductive arguments are subject to defeaters of at least two sorts: (i) undercutting defeaters (Pinto's underminers: "additional facts that undermine the inference") and (ii) rebutting defeaters (Pinto's overriders: "additional evidence that overrides the inference in question, by supporting the negation of its conclusion") (Pollock 1986, pp. 38-39) (Pinto 2001, pp. 102-103). Of this second type, Schum (1994, pp. 121 ff.) notes two different types: (a) contradictory evidence: where two pieces of evidence directly and positively support mutually exclusive events (or contradictory conclusions) and (b) conflicting evidence: where two pieces of evidence directly support different, mutually consistent sub-conclusions which in turn provide contradictory

---

[1] This is not to say that additional reasons (even additional deductively valid arguments) supporting the conclusion cannot be provided, only that they are unnecessary in that they neither contribute to the inferential strength of the initial argument, nor do they supplement it with any additional probative weight. Yet such redundant (Schum 1994, p. 126 ff.) evidence might be added as a kind of failsafe in the event that some otherwise sufficient reason is defeated or is not accepted by an audience.

Of course, the overall merit of a deductively valid argument can be strengthened by showing that it is sound, and the soundness of valid arguments can be affected by supplementation with additional premises when those added premises go to establishing the truth of the initial premises.

evidence for some main conclusion. Not only are non-deductive arguments subject to weakening and defeat, but they can be strengthened when supplemented with additional reasons where "a reason is the smallest self-standing unit of support for a position" (Blair 2000).

This paper concerns the strengthening function of corroborative evidence. Walton (2008c, 2009a), Walton and Reed (2008) recognized that corroboration poses a unique problem for the theory of evidence. So far as I can tell they are the first to attempt to address this problem, posing alternatives for the accurate modeling and proper evaluation of the evidential structure and operation of corroboration. Typically corroboration has been seen as having the same logical structure as convergence (Cohen 1977). While this posed some difficulties for probabilistic accounts, it failed to recognize a unique and uniquely problematic feature of corroborative evidence. Walton and Reed observe that this feature has to do with the double function corroborative evidence sometimes has, whereby in addition to acting as a reason for some conclusion it also somehow increases the probative weight of some other piece of evidence. This second, 'bolstering' function of corroborative evidence distinguishes it from merely convergent evidence.[2] Yet, it also raises the problem of the fallacy of double counting (Redmayne 2000). Walton and Reed address this problem and propose a number of alternative solutions for it.

This paper offers a critical examination of those solutions, and proposes an alternative account on which the double function is recognized and explained as non-fallacious.

15.2   Corroboration

Argument Strength and Strengthening

Informal logic provides three criteria for evaluating the strength of arguments: premise acceptability, premise relevance and inferential sufficiency or goodness (Johnson and Blair 1994b, Govier 2005b). Schum (1994, pp. 66 ff.) seems also to have adopted this three-part standard for the evaluation of evidence in law, using the criteria relevance, credibility (acceptability) and force (sufficiency). Following Johnson and Blair I will call this the R.S.A. standard of argument cogency.

Generally speaking, strengthening occurs when the overall probative force of a mass of evidence upon some conclusion is increased by additional evidence. Non-monotonic arguments can be strengthened in several ways.

1. *Premise support*: the acceptability of a premise used in the inference can be strengthened or established through the provision of additional reasons which bear upon the truth of that initial premise.[3]

2. *Convergence of primary reasons*: additional, independent reasons directly relevant to the main conclusion can be added to the argument.[4]

While still the subject of theoretical debate (Goddu 2007, 2009b) strengthening by premise support and convergence is largely well-understood in infor-

---

[2]For the remainder of this paper, I will use "corroboration" for the specifically narrow phenomenon that occurs when this 'boosting' effect is present. Corroborating evidence is the evidence that lends credibility or probative weight to corroborated evidence. Sometimes pieces of evidence can be mutually corroborating.

[3]This property is shared with deductive arguments, see fn. 1.

[4]This will strengthen the initial argument only to the extent that the additional reasons are not redundant (Schum 1994, pp. 126 ff.).

mal logic, argumentation and epistemology. A third, perhaps dialectical way in which arguments can be strengthened is:

3. *Preemptive rebuttal of defeaters*: additional reasons showing that a potential defeater does not apply or that some critical question can be answered satisfactorily can help to demonstrate that a defeasible warrant or generalization properly applies in some particular instance and can thereby strengthen the inferential connection in an argument.

While there is no standard way of modeling this third form of argument strengthening, Godden and Walton (2007) have demonstrated that critical questions function as commonplaces which test at least one of the R.S.A criteria.

Now, if corroborative evidence is to pose some unique or special problem to the analysis and assessment of arguments it must operate in some way different from these.

Weak Notions of Corroboration

There is a sense in which any piece of negatively relevant evidence, in counting against some conclusion, thereby fails to corroborate any positively relevant evidence (whether known or yet to be discovered) for that conclusion. Thus, there is a sense in which all positively relevant evidence for some conclusion corroborates all other evidence for that conclusion. Indeed, perhaps the weakest notion of corroboration is simply "is not inconsistent with" where any fact which does not count against the truth of some claim thereby corroborates the truth of that claim. Consistency-based notions of corroboration do not even involve argument strengthening. While positive relevance does involve argument strengthening, equating corroboration either with consistency or with positive relevance produces an extremely weak and not especially useful notion of corroboration.

Corroboration: Initial Accounts

Initial accounts of corroboration seem only to capture this loose notion of strengthening as just defined. For example, Cohen (1977) seems to equate corroboration with convergence, writing:

It is easy to describe in general terms the logical structure that is common to [testimonial] corroboration and [the] convergence [of circumstantial evidence]. If a conclusion, S, has its probability raised by each of two premises, R_1 and R_2, when these are considered separately, and R_2 is unconnected with R_1 unless through the truth of S, then the conjunction of R_1 and R_2 makes S more probable than does R_1 alone. (Cohen 1977, pp. 94-95)

Similarly, in explaining the legal idea of corroboration, Schum (1994, pp. 124 ff.) distinguishes two forms which correspond respectively to the convergent and premise support forms of strengthening just described.

It is quite true that these are entirely legitimate uses of the notion of corroboration, and accounts of the logical function of corroborative evidence in many ordinary cases. Yet, if corroboration only amounts to either convergence or premise support, then I submit that corroborative evidence poses no special problems for the theory of argument and can be dealt with by the analytical and evaluative tools already on hand. In this paper, I attempt to flesh out and specify a stronger, more restrictive sense of corroboration, which poses unique problems for the theory of evidence. This notion will turn on the bolstering effect associated with the fallacy of double counting.

15.3   Convergence and Corroboration

While it is clear that the ordinary and legal sense of corroboration includes convergent arguments, the mere convergence of reasons does not necessarily yield the bolstering effect that is uniquely corroborative.

Consider the following case (adapted from Cohen 1977, p. 94) in which evidence is convergent that is, as the mass of evidence accumulates it continually increases the likelihood of the conclusion's truth but is nevertheless not really corroborative in any strong sense. Recall that there are three essential components to any criminal case: motive, means and opportunity. Clearly, while showing that a suspect had either motive or opportunity goes some way towards demonstrating his guilt, showing that a suspect had both motive and opportunity raises the probability of his guilt substantially more than showing either on its own provided that the two are independent of each other except through the supposed fact that he committed the crime.

In such a case, each piece of evidence counts as an additional reason for the conclusion and thereby contributes to the overall argument. Yet no piece of evidence increases the probative strength of any of the others. That the suspect had motive as well as opportunity does not make the fact that he had motive count any more towards his guilt than it did on its own. The strength of the overall argument is increased with the addition of reasons, but adding reasons does not result in strengthening the probative force of any particular piece of evidence. Arguments having this evidentiary structure are usefully called convergent, but not corroborative.

Convergent, Non-Corroborative Arguments each new reason strengthens the
overall argument without increasing the strength of any single reason.

Contrast the above case with the following: several independent witnesses testify to some fact. In this case, not only does each new piece of testimonial evidence count as a primary reason for the conclusion, but it also somehow bolsters the probative value of some other piece of testimony. That several independent witnesses agree in their testimony is a reason to count each individual witness as more reliable than we would otherwise, in addition to providing additional reasons to accept the truth of the facts to which they testify. Arguments having this evidentiary structure are usefully called convergent and corroborative.

Convergent, Corroborative Arguments at least one reason (the corroborating
reason), in addition to strengthening overall argument by providing an independent reason for the main conclusion, also increases the strength of at
least one other reason in the argument.[5]

It is this apparent bolstering function of corroborative evidence that poses the real problem for the theory of argument and evidence. I now turn to a consideration of Walton's theory of corroboration specifically as it attempts to model and explain this bolstering effect of corroborative evidence.

---

[5]Schum (1994) recognizes this 'enhancing' or 'synergistic' effect of corroborative evidence, using the name "convergent evidence" for cases where "two or more items of convergent evidence may mean more to us, when considered jointly, than they do if we consider them separately ... [because] for example, that event F, if it occurred, would *enhance* the inferential force of event G on hypotheses {H,Hc}" (p. 126).

15.4   Walton's Model of Corroboration

Walton's (2008c) treatment of corroborative evidence is developed in Walton and Reed (2008). While recognizing the convergent component of corroborative evidence (Walton and Reed 2008, p. 538) (Walton 2008c, pp. 84 ff.), two hypotheses for modeling the bolstering function of corroborative evidence are proposed. Initially, (Walton and Reed 2008, pp. 539-540) supposed that corroborating evidence might be modeled as a kind of premise support for the corroborated evidence, such that "the corroborating argument supports one of the premises of the original argument" (Walton and Reed 2008, p. 552) (Walton 2008c, pp. 84 ff.). A variation of this first method was also considered whereby "the corroborating argument proactively rebuts a possible attack on the original argument by answering a critical question" (Walton and Reed 2008, p. 552). Second (Walton and Reed 2008, p. 544) (Walton 2008c, p. 300) was the hypothesis that corroborative evidence could be modeled as a special argumentation scheme which might be called the corroboration scheme. Walton (2008c, p. 302) is indecisive about which approach is best, and Walton and Reed (2008, p. 552) seem to favor the premise support approach.

Walton (2009a) postulates that there are two basic types of corroborative evidence: (a) *convergent corroborative evidence*, where the corroborating evidence gives direct and independent support to the main conclusion, and (b) *supportive corroborative evidence*, where the corroborating evidence supports the inferential link between the corroborated evidence and the primary conclusion.[6] An important development here is that corroborative evidence no longer works as a kind of premise support but instead as a kind of auxiliary strengthening of an inferential link. The strengthening function of corroboration is explained in type-(a) cases by the additional primary evidence, and in type-(b) cases by the preemptive or anticipatory rebuttal of a potential defeater.

Problems with Walton's Approach

Despite its pioneering advances in the treatment of corroboration, Walton's various approaches are not without their problems. Consider first the proposal that corroborative evidence should be modeled as its own argumentation scheme: the scheme itself is structured such that all the corroborating claims are linked together to support the conclusion that there is corroborative evidence for some claim. The corroboration scheme then acts in a convergent manner together with each individual piece of corroborated evidence in directly supporting the main conclusion (Walton and Reed 2008, pp. 544-545) (Walton 2008c, p. 303). In this way, the corroboration scheme attempts to capture the strengthening function of corroborative evidence by representing the specifically corroborative value of corroborating evidence.

---

[6] This roughly accords with Redmayne (2000, p. 150) who distinguishes the following three "ways in which one piece of evidence may support another":

   (i) *same fact corroboration*: "two pieces of evidence may be corroborative of the same fact, such as when two witnesses both report seeing the same event";

   (ii) *convergence*: "two pieces of evidence . . . point in the same broad direction"; and

   (iii) *credibility corroboration*: "one piece of evidence may support the credibility of another, such as when one witness testifies that another witness has a reputation for being truthful."

To avoid of the fallacy of double counting (discussed below) Redmayne discounts any boost effect of corroborative evidence in the first two ways and rejects any primary effect (i.e., direct evidence to the main conclusion) in the third way. Thus, while there are different ways that corroborative evidence can function in an argument, Redmayne maintains that any single piece of corroborative evidence has only one function.

This approach suffers a number of defects. First, it is not an explanatory model of corroborative evidence. It does not explain why or how corroborative evidence has the bolstering function it does and thereby it does not help the arguer or analyst to identify, understand or evaluate instances of corroboration. Further, as Walton and Reed (p. 552) recognize, convergence with corroborated reasons does not seem to accurately represent the logical operation of the bolstering function provided by corroborating evidence. The specifically corroborative value of corroborative evidence is not a separate and independent line of support for the main conclusion, but is instead an increasing of the degree of support given by each corroborated reason. Nor does the corroboration scheme capture the idea of their second hypothesis that the inferential strengthening of corroboration occurs through a preemptive foreclosure on some potential defeater of the initial presumptive argument. Finally, given the existing tools for operationalizing the transmission of probability or plausibility through a convergent argument, it is not clear that the boosting effect of corroboration will be properly calculated using the schematic method.

Modeling corroborative evidence as premise support seems to fare no better. Consider a paradigm case of corroboration where the bolstering effect is present: multiple independent witnesses testifying to the same fact. Walton has proposed two premise-types in the corroborated argument that are candidates for the support offered by corroborating reasons: (i) the premise to the effect that Witness 1 has testified that P (Walton and Reed 2008, p. 540), and (ii) where each testimonial reason is modeled using a position to know scheme, a premise to the effect that Witness 1 is telling the truth in testifying that P (Walton 2008c, p. 298). (Type-(ii) premises would frequently be implicit but unstated in an initial argument.)

The first proposal clearly fails since the testimony of some additional witness to the effect that P does not provide any evidence for the claim that some initial witness also testified that P. Indeed Walton (2008c, pp. 85-86) seems to claim that circumstantial evidence showing that a bullet found in the body of a victim matches a suspect's gun could give premise support to a witness's testimony that he saw the suspect shoot the victim. Yet, the additional testimonial or circumstantial evidence is plainly irrelevant to type-(i) premises. Instead, such additional evidence goes towards showing that P, not that someone reported that P or claimed to have witnessed it.

The second proposal seems to address this problem. Further, it captures the idea that the credibility of a witness that is the reliability or veracity of a source is positively affected by corroborating evidence. Thus, there is a sense in which the reliability of testimony or other evidence is strengthened through corroboration. Yet, this proposal still fails to provide any substantive answers concerning the how and why of corroboration. Walton's proposal that a typically enthymematic type-(ii) premise is supported by corroborating evidence fails to explain why such strengthening occurs. What is the evidential rationale for the boost-effect? Also, the operation (or mechanism) by which this strengthening occurs is similarly unexplained. It remains unclear whether the type of strengthening provided by the boost-effect of corroborative evidence is properly modeled as a kind of premise support or, as Walton later suggested, the strengthening of an inferential link.

Virtues and Directions from Walton's Account

Walton's treatment of corroboration recognizes several of its distinguishing features, including: its difference from convergent evidence (Walton and Reed 2008, p. 533), the requirement that corroborative evidence come from independent sources (Walton and Reed 2008, pp. 536/547), and the boosting effect that corroborating evidence has on corroborated evidence (Walton and Reed 2008, p. 543).

Further, Walton's hypothesis that this strengthening occurs because of a pre-emptive answering of a critical question applying to the initial, defeasible inference offers an initial explanation of why such a boosting effect occurs. Thus, Walton's account offers the following general picture of how corroboration works: "It fills gaps by anticipating objections, thus making the original argument more plausible than it was before" (Walton and Reed 2008, p. 539). This, I suggest, is a promising hypothesis to which I return at the conclusion of the paper.

15.5   Double Counting and the Corroboration Effect

Some have challenged the legitimacy of the bolstering effect of corroborative evidence on the grounds that it gives rise to the fallacy of double counting.

In its simplest versions, the fallacy of double counting involves over-representing the likelihood of some outcome or event. Walton and Reed (2008, p. 532) explain it with the following example: suppose you roll two fair dice, what is the probability that at least one 5 will turn up? One might think that the probability is 1/3 (1/6 + 1/6). But this answer is mistaken because it double counts the possibility that each die will land showing 5. Thus, in fact, the correct answer is that the probability is only 11/36.

As it pertains to corroborative evidence, the fallacy of double-counting involves over-valuing some piece of evidence by counting it twice. Thus, it specifically arises from the purported boosting effect of corroborating evidence.

Redmayne's Argument Against Double-Counting Corroborative Evidence

According to Redmayne (2000) the fallacy of double counting occurs whenever some auxiliary, bolstering function is attributed to corroborative evidence in addition to its primary function in the argument.[7] His solution to the problem is to subtract the value of the 'double-counted' evidence from the total probative weight of the argument.

Redmayne's (p.151) argument against double-counting seems to work as follows:

P1. Corroborative evidence has a double function in an argument.
P2. In accounting for this double-function the probative value of corroborative evidence is double-counted.
P3. This double-counting is fallacious and overvalues the probative weight of the evidence.
C. Therefore, the value of double-counted evidence must be subtracted from the overall probative weight of the argument.

Given the case I have attempted to make for the unique bolstering function of corroborative evidence, Premise 3 is the contentious premise in Redmayne's

---

[7]In contrast to Redmayne's account, Schum (1994, p. 128) offers a different, and more agreeable, account of the fallacious double counting of evidence whereby it occurs as a kind of evidential redundancy. Here super-sufficient evidence is offered for a claim and is fallaciously counted as making that claim more certain than is actually supported by the evidence. For example (Schum 1994, p. 129), suppose that there is an abundance of circumstantial evidence that an accused was present at a crime scene. This may result in a 'cumulative redundance' which may tend to lead an audience to overvalue the presence of the accused at a crime scene as evidence that she committed the crime. Schum's account of double-counting as evidential redundancy, then, does not necessarily discount the dual-function of corroborative evidence.

argument. While he seems to take it as manifestly acceptable, he supports it with the following argument:

> If you doubt the double-counting problem, consider the following reduction (inflation?) *ad absurdum*. C reports a recovered memory of abuse by D. Witness W also reports recovering a memory that she saw D abuse C. If you think that W's report increases the probative value of C's report without at the same time reducing the probative value to be drawn from W's report, you can obtain infinite probative value from the two reports. First, use W's report to increase the probative value of C's report. Then, use C's report (which now has enhanced probative value) to increase the probative value of W's report (which will now be boosted beyond its original probative value). Next, use W's report to further enhance C's report. And so on, until your degree of belief in the proposition "D abused C" is as high as you like. Redmayne (2000, p. 151 fn. 14)

Thus, Redmayne sees any double counting as producing an iterative regress whereby the probability of any corroborated claim, no matter how initially unlikely, becomes certain.

A Rebuttal of Redmayne's Argument

There are at least two problems with this argument. First, it assumes an iterative application of corroborative evidence. Redmayne's argument works not by making the plausible supposition that there is a single occurrence of the boosting effect. (Surely it is at least plausible that the mutual agreement of the testimony of several independent witnesses boosts the credibility of each individual witness and thereby the probative value of their individual testimonies.) Instead, it works by supposing that this newly-bolstered piece of old evidence can be used again to re-bolster other pieces of corroborated evidence. Redmayne's argument claims that the new credibility-value of each witness's piece of testimony should, in view of its retained agreement with the testimony of the other witnesses, *again* boost the credibility of each individual witness. This is, it assumes that having judged that one witness's testimony is more reliable than we initially took it to be because it accords with the testimony of other witnesses, we should now say that the testimony of those other witnesses is even more reliable because it accords with a more credible (set of) witness(es). Yet surely this is implausible, and is to double count fallaciously. Indeed, it seems to be a case of circular reasoning. In the first instance, the boosting effect is justified because new information (the corroborating evidence) has been discovered, and this new information prompts a revision of the probative value attached to information already possessed. The reason that I found A's testimony to be more credible than I did initially is that I learned about B's testimony and that A's testimony accorded with B's. Redmayne's assumption of iterative application cannot be similarly justified. Following the first occurrence of the boost effect, no new information is discovered or enters into the argument. Instead, the iterative assumption allows an informational echo (rather than a new informational voice) to continue to affect the weighting of evidence. Ultimately, this iterative echo effect allows a piece of evidence to bootstrap its own probative value in a way that is fallaciously circular. Thus, Redmayne's iterative assumption does not merely allow a single piece of evidence to perform several plausibly legitimate functions in an argument; instead it conspicuously allows a single piece of (revalued) evidence to perform *the same* function in an argument several times. So the first problem with Redmayne's argument is that it unacceptably assumes an iterative application of corroborative evidence.

As much as the iterative assumption should be denied, it is not clear that Redmayne's conclusion follows even if it is allowed. Redmayne concludes that iterative double counting will always produce the certainty of the claims involved given enough iterations. A second problem with the argument is that it assumes that the full probative weight of bolstered evidence to be re-counted at each iteration, instead of only the specifically bolstered value (i.e., incrementally increased) of that evidence. Yet if it is only the bolstered value of the evidence that is counted, this will become incrementally smaller at each iteration. While the resulting probability will admittedly be overvalued, this value will, nevertheless reach a limit at something less than certainty depending on what the initial probative value of the evidence was. Thus there is no reason to think that even the iterative application of the boost effect of corroborative evidence necessarily results in the certainty of the initially uncertain claims involved.

Redmayne's Solution and its Failure

Having concluded that allowing the bolstering function of corroborative evidence is a fallacious instance of double-counting, Redmayne (p. 151; quotation amended for generality) proposes the following as a general solution: given any two pieces of evidence, R1 and R2 where one tends to corroborate the other, "if we have taken some probative value from [R1] to add to [R2], we must remember to subtract that same amount when we consider [R2]." Thus, Redmayne's solution to the problem of double-counting is to subtract the value of any double-counted evidence in the final tallying of the probative weight of a mass of evidence. In doing so, it rejects the idea that corroborative evidence performs a double function in argument, and actively negates any value this double function might add to the probative merits of an argument.

On the final analysis, then, when corroborative evidence is properly modeled on Redmayne's account it has the structure either of premise support or of convergence. Redmayne (p. 151) writes:

> The logic of corroboration, then, is this. With the exception of credibility corroboration [which acts as a kind of premise support], the corroborating evidence supports the corroborated evidence by increasing the probability of the hypothesis [i.e., conclusion] that both are being used to prove. The corroborating evidence does not increase the probative value of the corroborated evidence. Metaphorically, it adds another strand to the rope; it does not increase the strength of the existing strand[s].

The failure of Redmayne's proposed solution is that it actively negates any bolstering effect that corroboration might produce. That is, it says that we should not count a witness's testimony as more credible than we would otherwise because it accords with the testimony of other independent witnesses, while at the same time acknowledging the convergence of reasons offered by multiple independent pieces of testimonial evidence speaking to the same fact. But surely this is wrong. That several independent witnesses agree makes each witness's testimony more credible, as well as lending cumulative support to the final conclusion. The solution is not to ignore, discount or subtract the bolstering effect of corroboration, but instead to find a model that successfully explains it and incorporates it into a working model of argument analysis and evaluation.

A final problem with Walton's account of corroboration, I argue, is that it accepts Redmayne's flawed arguments against the dual function of corroborative evidence, and accepts that double-counting corroborative evidence is always fallacious. For example, Walton (2008c, p. 299) writes that "Each of these two methods of diagramming the argument [i.e., as premise support and as a convergence of primary reasons] in this case [of convergent testimonies] makes the argument inherently reasonable by itself. The problem of double counting comes when the two ways of analyzing the evidence are combined" – i.e., when the dual function of corroborative evidence is modelled.

In the final section of this paper, I offer an explanatory model of corroborative evidence which accounts for its dual function. In doing this, I hope to make a positive case for the thesis that, in addition to its primary effect, the bolstering effect of corroborative evidence is legitimate rather than fallacious.

15.6   Explaining Corroboration: Inference to the Best Explanation

As it turns out, the structure of the problem presented by corroborative evidence is remarkably similar to one more familiar in epistemology the problem of coherence as a source of justification. I suggest that the solution to this latter problem may serve as a model for a solution to the former.

Against foundationalism, which claims that some, basic beliefs are inherently justified and that these basic beliefs provide the ultimate justificatory foundations for all other beliefs in the system, coherentism claims that at least sometimes justification arises from the coherence of a set of beliefs with one another (Boujour 1985). A significant problem for coherentist epistemologists is explaining how coherence can be a source of justification thereby demonstrating its legitimacy. The problem is an especially acute one, since it is well-known that agreement does not entail truth. As Elgin (2005, p. 159) aptly puts it, "coherence can readily be achieved through epistemically illicit means." So, how can coherence be indicative of truth?

Coherence occurs in many epistemically relevant contexts, such as when information from our various sensory mechanisms coheres to present a unified multi-sensory picture of the world, or when our memories cohere with circumstantial artifacts (e.g., photographs). Lewis (1946) considers an example of coherence which closely resembles a stereotypical situation of corroborative evidence. In this case a group of independent but individually unreliable witnesses all give substantially similar accounts of some event, e.g., a theft. Here, each piece of testimony coheres with the others. "Given the unreliability of the witnesses," Elgin (2005, p. 157) writes, "we might expect them to be wrong about the thief. But we would not expect them to all be wrong in the same way. The fact that they agree needs an explanation." Now, there are a variety of possible explanations for such agreement. It might be the case that the witnesses are in cahoots and have jointly concocted a fabricated story. Yet knowing that the witnesses are independent eliminates this as a possible explanation. It might be the case that each of the witnesses is somehow mistaken or lying. But this seems unlikely given the similarity of their accounts; it is implausible that they would all be mistaken or lying in the same way. Indeed, there are many possible explanations of the similarity of the testimonies of the witnesses. One of these is that their testimonies are indeed true that the reason the witnesses have testified as they have is that they did indeed witness what they claim to have, and that things occurred as they witnessed them to have.

Elgin (2005) argues that in this last eventuality, the agreement of the different testimonies is evidence of their individual truth. Thus, she (p. 160) proposes the thesis that "coherence conduces to epistemic acceptability only when the best explanation of the coherence of a constellation of claims is that they are (at least roughly) true."[8]

Given the analogousness of corroboration and coherence as problems in the theory of evidence, I propose that inference to the best explanation (IBE) can also provide a solution to the problem of corroborative evidence. Corroborative evidence has a dual function: primarily it provides direct support for some conclusion. In doing so, it makes the conditional probability of the conclusion (given the corroborative evidence) higher than its antecedent probability. This, in turn, gives rise to the auxiliary function of corroborative evidence whereby it increases the probative value of some other piece of corroborated evidence. This bolstering effect is explained by the conditional probability of the corroborated evidence being higher given the truth of the conclusion than the conditional probability of the evidence given the negation of the conclusion. The truth of the conclusion is not evidence for the truth of the corroborated premise, rather it is the best explanation of it. So long as the best explanation of the corroborated evidence is the truth of the conclusion, the bolstering effect of corroborative evidence is entirely legitimate. Since corroborating evidence provides an independent reason to accept that the conclusion is true, the inference to the best explanation provides an additional reason to accept that the corroborated premise is true. Importantly, then, while corroborative evidence can strengthen an argument by lending credibility to a corroborated premise, it does not do so by offering premise support at least not directly. Rather it is through an entirely separate inference that the corroborated argument is strengthened.

The IBE account of corroborative evidence has a number of explanatory virtues. First, it avoids the problem of double counting while allowing for and explaining the dual effect of corroborative evidence. The same evidence is not counted twice on the IBE theory. Rather, an explanation is provided as to how the additional reason for the main conclusion also indirectly strengthens the corroborated reason. Second it explains why the independence of corroborating evidence is such a crucial feature of it. If it is not known that corroborative evidence is independent, then a viable explanation of the convergence of several pieces of putatively corroborative evidence is their causal inter-dependence. Unless this explanation can be eliminated, the evidential value of corroborative evidence is nullified because the IBE condition fails to be satisfied. Third, it explains why some evidence, while convergent, is non-corroborative namely when the truth of the conclusion is not the best explanation of the putatively corroborated evidence. For example, the putative fact that A killed B (as evidenced by A's opportunity to kill B) is *not* the best explanation of A's having the motive or means to kill B.

15.7   Conclusion

I have argued that corroborative evidence has the following dual function: while offering direct support to some conclusion C, corroborative evidence has an auxiliary function of bolstering the probative weight of some additional piece of corroborated evidence. This bolstering effect is explained by using inference to the best explanation. So long as the truth of the conclusion provides the best explanation of the corroborated evidence, and the truth of the conclusion is supported by the corroborating evidence, there is additional reason to find that the corroborated evidence is also true. This account, I argue, recognizes and explains the

---

[8] As van Cleve (2005) observes, there remains the issue of whether each independent source must have some positive initial credibility which can be amplified by coherence, or whether coherence alone can produce positive credibility where initially there was none. A moderate foundationalism holds that coherence has an ampliative but not a productive function. While van Cleve summarizes arguments on each side of this issue, I do not here attempt to resolve it.

dual function of corroborative evidence without giving rise to the fallacy of double counting. Furthermore it suggests a way that corroborative evidence can be modeled and evaluated in theories of argument. I conclude by relating the account proposed herein to Walton's account.

Regrettably, Walton and Reed (2008) seem to reject the IBE explanation of corroborative evidence. Considering the example of a jury deliberating upon the mass of evidence in a case, they write:

> The jury shouldn't be making up their minds by arguing as follows: what the witness said has now turned out to be true, or at least highly plausible as shown by this new [corroborating] evidence, therefore his [original, corroborated] argument from witness testimony must be correct, or should be taken as stronger than it was before. What seems to be wrong with this argument is that it judges whether an argument is correct based on the conclusion alone rather than on the inferential link between the premises and the conclusion. It reasons backwards that since the conclusion has turned out to be true, the argument must be a good one, either because of the inferential link between the premises and the conclusion, or because one of the premises has been shown to be true. (p. 551)

Walton and Reed's point here is well-taken. Reasoning from the acceptability of a conclusion to the acceptability of a set of reasons for that conclusion is to entirely neglect the evidentiary structure of argument. Yet, Walton and Reed do not consider the evidential structure at work in inference to the best explanation. The reasoning does not proceed from the truth of a conclusion to the acceptability of some premise, but rather from the elimination of all other possible explanations of the truth of the premises except the truth of the conclusion.

In having this structure, the IBE explanation of corroborative evidence bears a close resemblance to Walton and Reed's own thesis that corroborative evidence works by providing pre-emptive rebuttals to potential defeaters of a plausible argument. Walton and Reed suggest that the potential defeaters have the form of critical questions which are pre-emptively answered. Yet because critical questions provide only a commonplace rather than comprehensive inventory of potential defeaters (Godden and Walton 2007, p. 269), a more general account of pre-emptive rebuttal is desirable. On the IBE account, potential defeaters take the form of alternative explanations of the truth of some premise which do not involve the truth of the conclusion (i.e., counter-examples). Corroborative evidence, in providing independent evidence supporting the truth of the conclusion, helps to show that no counter-examples apply in this instance. By showing that no counter-examples apply, corroborative evidence strengthens the link between the corroborated premise and the conclusion.

So, there is an important congruence between Walton's account and the IBE account proposed here, which helps to isolate a distinctive feature of corroborative evidence, and to explain its unique bolstering function. Indeed, the IBE account accords well with Wigmore's (1913, p. 751) description of the function of corroborative evidence in law as "closing up other possible explanations." Perhaps then a final virtue of the inference to the best explanation account of corroborative evidence is its coherence with other, existing accounts which thereby lend it a degree of corroboration.

CHAPTER 16

WALTON'S EQUIVOCAL VIEWS ABOUT INFORMAL LOGIC

Ralph H. Johnson

16.1   Introduction

Over the years Douglas Walton has been one of the leading voices of informal logic. He presented a paper at the First International Symposium on Informal Logic in 1978. Walton has published many articles in *Informal Logic* that devoted a special issue to his work (Vol.27, No.1). He is the author of a book from Cambridge University Press by that name (Walton 1989, 2008a), and has co-authored a history of informal logic with Alan Brinton (Walton 1997c). He also wrote the entry on "Informal Logic" for *the Cambridge Dictionary of Philosophy*.

It is, then, interesting to note that Walton has expressed doubts about whether there is any such an inquiry as informal logic, and if there is, whether "informal logic" is the best name. In his (Walton 1998b), Walton referred to "an identity crisis" in informal logic, citing a 1997 review by Binkley of *New Essays in Informal Logic*. There Binkley had mentioned "an identity crisis of informal logic, arising from the question of whether this new subject should be thought of as a branch of logic, or as a distinct discipline in its own right" (p. 853).

In this paper, I want to discuss this "equivocal" relationship Walton appears to have to informal logic. That is, in some writings, he appears to be a strong proponent of it, while in others he seems more sceptical. I begin by outlining Walton's conception of informal logic, or better, his various conceptions of informal logic. Then I review the reasons for the claim, which Walton endorses, that informal logic is undergoing an identity crisis. Are the reasons given good reasons for this claim? Further, Walton has recently proposed that the enterprise might be better called "semi-formal" logic. Is that a better name? Finally, I want show that in spite of this rather ambiguous relationship with it, Walton has been instrumental in introducing the "wares" of informal logic into what I refer to as "the new digital environment."

16.2   Walton's various conceptions of informal logic

As far as I am able to tell, there is no thematic treatment in Walton's work of informal logic, and certainly no attempt to define "informal logic," until 1989. Before that, there is just a 1974 paper, co-authored with John Woods, "Informal Logic and Critical Thinking" which appeared in Education. That paper contains no definition of "informal logic," and the best inference, given the contents of the paper, would be that its authors take informal logic to have something to do with the informal fallacies (Woods and Walton 1974, p. 84).

But beginning in 1989, Walton has offered a number of characterizations of informal logic, as well as some important claims about it that I investigate in this paper. It should be mentioned that this descriptor — informal logic — functions minimally in Walton's work. There are a number of descriptors that are much more closely associated with his work: e.g., argumentation, theory, fallacies, fallacy theory, practical reasoning — all these of which occur as descriptors more regularly. Still, as Walton is, for good reasons, associated with informal logic, his views about informal logic are to be taken seriously.

(1989) Informal Logic: A Handbook for Critical Argumentation

This work is important because it was the first monograph dealing with informal logic to be published by one of the leading publishing houses in the world — Cambridge University Press.[1] Walton had already established a reputation as one of the leading theoreticians associated with informal logic. His approach here is largely descriptive and discursive. Noting the importance of what he calls "logical semantics", he states that "the other eight chapters are mainly about the pragmatics of argumentation" (p. ix), since "applying critical rules of good argument to argumentative discourse in controversial issues in natural language is an essentially pragmatic endeavor" (p. ix). Walton is not the first to associate informal logic with pragmatics,[2] but was one of the first to conceive of informal logic as dialogical and rule-based. Walton writes that the "basic requirement of critical argumentation is that any argument that a critic attempts to evaluate must be set out as sympathetically appreciated in the context of dialogue in which the argument occurs" (pp. ix-x). Most approaches to informal logic up to this point had been criteria-based, focussing on argument conceived as a certain kind of structured discourse.[3] Instead Walton here sets the focus on dialogue. (One senses the influence of Hamblin, who in *Fallacies* criticizes formal logic and winds up endorsing formal dialectic (Chapter 8).) Walton later refers to this as "the dialectical approach" (p. x), in the question-answer context of an argument and concludes: "Thus generally the theory of informal logic must be based on the concept of question-reply dialogue as a form of interaction between two participants, each representing one side of an argument, on a disputed question" (p. x). Though no definition of "informal logic" is given, it is conceived of as a pragmatic undertaking[4] in the setting of a dialogue.

(1990) What is Reasoning? What is an Argument?

In this paper, Walton presents a slightly different view of informal logic, based on contrasting formal with informal logic. Walton writes:

> Formal logic has to do with the forms of argument (syntax) and truth values (semantics).... Informal logic (or more broadly, argumentation, as a field) has to do with the uses of argumentation in a context of dialogue, an essentially pragmatic undertaking. (pp. 418-419)

Note the suggestion made in passing that Walton considers that informal logic is, more broadly, the field of argumentation. This is a recurring theme in his work, which I will comment on below. In this article, Walton invokes many of the same

---

[1]I do not mean to ignore Govier's *Problems in Argument Analysis and Evaluation* (1987). But it is a collection of individual papers some of which had already been published on different occasions.

[2]The idea is there in Fogelin (1978) but not thematically developed. See also (Scriven 1980).

[3](Scriven 1976, Johnson and Blair 1994b, Fogelin 1978, Govier 1985, Fisher 1988)

[4]In *Manifest Rationality*, I addressed how informal logic fares with respect to the semantics, syntax, pragmatics distinction. See pp. 368-69.

ideas as in 1989, but in the context now of discussing the relationship between formal and informal logic. Walton writes:

> Hence the strongly opposed current distinction between informal and formal logic is really an illusion, to a great extent. It is better to distinguish between the syntactic/semantic study of reasoning, on the one hand, and the pragmatic study of reasoning in arguments on the other hand. The two studies, if they are to be useful to serve the primary goal of logic, should be regarded as inherently interdependent, and not opposed, as the current conventional wisdom seems to have it. (p. 419)

When properly understood, Walton thinks, these two logics are not in competition but rather are complementary. (The further and interesting claim that they are "interdependent" is not elaborated upon.) To spell out their complementary nature, Walton relies on the traditional distinction between syntax, semantics and pragmatics, assigning to formal logic the syntactical and semantical aspects of the study of argumentation, and to informal logic the pragmatic aspect. Hitherto there had been an underlying current that saw informal logic as antithetical to formal logic (see (Scriven 1980, Johnson and Blair 1980)) but by the time this paper was written, that antipathy had largely disappeared.

(1995) Cambridge Dictionary of Philosophy

Five years later, Walton wrote the entry on "informal logic" for the *Cambridge Dictionary*. We shall see that it differs markedly from previous definitions:

> Informal logic, also called practical logic, the use of logic to identify, analyze and evaluate arguments as they occur in the contexts of discourse in everyday conversations. In informal logic, arguments are assessed on a case by case basis, relative to how the argument was used in a given context to persuade someone to accept the conclusion, or at least give some reason relevant to accepting the conclusion.

His articulation here comes closer to the sort of definition given by (Munson 1976).[5] Walton also here must be understood as focusing on the practice of informal logic rather than its theoretical role. Still this entry seems something of an oddity. The focus on conversation and context are in line with earlier discussion. But the claim that "informal logic" is "also called practical logic" is new and, I believe, unwarranted; almost no-one (to the best of my knowledge) refers to informal logic as practical logic.[6] "Practical logic" is associated with practical reasoning, which for a long time has been identified with a certain kind of reasoning, usually deliberative and normative — a much narrower focus than informal logic has. Walton continues: "The use of logic to identify, analyze and evaluate arguments as they occur in the contexts of discourse in everyday conversations. In informal logic, arguments are assessed on a case-by-case basis." But informal logic, insofar as it is a general theory, does not itself take up and evaluate any specific arguments, except as an illustration. As an inquiry it is perfectly general, setting out criteria, standards, rules, procedures etc.

---

[5]"Informal logic is the attempt to make explicit the principles or standards that are involved in ordinary everyday activities of establishing and evaluating claims and using language effectively in the process of communication and rational persuasion."(Munson 1976, p. 3).

[6]Thomas did call his important textbook *Practical Reasoning in Natural Language*, and Govier titled her text *A Practical Study of Argument*.

(1997) Brinton & Walton, Historical Foundations of Informal Logic

Yet another, slightly different, view emerges in this book, the first attempt to write a history of informal logic, about which the authors say:

> Informal logic has yet to come together as a clearly defined discipline, one organized around some well-defined and agreed upon systematic techniques that have a definite structure and that can be decisively applied by users. Nothing analogous to the great flowering of formal logic of the past hundred years has occurred or appears quite ready to occur in informal logic. (p. 9)

In the background of this claim that informal logic has yet to come together as a discipline, one can sense the presence of formal logic as paradigm in the use of such phrases as "well-defined and agreed upon systematic techniques," "definite structure," "decisively applied." This is unfortunate for it amounts to complaining that informal logic is not formal deductive logic, which is somewhat at odds with the division of labour he assigned in (1990). Walton and Brinton stress informal logic's lack of definition and go on to cite "the diversity of thinking" (p. 9) about the issues and problems that fall under its study. They cite as a reason that informal logic will not attain the status of formal logic is that it is a practical subject that requires "the interpretation of natural language argumentative discourse" and "attention to the role of participants and the feature of persuasion" (p. 9). But "its aim with respect to argument, like that of formal logic, is to provide the basis for judgements about the correctness or incorrectness of inferences" (p. 10) and they understand that this requires attention to "much more than the form of these inferences."

I want to make a number of comments. First, Walton and Brinton seem to conflate argument and inference, as many do. In our attempts to define "informal logic," Blair and I have been careful to specify argument/ation as the focus of the inquiry, and I have argued that to conflate argument and inference is to invite conceptual confusion.[7] Second, Walton and Brinton are right when they say that the informal logician will have to grapple with much more than the form of these inferences. The way that I would put this is to say that informal logicians must find ways to appraise arguments as to their correctness that does not rely on their being reduced to their logical form. Third, if one chooses to regard argument as an inference from premises to conclusion, then one way that informal logic has been characterized from the start is as an inquiry that pursues the issue of whether there is not a third type of inference (neither deductive nor inductive (Blair and Johnson 1980)). Fourth, it is true that informal logic has not matched "the great flowering of formal logic of the past hundred years." But there are reasons. If "the great flowering" refers to the development of axiomatic methods that yield decidable outcomes, then it seems clear that informal logic will never be able to achieve this kind of success — because its very subject matter precludes it — as indeed Walton and Brinton themselves seem to recognize.

Finally, in our chapter in this volume, we offered the following definition:

> By informal logic we mean to designate a branch of logic which is concerned to develop non-formal standards criteria [and] procedures for the analysis, interpretation, evaluation, critique and construction of argumentation in everyday discourse (Johnson and Blair 1987, p. 161).

---

[7]See (Johnson 2000, pp. 92-95)

There is no incompatibility between Walton's various conceptions and ours. Both stress the focus of informal logic on argumentation in everyday discourse. Both stress analysis and evaluation as primary activities which informal logic supports. I do find it odd that in none of his attempts to define "informal logic" has Walton made mention of our 1987 definition — even if only to say why he finds it inadequate.

We come to Walton's 1998 paper where yet a different view appears, as I shall discuss below. Let me summarize the findings thus far. First, Walton associates informal logic closely with natural language argumentation, and with pragmatics rather than syntax or semantics. Second, Walton does not have one consistent definition of "informal logic" but rather several different of ways of characterizing it. Sometimes it is identified with practical reasoning, sometimes with the field of argumentation. To that degree his conception appears equivocal.

Walton's position on informal logic is equivocal in another sense; to the degree that "equivocal" suggests uncertain, we will find that Walton appears to be uncertain about the very existence of informal logic, as we see in the next section.

16.3   The identity crisis in informal logic

In the 1998 paper, Walton takes what seems to be a new tack. Here he leans on Binkley's review of *New Essays in Informal Logic* (1995) in which Binkley has suggested that informal logic was undergoing an identity crisis. In what follows, I first briefly summarize Binkley's position in his paper and then discuss Walton's deployment of it.

In his review of *New Essays in Informal Logic*, Binkley detects a number of themes, one of which has to do with the identity of informal logic: "What exactly is this logic? Is it really a branch of logic, in the sense in which that subject is seen to be a branch of philosophy?" (p. 261). Then Binkley proceeds to indicate how some of the essays in the volume may be seen as a response to this question. To some degree, the crisis may be seen as wondering just how informal logic can position itself in relationship to formal logic, on the one hand, and epistemology on the other — and as well, to differentiate itself from critical thinking and argumentation theory. But, as I read him, Binkley does not himself assert that there is an identity crisis; rather he is picking up on what he takes as one theme in the volume he is reviewing, and, in my judgement, there is some legitimacy to such an interpretation.

The issue of just what informal logic is and what its various tasks are has been with us from the very start. In that sense, its perennial pursuit of identity might perhaps be taken to reflect the disproportionate number of Canadians (Woods 2000, p. 161) who have become identified with informal logic and who perhaps have brought to the inquiry, the well-known quest for a Canadian identity. But the posing of this question need not be taken as any indication of a crisis — and indeed Binkley does not himself assert that there is such.

Walton takes a much stronger position, claiming that "there is considerable ambivalence and uncertainty, even with the exponents of informal logic themselves, how it should be titled and defined, on how it ought to be presented,..." (p. 853). But Walton gives no evidence to support this rather crucial claim. Referring to Binkley's review, Walton asks: "This identity crisis is implicit even in doubts about the exact terminology that should be used to label the new subject. Should it be called argumentation or critical thinking, without the term logic being used at all?" In response to this question, I would say that no one should propose that informal logic be equated with either argumentation or critical thinking. To be sure, the connection between informal logic and argumentation is strong, since historically informal logic portrayed itself as a logic of (everyday, ordinary) argumentation (but see (Woods 2000, pp. 156-157), (Goddu 2009a), and the close ties to critical thinking are explicable from the fact that informal logic emerged initially out of concerns having to do with the teaching of logic. These coincided with the thinking skills movement that emerged in the 70s (Resnick 1987) which ultimately led to the connections between informal logic and critical thinking. On several occasions Blair and I (Johnson and Blair 1994a, 1996, 2002) argued that informal logic and critical thinking are quite different:

> As contrasted within formal logic as a branch of a particular discipline (logic) and partly defined by its subject matter (arguments), critical thinking refers to an attitude of mind whose application knows no disciplinary boundaries (1996:p. 165).

Walton continues: "Or should it be called applied logic or practical logic, if the word 'logic' is appropriate"(p. 853). In response to this alternative, I suggest there are good reasons for avoiding both of the above terms. I have already said why 'practical logic' strikes me as inapt. 'Applied logic' suggests applied formal logic, which is not what informal logic, is about. (For an example of this kind of use of logic, see Pospesel & Lycon (1997). As to whether the word 'logic' is appropriate, that depends on how one understands that term. If logic is taken to mean the normative study of reasoning, then I would say that it is appropriate. There are various logics associated with the various types of reasoning. We can make the point that informal logic is concerned with argumentation, which cannot be identified with either entailment (the subject of formal logic) or even inference. Walton then writes: "The very phrase 'informal logic' looks like an oxymoron." It looks like an oxymoron because, one presumes, logic is thought by its very nature to be formal. But this objection disappears if one works with a generic conception of logic like the one above.

Moreover, the objection can be defused if one understands the sense of "formal" which the "in" of "informal logic" negates. Here I offer my explanation – essentially repeating what I wrote in (Johnson 2000) in which I invoked the distinction of various senses of 'form' developed by Barth and Krabbe (1982).

> An obvious point is that "informal" takes its meaning in contrast to its counterpart — "formal." And yet this point manages not to be made for a very long time, and hence the nature of informal logic remained opaque, even to those involved in it, for a long period of time. Here it is helpful to have recourse to Barth and Krabbe (1982: p. 14f) where they distinguish three senses of the term "form."
>
> By "form$_1$," Barth and Krabbe mean the sense of the term which derives from the Platonic idea of form, where form denotes the ultimate metaphysical unit. Barth and Krabbe claim that most traditional logic is formal in this sense. That is, syllogistic logic is a logic of terms where the terms could naturally be understood as place-holders for Platonic (or Aristotelian) forms. In this first sense of "form," almost all logic is informal (not-formal). Certainly neither predicate logic nor propositional logic can be construed as term logics. However, such an understanding of informal logic would be much too broad to be useful.
>
> By "form$_2$," Barth and Krabbe mean the form of sentences and statements as these are seen in modern logic. In this sense, one could say that the syntax of the language to which a statement belongs is very precisely formulated or "formalized"; or that the validity concept is defined in terms of the logical form of the sentences which make up the argument. In this sense of "formal," most modern and

contemporary logic is "formal." That is, such logics are formal in the sense that they canonize the notion of logical form, and the notion of validity plays the central role normatively. In this second sense of form, informal logic is not formal, because it abandons the notion of logical form as the key to understanding structure and likewise abandons validity as constitutive for the purposes of the evaluation or argument(ation).

By "form₃," Barth and Krabbe mean to refer to "procedures which are somehow regulated or regimented, which take place according to some set of rules." Barth and Krabbe say that "we do not defend formality₃ of all kinds and under all circumstances." Rather "we defend the thesis that verbal dialectics must have a certain form (i.e., must proceed according to certain rules) in order that one can speak of the discussion as being won or lost" (p. 19). In this third sense of "form", informal logic can itself also be formal. There is nothing in the informal logic enterprise that stands opposed to the idea that argumentative discourse should be subject to norms, i.e., subject to rules, criteria, standards or procedures.

This last point requires emphasis, because even though both Blair and I have made this point several times, it does not appear to have sunk in. *There is nothing in the informal logic initiative that prohibits or is antithetical to the use of formal methods or procedures.* The point is that any logic must be about normativity, and informal logic differs from formal logic in that the normativity it pursues is not that associated with validity. Regarding rigour, as Blair and I (Blair and Johnson 1991) and Govier (Govier 1987, 1999) have argued, much depends here on how one understands the ideal of rigour. A rigorous proof of a mathematical theorem is one thing; a rigorous police investigation quite something else. Formal logic understandably aims at something like the former type of rigour; informal logic, understandably, aims rather at something like the latter.

In any event, I have given my reasons for thinking that the objection that informal logic is an oxymoron is not a strong one.[8]

Walton then states that "Informal Logic is a confrontational phrase that seems particularly off-putting, especially now that so many in the field of computer science have taken up with argumentation as a much-needed component in computer programming" (p. 853). Walton is here referring to an important development that I will be commenting on further below — how it is that the subject of argumentation came to be a focus for computer studies. Walton's point that the phrase "informal logic" is confrontational seems predicated on the old idea that there is some inherent tension between formal and informal logic. But his own (1990) gives the lie to any such suggestion, so it is somewhat surprising to find that idea being put in play here.

I have argued thus far that Walton's views about informal logic are equivocal and that his misgivings about the name "informal logic" seem to be based on some misunderstandings. Next I discuss his recent suggestion about a better name.

16.4   Would 'semi-formal logic' be a better name?

From the beginning, there has been a theme in the literature in which the name — informal logic — is lamented and castigated, not only by Walton but by others. Hitchcock who dislikes this name because of "its unfortunate connotation

---

[8]For further comments see (Johnson 2000, pp. 252-260).

of sloppiness and lack of rigour" (2000, p. 130). In a recent article (2009), I commented on this matter:

> My rejoinder to this criticism is to wonder, first of all, if philosophers really are that much influenced by the name. My next thought is to wonder what name might have worked better. Other names have been suggested: *Natural logic*: but this is already in use by (Grize 1982) for a very different project. *Practical logic*: possibly but perhaps too connected to the idea of practical reasoning, which would cover some of the sorts of argument dealt with by informal logic, but not all. *Applied logic*: Blair and I used this name to designate the course we created at the University of Windsor in the early 1970s (see (Johnson 2009)) but later, we set this term aside for its implicit suggestion that what we were doing was applying formal logic. *Ordinary logic*: this was already used by Ennis for the title of a logic text — but I wonder if this is any better. Recently some have suggested *Philosophy of Argument* (Govier 1999, Blair 1999, 2003). But see (Blair and Johnson 2009) for his more recent, more cautionary view.

In his new introduction to the second edition of *Informal Logic*, Walton suggests that maybe "semi-formal logic" would be a better term. Walton leads into this by writing: "At this point in the history of the subject, it is timely to raise the terminological question of whether it should still be called informal logic, or something else, such as argumentation" (2008, p. xiii). It seems to me that this suggestion will fare no better, and maybe even worse, than "informal logic." It runs up against a semantic obstacle in that 'argumentation' is quite a different type of term. Informal logic is meant to refer to an inquiry, a discipline; the term 'argumentation' refers to subject matter, a practice or a process which may be studied. Such an inquiry might better be called Argumentation Theory. Walton, however, seems to want to retain the term 'logic,' saying: "It is good that the term 'logic' should be retained, but it is a problem that for the purpose of computing, an exact science, any useful system of analysis and evaluating arguments must be at least partly formal" (p. xiii). Here it seems to me we come to the crucial point. The reference to "computing" may catch some by surprise, so let me provide some important background information.

Walton began his study of argumentation in the 1970s, using the methods and ideas he had learned in his graduate education in philosophy: namely logic, epistemology. Though Walton has never repudiated a role for formal logic, it appears he came to understand that informal logic had an important contribution to make to the study of argumentation. As his research developed, Walton became aware of the possibilities that this new focus — argumentation — had for other disciplines. One of these was the area of computer studies/computer science, sometimes referred to as AI. While traditional AI was oriented toward the attempt to understand human knowledge better by seeking to replicate it using computer routines, an important shift occurred in the late 80s, which I say more about in the next section. Walton was one of the first working in informal logic to see the possibilities that work done under the rubric of "informal logic" had for this new focus in Computer Studies. In 2003, he began a working relationship with Chris Reed and Henry Prakken.[9] I believe that the role assigned to computational approaches to the study of argumentation is one of the reasons

---

[9] Articles on argumentation and computation by Douglas Walton include (Walton 1997c), (Walton 1999b), (Walton 2000b), (Bex et al. 2003), (Walton 2005f), (Walton and Reed 2005), (Walton 2005e), (Walton 2005b), (Reed and Walton 2005), (Walton and Godden 2006), (Rowe et al. 2006), (Walton 2006e), (Walton 2006d), (Walton 2006a) and (Walton and Godden 2005).

for Walton's resistance to the name "informal logic." Computational approaches must to some degree be formal. But the idea that useful systems or methods of analysis must be "at least partly formal" is not a problem for informal logic. What would be a problem is the insistence that such methods be committed to the concept of logical form and be normativized by reference to the standard of validity. But I see no such insistence.

I return now to the suggestion Walton has made that 'semi-formal logic' may be a better term. Walton is here basing his views on a paper by Verheij, (Verheij 2003a, p. 172) which refers to "semi-formal rules of inference" or "semi-formal argument templates." Yet it seems to me that this suggestion will fare no better. At the very least, the term "semi-formal" will need clarification in this context; and it seems that any objections that one would have to "informal logic" based on its connotations would readily transfer to "semi-formal logic."

16.5   Informal logic and the digital environment

I turn now to a point touched on in the recent discussion about the best name for this inquiry. I offer some reflections on the importance of the developments that Walton alluded to when he wrote: "Argumentation, also called informal logic, based on this new notion of argument, is now being widely used as a new model of practical reasoning in computing, especially in agent communication in multi-agent systems. (2008, p. 1)." To appreciate the significance of this statement, I offer the following explanation.

It appears that we are in the process now of moving from one cultural environment to another quite different one and that this move will have important repercussions both on the practice of argumentation and our attempt to understand it and theorize about it, whether that attempt is called informal logic semi-formal logic or something altogether different. Many think of this new environment as the *digital environment*,[10] by contrast to the older analogical environment largely dominated by print media: by books, encyclopaedias, papers, journals, magazines. In the analogical environment, a person makes an argument in an article that she publishes in a journal, someone reads it and makes criticisms, the other responds — positions gradually change, become more developed and these developments can be mapped. The new digital environment is of course a direct offspring of the computer revolution and the creation of what is sometimes called cyberspace (the world-wide web, virtual reality, etc). Some of the old forms have been transported into the digital environment — i.e., scholarly journals go online and online encyclopediae have developed: viz., the *Stanford Encyclopaedia of Philosophy*. Scholarly conferences now routinely allow for virtual participation, and indeed some newer forms would not be possible without the Internet and Cyberspace. (See the Appendix).

The internet has made new forms of scholarly interaction possible. But it is not just that the older forms have been incorporated into the important shifts that are taking place in this new digital environment. For example, Google has had an enormous influence on how many people approach research. Much of the journal literature can be retrieved online and thus library searches and the way in which research is carried out have changed. When you go into a university

---

[10]The phrase is not new: "Since the late 1980s it has become accepted wisdom that we are living in a 'digital environment', one that embodies a 'law of code' (a law founded on US libertarian values regarding free speech and private enterprise) that will bring nations - and individuals - together in a global market featuring ubiquitous access to information via electronic media. That environment is supposedly both unprecedented and inevitable, although delayed in parts of the world which that have not inhaled the zeitgeist." http://www.caslon.com.au/digitalguide.htm

library these days, the first thing that will strike you is the work areas that provide students with computer access, or with the needs for computer use (WiFi). The old file card catalogue have been replaced by computers which allow the student to search one's own library holdings, as well as those in other libraries all over the world. The process of publication has changed as well. A scholar can now "publish" or make available his articles on his website, or via a weblog. This new digital environment is far more inclusive and decentralized; access to it is largely unrestrained, though the library has to pay significant fees. In the old environment, one could only have access to publication by passing through a set of filters. To publish an article one had to submit it to a journal and pass through the editorial process. I could go on and on detailing changes but the proposition that I want to put forth is that in this new digital environment, the "wares" of informal logic are prominent. How did this happen?

To set the stage, I need to make reference to the shift that took place in the late 80s when many computer studies researchers began to focus their attention on modelling argumentation (rather than knowledge). I cannot do better than quote the explanation given by Dix, Parsons, Prakken and Simari (researchers in AI and CS) in their paper, "Research Challenges for Argumentation":

> The first articles on argumentation in computer science appeared circa 20 years ago. Since then we have seen great advances, establishing a solid theoretical basis, a broad canvas of applications, and most recently, some realistic implementations.
>
> Modeling commonsense reasoning in AI is a difficult task given that it almost always occurs in the face of incomplete and potentially inconsistent information. Argumentation formalisms are defeasible reasoning systems which work by considering the reasons that lead to a given conclusion (or claim) through a piece of reasoning (the supporting argument) and the potential challenges (or counterarguments) for accepting that conclusion In this manner, the mechanisms proposed model reasoning as a dialectical process, i.e., the exchange of arguments and counter-arguments respectively advocating and challenging the claim of the initial argument. This process offers a remarkable tolerance to the problems introduced by the potential inconsistency and/or incompleteness of the knowledge source. Within the study of argumentation, we can distinguish a number of strands, each of which is an active area of research, and between which there is unfortunately little crossfertilization. Argumentation's roots in models of disputation, for example, mean that it is well suited for the study of models of dialogue. Its ability to capture and resolve different points of view means that argumentation has found many applications in the area of multi-agent systems, since the different systems, since the different agents in such systems frequently have mutually inconsistent information and goals. [Retrieved from http: //drops.dagstuhl.de/opus/volltexte/2008/1577/pdf/08042. Dagstuhl_Manifesto.pdf]

Walton has played a pivotal role as a mediator between research initiatives that stem from his work on argumentation and informal logic, and this evolving agenda for Computer Studies. For example, argumentation schemes have turned out to play an important role in the attempt on the part of those in computer studies to model argumentation. These schemes had been developed by informal logicians, including Walton and other argumentation theorists in the mid-90s. Argumentation schemes are not formal$_2$ (in the sense defined above) devices; they do not rely on the notions of validity, logical form, or entailment and yet they are formal₃ and they have found increasingly wide application in computer studies approaches to argumentation.

In addition, the wares of informal logic are clear in the work that Walton and others have done on argument diagramming. Araucaria is a software tool for analyzing arguments developed by Chris Reed and Glenn Rowe. The authors describe it as follows:

> [Araucaria] aids a user in reconstructing and diagramming an argument using a simple point-and-click interface. The software also supports argumentation schemes, and provides a user-customisable set of schemes with which to analyse arguments. Araucaria has been designed with the student, instructor, and researcher in mind. It is sufficiently straightforward to be useful to students learning how to reconstruct arguments, diagram them, and apply argumentation schemes. It is sufficiently flexible for instructors to provide their own examples, sample analyses, and alternate sets of argumentation schemes. Finally, it is also sufficiently powerful to be of use in research, particularly in providing examples of argument analyses to support claims. (Retrieved from http://araucaria.computing.dundee.ac.uk/)

One can see in their description the importance of argumentation schemes, which play a central role in this approach.

To summarize, in the newly emerging digital environment, the wares of informal logic are well represented, and this is in no small measure due to the work of Douglas Walton who has been one of the principal exporters.

16.6 Conclusion

In this paper I have attempted to set forth Walton's views about the nature of and the viability of informal logic as an inquiry. I have argued that his views about the nature informal logic are equivocal, both in the sense of that there is some ambiguity in how he understands the term, but as well in his doubts about its very existence. Perhaps, as one referee has suggested,

> "[in] fairness to him [Walton], he has grappled with the matter over a period of time in which his own allegiances may have shifted from a more logical to a more dialectical approach. Further, Walton is not happy with the name "informal logic," having expressed some scepticism about its suitability. However, I tend to side with one astute reader who put the matter this way: "While the name 'informal logic' might not be perfect, it does not appear that there are any better alternatives in the offing. To replace it with an equally imperfect alternative would be useless labour."

In spite of his uneasiness about the name and his concerns about an identity crisis, Walton has done as much as anyone to help develop the methods, procedures, and approaches associated with informal logic, and that through his efforts, these have become increasingly prominent in what I have called the new digital environment.

Appendix

Phenomenon #1 (digital conference)

ESSENCE [E-Science/Sensemaking/Climate Change] is the world's first global climate collective intelligence event designed to bring together scientists, industrialists, campaigners and policy makers, and the emerging set of web-based sensemaking tools, to pool and deepen our understanding of the issues and options facing the UN Climate Change Conference in Copenhagen in December 2009. The event, starts online in January 2009 and culminates in a conference at the National e-Science Institute in Edinburgh, in April 2009. During the pre-launch phase, we are beginning to identify and assemble teams of scientists, industrialists, campaigners and policy makers to work with the tool developers on specific aspects of the complex set of issues around climate change.

The aim is to develop a comprehensive, distilled, visual map of the issues, evidence, arguments and options facing the UN Climate Change Conference in Copenhagen, that will be available for all to explore and enrich across the web.

The project is founded on principles of openness, transparency, and discovery; with no preconceptions about the conclusions that will emerge from the event. If you are scientist, industrialist, campaigner, policy maker, tool maker or someone with other ideas and resources to contribute and are interested in learning more about and participating in ESSENCE, please get in touch.

Ready?... / ...Engage

Organised by: www.GlobalSensemaking.net, Open University UK, Debategraph, MIT Center for Collective Intelligence, UN Millennium Project, UK e-Science Institute

Phenomenon #2 Democracy in a digital society (digital research project)

Dear Colleagues in particular in EU countries:
Who would be interested in being involved in a project on Democracy in the future digital society? A consortium based here in Denmark, led by the Danish Broadcasting Corporation, is preparing an application to the EU. If given, the grant will be 2,7 million Euros.

Some of my colleagues from the Department of Media, Cognition and Communication at the University of Copenhagen will participate. We are very interested in quick response from researchers and institutions that might be interested in participating, particularly from EU countries, and in particular from Central and Southern Europe. Please send me a reply email as soon as possible to indicate your interest. It will not be binding, but we will get in touch with you to discuss your possible involvement. The final application is due on January 17.

Below is some information about the project.

Sincerely,

Christian Kock
Professor of Rhetoric
Department of Media, Cognition and Communication
University of Copenhagen
Denmark

Scenarios for democracy in the future digital society

The digital information and communication technology (ICT) will change most inter-human relations in what you could call the future digital society. And so far we have only seen the beginning of this complete change.

The digital technology has established global technical networks providing new business opportunities and worldwide professional collaboration. The electronic media have become global and multitudinous due to digital technology and all over the world mobile phones are now more common than the old wired device. With the popularization of the internet communication at all level of society the social life are changing and will continue to change. ICT is not only providing new facilities, but is a catalyst for transformation of our socio-economic and political system.

How will this transformation affect the democracy in Europe? How will ICT influence the inter-human relations? How will the political management of society be organised with the manifold communication channels available for the people? And how will the role of independent public service media be in the future digital society? The project will develop scenarios for the development of the democratic society in the digital future with special focus on the role of the independent public service media in relation to the democratic political process, peoples participation and daily life.

The project partners include research institutions in various relevant fields, public service media institutions in small and big European countries and CSOs engaged in media, freedom of speech and political participation.

Phenomenon #3 COMPUTATIONAL MODELS OF ARGUMENT

FIRST CALL FOR PAPERS

CMNA IX - THE 9TH INTERNATIONAL WORKSHOP ON COMPUTATIONAL MODELS OF NATURAL ARGUMENT Organizers: Nancy Green (University of North Carolina Greensboro), Floriana Grasso (University of Liverpool), Rodger Kibble (Goldsmiths College, University of London), Chris Reed (University of Dundee)

The series of workshops on Computational Models of Natural Argument is continuing to attract high quality submissions from researchers around the world since its inception in 2001. Like the past editions, CMNA09 acts to nurture and provide succor to the ever growing community working in "argument and computation". AI has witnessed a prodigious growth in uses of argumentation throughout many of its subdisciplines: agent system negotiation protocols that demonstrate higher levels of sophistication and robustness; argumentation-based models of evidential relations; groupwork tools that use argument to structure interaction and debate; computer-based learning tools that exploit monological and dialogical argument structures in designing pedagogic environments; decision support systems that build upon argumentation theoretic models of deliberation to better integrate with human reasoning. The workshop focuses on the issue of modelling "natural" argumentation. Naturalness may involve the use of means which are more visual than linguistic to illustrate a point, such as graphics or multimedia. Or to the use of more sophisticated rhetorical devices, interacting at various layers of abstraction. Or the exploitation of "extra-rational" characteristics of the audience, taking into account emotions and affective factors. In particular, contributions will be solicited addressing, but not limited to, the following areas of interest:

- The characteristics of "natural" arguments: ontological aspects and cognitive issues.

- The computational use of models from informal logic and argumentation theory,

- The linguistic characteristics of natural argumentation, including discourse markers, sentence format, referring expressions,and style.

- The generation of natural argument

- Corpus argumentation results and techniques

- Models of natural legal argument

- Rhetoric and affect: the role of emotions, personalities, etc. in argumentation.

- The roles of licentiousness and deceit and the ethical implications of implemented systems demonstrating such features.

- Natural argumentation in multi-agent systems.

- Methods to better convey the structure of complex argument, including representation and summarisation.

- Natural argumentation and media: visual arguments, multi-modal arguments, spoken arguments.

- Evaluative arguments and their application in AI systems (such as decision support and advice giving).

- Non-monotonic, defeasible and uncertain argumentation.

- Computer supported collaborative argumentation, for pedagogy, e-democracy and public debate.

- Tools for interacting with structures of argument

- Applications of argumentation based systems

For information on IJCAI-09 see: `http://ijcai-09.org/index.html`

CHAPTER 17

WALTON AND THE TRADITION OF PLAUSIBILITY ARGUMENTS

Christopher W. Tindale

17.1 Introduction

Plausibility arguments, or what I will here call *eikos* or likelihood arguments, have a long tradition in the history of logic, although not necessarily a favourable one. Douglas Walton, a major contemporary proponent of such reasoning, judges that Plato's denunciation of the Sophists created a strong prejudice against plausible reasoning that continues to be felt. Plausible reasoning, again as expressed by the Sophists, has become an unintelligible part of logic, along with a general shift on the part of logicians away from both rhetoric and dialectic (Walton 1998b, p. 16). Yet in spite of this acknowledgement of the role that plausible reasoning played in the argumentation of the Sophists, his own history, that tracks through Cicero, Locke (whose famous case of the Dutch ambassador and the King of Siam uses 'probability' in the sense that the ancients used 'plausibility') and Bentham, finds its origin in the works of Carneades of the later Academy (Walton 2002a, chap. 4). In fact, when he establishes his new theory of plausibility in chapter 5 of that work, it is a skeptical theory that has its acknowledged origins only in the sayings of Carneades rather than those of the Sophists three centuries earlier. What is called for, then, is a recovery of that earlier history to measure the degree to which there is a coherent theory of plausible arguments that can both inform later models and serve as part of a fuller appreciation of the types of argumentation that preceded both Plato and Aristotle. It is not my intention to dispute the importance of what Walton has uncovered in Carneades; but only to demonstrate a way in which we have much richer resources in our intellectual history than is usually appreciated. Again, it is not my intention to examine in too much detail the treatment of *eikos* arguments provided by Aristotle, except as this becomes relevant to the central discussion.

By a plausible argument Walton means one that "proceeds from premises that are more plausible to a conclusion that was less plausible before the plausible argument was brought to bear on it" (Walton 1998b, p. 242).[1] If these are to be the modern instantiations of *eikos* arguments, then this definition will not exactly accord with what we are to learn from the study of ancient *eikos* arguments. Elsewhere, however, Walton (2001c, 2002a) clearly equates plausible reasoning with *eikos* arguments in a way that conforms to what we find in the extant texts: as a kind of common knowledge that reflects general human experience as this is instantiated in a particular group of reasoners. The first thing to note, though, is that this is not the understanding that Plato attributes to the *eikos* arguments of the Sophists (and on which the noted repudiation depends). So the study must begin with Plato's account.

---

[1] Another contemporary engagement with plausibility theory is that of Rescher (1976), although

17.2   Plato's Story

In the *Phaedrus*, Socrates and Phaedrus discuss the advice that should be offered to students of speechmaking. The speeches they have been considering up to that point are noteworthy for the way they collect and divide certain kinds and parts, and those who do this correctly are identified as "dialecticians" (266c). Here we have Plato's own interest in the organization of ideas within discourses. But then the contrast is marked with the "masters of speechmaking," "Thrasymachus and the rest of them," who are purveyors of rhetoric rather than dialectic. Still, it is suggested, there may be something of value in what is written up in books on the art of speaking:

> Socrates: First, I believe, there is the preamble with which a speech must begin. This is what you mean, isn't it–the fine points of the art?
> Phaedrus: Yes.
> Socrates: Second come the narrative of facts and the testimony of witnesses; third, the evidence (*tekmêria*); fourth, the probabilities (*eikota*). And I believe that that excellent Byzantine word-crafter adds proof (*pistôsis*) and counter-proof (*epipistôsis*).
> Phaedrus: You mean the worthy Theodorus? (266d-e)

The itemizing of details continues, drawing in the contributions of other major Sophists like Gorgias, Prodicus, Hippias and Protagoras. But our primary interest here is the attention given to probabilities or likelihoods (*eikota*). This, at least, is the key idea that they set in opposition to truth later in the dialogue (273a). Here, the attribution is to Tisias who, along with Corax, is credited with the first handbook on rhetoric. The nature of such a handbook and the degree to which it would deserve the title "rhetoric" is a matter of some dispute. But Plato here has a clear message he wants to communicate on how "likely" should be understood:

> Socrates: Then let Tisias tell us this also: By "the likely" (*eikos*) does he mean anything but what is accepted by the crowd?
> Phaedrus: What else?
> Socrates: And it's likely it was when he discovered this clever and artful technique that Tisias wrote that if a weak but spunky man is taken to court because he beat up a strong but cowardly one and stole his cloak or something else, neither one should tell the truth. The coward must say that the spunky man didn't beat him up all by himself, while the latter must rebut this by saying that only the two of them were there, and fall back on that well-worn plea: "How could a man like me attack a man like him?" The strong man, naturally, will not admit his cowardice, but will try to invent some other lie, and may thus give his opponent the chance to refute him. And in other cases, speaking as the art dictates will take similar forms. Isn't that so Phaedrus? (273a-c).

---

his treatment is far more rigorous. Tracing the theoretical or systematic interest in plausibility back to Aristotle's *Topics*, Rescher proposes a theory that will effect a transition from the reliability of sources to the plausibility of their utterances. We can agree with his point of origin for the theory since Antiphon's treatment that we will explore, while deemed foundational (Enos 1980, p. 182) is hardly systematic.

Rescher himself marks a sharp distinction between plausibility and probability. The former is concerned with the reputation of the sources of claims, and thus is external; probability is by contrast internal, involving the content of the claims (p. 28).

The example that Plato uses here is the same one that Aristotle employs when attacking the probabilities or likelihoods of Protagoras, and accusing him of making the weak argument strong (Rhetoric II, 24).[2] Like Aristotle, who sets the Sophistic probabilities against "real" probabilities, Plato here opposes likelihood to what is true. But unlike Aristotle, Plato has what is *eikos* defined as what the crowd believes, suggesting it is a matter of opinion.[3] On such terms, part of Plato's disdain for the Sophist's ability to persuade an audience becomes clear: if opinion has no firmer foundation than verbal persuasion, then what is likely becomes a mere topic for manipulation. After all, should we believe Plato's suggestion that Tisias' discovery of the technique led him to the example of the weak and strong men, when we are told no more than that it is "likely"? Certainly, though, for Plato, *eikos* is a pale shadow beside the truth. But there may be reason not to equate what is likely with the "opinion" of the crowd, nor to see such issues in terms of an opposition between false and real probabilities or between what is likely and what is true.

Again, in his version of the example Aristotle sets what is probable *generally* against what is probable *in the circumstances*.[4] It is only the former that seems to be judged "real," presumably because of its wider applicability. Thus, Aristotle's understanding of what is *eikos*, in the Sophistic sense, seems different from Plato's and to involve the specificities of particular cases. If what is likely is restricted to the uniqueness of a case, then is has less interest for the kind of system that Aristotle is building. What is not at stake in the *eikota* of Sophistic arguments is the more rigorous sense of probability that we now associate with induction (but which Aristotle's interest in the "general" seems to point toward).[5] There is some distance between deciding the probability that an eclipse will appear in the next month and in deciding the probability that a gaily-dressed man abroad after dark is an adulterer.

On terms drawn from earlier thinkers like Protagoras and Antiphon, what is likely is a case-by-case determination, and is accepted by the crowd on the grounds not of their opinion alone, but their *experience*. That is, while they may indeed have formed an opinion about what is likely to be the case in a situation, the source of that opinion is not simply a Sophist's discourse but their own experience of such situations (Kraus 2007, p. 7). Hence, their own understanding of their humanity is the standard by which they judge what is the case. In the same way, what is false or real assumes a perspective that is foreign to the thinking of figures like Protagoras or Antiphon, who do not believe such things can be decided. In situations where we cannot recover the "facts" or know what happened, we must decide what probably happened using less than first-hand experience. This is not to deny the value of direct experience; Antiphon's ped-

---

[2] Aristotle attributes the example to Corax rather than Tisias, but it seems the two are so closely linked that what holds for one may be transferred to the other. In fact, Socrates continues in the *Phaedrus* passage to suggest that the art that disguises may also disguise the author ("whatever name he pleases to use for himself" – 273c).

[3] Kraus (2007) draws attention to the 'similarity' interpretation of *eikos* that seems pronounced in Plato's account. At 273d, for example, Socrates imagines he is addressing Tisias and suggests that people accepted his *eikos* because of its likeness to truth. In this, Plato seems to be shifting the discussion onto his own terms. Kraus himself adopts a reading of *eikos* more in keeping with the view developed in this paper.

[4] "Both alternatives seem probable, but one really is probable, the other so not generally, only in the circumstances mentioned" (Rhetoric 2.24.11: Kennedy translation).

[5] Whatever *eikos* is taken to mean, it would be misleading – as Walton has noted – to give it the sense of mathematical probability (Gagarin 2002, p. 29 fn57). (Solmsen 1968, p. 317) calls *eikos* one of the "traditional" types of evidence that will receive a more logical foundation when reinterpreted by Aristotle. The sense of that tradition can be witnessed in the apparent presence of *eikos* arguments in Sophocles' *Oedipus the King* (Kennedy 1963, p. 30). To avoid confusion as much as possible, I will use "likelihood" for *eikos*, rather than "probability."

agogic speeches make the point that direct evidence has greater value. But in certain circumstances, likelihood is not simply a poor substitute for truth, it is the only resource available. Hence, the importance of setting out the likelihoods in a case. These points will become clearer as we explore some examples from Antiphon. His work will provide both the core and the bulk of my discussion of early *eikos* arguments because it is the richest resource that we have in this respect.

17.3   Examples from Antiphon[6]

Antiphon's *Tetralogies* are generally recognized to be models of argumentation devised for the instruction of students.[7] Within the course of three set pieces, each comprising opposing arguments between litigants, Antiphon rolls out arguments of direct and indirect evidence, arguments based on the testimony of witnesses, strategies of turning the tables and opposing arguments, ethotic appeals to character and conduct, counter-factuals or hypothetical antitheses, and above all the arguments based on likelihood (or, in some cases, unlikelihood). Nowhere else can we find such a concentrated study of their nature and value.

Tetralogy I is the most accomplished in this regard. The circumstances involve a dispute between two enemies, one of whom is about to be taken to court by the other. Before this can occur, the one bringing the charges is murdered. The only witness is his dying slave, who identifies the other man as the murderer. There would seem, then, to be a very strong case against the defendant, who is being prosecuted by the deceased man's friends. Each side offers two speeches, or arguments, the second of which aims to rebut the other's first attempt in each case. The primary strategy of each side is to establish what is most *eikos*, and so win the jury over to their side. Accordingly, this is a useful text for considering how arguments from likelihood were to be evaluated and which decided the stronger.

The prosecutors in their first speech set the tone by indicating how difficult it is to detect and expose crimes that have been carried out by natural criminals who plan their acts carefully with strict attention to not getting caught. Because of this, the jury "must give the utmost weight to any indication whatever of likelihood (*eikos*) that is presented" (DK 87 B1:2.1.2).[8] They then issue a series of *eikota*, or in the initial move, what is not *eikota*:

> It is not likely that professional criminals killed this man, as no one would give up an obvious and achieved advantage for which he had risked his life, and the victims were found still wearing their cloaks. Nor again did anyone who was drunk kill him, since the murderer then would be identified by his fellow guests. Nor would the victim be killed because of a quarrel, since people would not quarrel in the dead of night and in a deserted spot. Nor was it a case of a man aiming to kill someone else and killing the victim, because then his attendant would not have been also killed (2.1.4).

---

[6] A debate from the ancient tradition through to contemporary commentators concerns the number and identity of Antiphons involved, with the two principal candidates being Antiphon the Sophist and Antiphon of Rhamnus (see (Gagarin 2002, Chapter 2); (Pendrick 2002, Introduction)). While current considerations favour a single author, the resolution of this debate is not really crucial to the present discussion.

[7] For positive comment on Antiphon's general place in the development of forensic argument see (Enos 1980).

[8] The numbering for citations here (and later from Gorgias) is the standard paragraphing from Diels and Kranz (1952).

Each unlikelihood is accompanied by one or more supporting reason, all of which, of course, is defeasible, but the sum of which shifts the weight of proof in the prosecution's favour. Clearly, people *could* quarrel in the dead of night and at a deserted spot; the arguers do not deny this. But given the jurors' experience of how people behave, how likely is this particular action?

Following these negative suggestions, attention shifts to what is likely, in the prosecution's opinion: "Who is more likely to have attacked him than an individual who had already suffered great injuries at his hands and could expect to suffer greater ones still?" (2.1.5). Details of the past history between the two are then given to support this likelihood. Included in this is that, since the defendant was an old enemy of the murdered man and had brought several unsuccessful cases against him, and since he had been indicted by the dead man in several cases, all of which he lost at the cost of much property, and since he bore a grudge for this, then it "was natural for him to plot against him, and it was natural for him to seek protection from his enmity by killing his opponent" (6-7). Here, the likelihood is supported by claims about what a man of this nature would do.

The prosecutors then refer to the absence of direct witnesses, beyond the slave who subsequently died. But they will produce these indirect witnesses who heard the slave's testimony. They then sum up by insisting the jury cannot acquit the defendant because "conclusions from likelihood (*eikota*) and from eyewitnesses have alike proved" his guilt (10).

In his own opening speech, the defendant matches the likelihoods that have been brought against him:

> ...they assume me to be a fool. For if now, because of the magnitude of my enmity, you find me guilty on the grounds of likelihood (*eikotôs*), it was still more natural for me to foresee before committing the crime that suspicion would devolve upon me as it has done, and, if I knew of anyone else who was plotting the murder, I was likely to go as far as to stop them, rather than to deliberately fall under obvious suspicion (2.2.3).

The strategy here is to rebut one likelihood by advancing another that is deemed more likely, given the same claims of enmity between the two men.[9] In terms that Walton will employ centuries later, the case builds to a conclusion that was less plausible before the plausible argument was brought to bear on it. The defendant continues: it is not unlikely but likely that a man wandering about in the night would be murdered for his clothes (that he still had them, simply indicates the murderers were frightened off); and others who hated him (though less than the defendant) are more likely to have murdered him, knowing suspicion would fall on the defendant. Then he turns to the evidence of the dead slave: since he would be terrified by the situation "it is not likely (*ouk eikos*) that he would recognize the murderers, it is likely (*eikos*) rather that at the instigation of those who were his masters, he would assent to what they suggested" (7). Moreover, he argues, if likelihoods are to be treated like facts, then they should consider it more likely that he would stay out of sight so as to avoid recognition. Later, the defendant again opposes likelihood to fact or actuality (10) to insist that if the likelihoods condemn him, then he must have been provoked and acted in self-defense, otherwise he would not have seemed a likely murderer.

The subsequent second speeches of the prosecution and defense continue the pattern of rebutting likelihoods previously argued and advancing alternatives.

---

[9] And we might note here that the logic used is a variant of the strong man's logic in the case discussed earlier from the *Phaedrus* and Aristotle's *Rhetoric*: the likelihood that he would be a suspect favours the likelihood that he did not do it, because he would have known he would be a suspect.

They also set these *eikos* arguments in the context of other types of argumentation. The prosecution concedes that those who actually (in fact) committed the
murder are unknown. But given this, proof must be based on what is *eikos*, since
crimes of this kind are not committed in the presence of witnesses. This suggests
that Antiphon recognizes the relative value of *eikos* arguments. Matters of fact,
were they established, would carry greater weight. And a better way to establish
such facts would be on the testimony of witnesses (2.3.8). But some cases, like
the present one, cannot have recourse to such evidence and must therefore be
decided on the likelihoods.

Likewise, the defendant attempts just such a shift in strategy by proposing
that he will not rely on likelihood to prove he was not present (as the prosecution has done), but on fact – he is prepared to surrender all the slaves in his
household for torture in order to show that he was at home in bed that night
(2.4.8). (Gagarin 2002, p. 116) suggests that the reason Antiphon introduces such
a startling piece of evidence so late in the exchange (that is, that the defendant
had been home that night) is because he has the secondary interest of demonstrating how different kinds of argument or proof (*pisteis*) can be weighed against
each other. Arguments from likelihood are necessary in cases where no better evidence is available, but first-hand factual evidence (*pragmata*) and the testimony
of witnesses (even slaves, after torture) would be stronger. Gagarin goes so far
as to propose that Antiphon is both showing the value of *eikos* arguments and
challenging them in general, because the defendant argues that likelihood does
not have the status of facts against him (2.2.8). But, given what we see Antiphon
arguing elsewhere about facts, it is not clear that he would make this second,
more general, point, except perhaps to teach students to exploit the beliefs that
others have about reality. After all, how would one establish "the facts" of a case.
Even the opposition between indirect evidence (*eikota*) and direct evidence in the
form of the testimony of witnesses is a matter of competing discourses. Gagarin
himself acknowledges something of this in his analysis of Tetralogy 2:

> Tetralogy 2 thus moves the discourse about the relationship of lan-
> guage and reality to another level compared to the discussion in Tetral-
> ogy 1, where the issue was one important types of *logos*, the argument
> from likelihood, and its relation to factual evidence. There *logos* and
> *pragma* are opposed, but Tetralogy 2 presents a complex symbiosis of
> the two in which facts control words, but words also control facts,
> since the truth of these facts depends on the words that represent
> them. Language corresponds to reality, but since reality is complex,
> that correspondence does not necessarily involve a single *logos* and a
> single reality (Gagarin 2002, p. 126).

In fact, it seems to overstate the case for Antiphon to suggest that language
corresponds to reality. For him, as for Protagoras, reality was beyond the direct
grasp of humanity and could only be conveyed in a series of *logoi*, the merits
of which need to be tested against the standard of human experience rather than
against reality as it "actually" is. Likewise, the *pragma* of the case being examined
in Tetralogy I are subject to the *logoi* that express them. There is a murdered
man (the case would not exist otherwise), but the identity of the murderer or
murderers (which is the fact that the defendant insists must be established) is
simply not accessible.[10] *Eikos* arguments, then, have merit alongside other types
of evidence that might be used to decide a case (rather than establish the "truth"
about it), and here the exercise of Tetralogy I amply demonstrates that merit.

---

[10]Outside of a reliable confession, which would undermine the point of the exercise and still be no
more than one *logos* to be weighed against others.

17.4   The range of *eikos* arguments

The examples from Tetralogy I represent only one speech in which Antiphon employs *eikos* arguments, although they are particularly fine illustrations of what is involved. The third and last Tetralogy also makes extensive used of this argumentative strategy, as indeed Antiphon does elsewhere. In his own defense trial in Athens, he resorted to at least one *eikos* argument, thus perhaps indicating the value of this strategy to him.[11] A member of the oligarchic government of the Four Hundred who were overthrown in 411, Antiphon was one of the few who did not go into exile and was executed by the Thirty. In one of the remaining fragments of his speech, he seemingly argues that he wanted only the best for Athens by asking "What likelihood (*eikos*) is there that I should want an oligarchy?" (Sprague 1972, frag 9.19). Since, as the prosecution claimed, he composed defenses for other people, he was not likely to be of any importance under an oligarchy compared to a democracy. This is the only speech of Antiphons of which we do know the outcome – like at least one other famous Athenian defendant,[12] he was not successful in persuading the jury.

Another particularly strong example of *eikos* arguments arises in a speech Antiphon wrote for a court case, "On the Murder of Herodes." (Usher 1999, p. 40) judges this speech responsible for Antiphon's reputation as a pioneer of *eikos* arguments. The defendant, Euxitheus, (a citizen of Mytilene) has been charged in a questionable way with the murder of an Athenian, Herodes. The two men had been together with others on a boat bound for Thrace. After a night of drinking while in harbour, Herodes goes missing. Euxitheus helps in the search, but in the end continues on the way to Thrace. After he has left, he is accused of the murder (although no body is ever found), supposedly on the testimony of a slave who the prosecuting parties subsequently executed before he could be questioned.[13]

The bulk of Euxitheus's case involves the likelihoods (*ta eikota*), and they are considerable (DK 87 B5: 25-63): that he was not accused immediately is likely because the prosecutors needed time to fabricate their evidence; if he had left the vessel (contrary to his insistence), it is likely that some evidence or clue of the murder would be found around the harbour; if a boat was used, it is likely that some sign would be present on that boat; it is likely that the slave was promised his freedom if he accused Euxitheus (the jury should be aware that this happens); if the defense had had access to the slave to torture him (as was the accepted custom), then it is likely he would have retracted his accusation, as indeed he did when he realized he was going to die; it is also judged unlikely that the murderer would have used an accomplice (as the slave was deemed to be); it is likely that such a crime could not have been committed without attracting attention, nor without leaving tangible evidence. In sum, Euxitheus asks: "men of the jury, how can these things be likely?" (45). To emphasize the import of this, Euxitheus indulges in repetition of the slave's treatment and the likelihoods to be drawn from the prosecution's behaviour with regard to it.

Euxitheus uses other argumentative strategies, indulging in a pathotic appeal for the jury's pity (73) and defending his father's character in an ethotic argument

---

[11] I say at least one because, in spite of the high reputation that this speech seems to have held among Antiphon's contemporaries and in the subsequent tradition, we have only three short fragments of it left (much of this only discovered on papyri in 1907). Among these, only the one *eikos* argument is evident.

[12] Gagarin (Gagarin 1997, p. 249) sees parallels between this speech and the one Plato attributes to Socrates in the Apology.

[13] The unorthodoxies of the case, of which Antiphon makes much in the speech, include the fact that Euxitheus was accused of murder, but arrested by a procedure intended for common criminals. After his arrest he was kept in custody until the trial, in spite of his being willing to provide the usual sureties. Moreover, the slave whose testimony seems pivotal in the charges against him was executed before the defense had access to him.

(74-79), and, as we have seen, arguing that the normal rules of procedure in such cases had been violated. The latter may well give strong support to the likelihood that the prosecutors have fabricated evidence and, as Gagarin suggests (Gagarin 2002, p. 160) the reputation of the speech in antiquity makes it likely that the strategy of juxtaposing these points was successful.[14]

While an undoubted master of the strategy, Antiphon was not alone in his use of *eikos* arguments, and a balanced consideration of their import requires the examination of a few additional cases. An interesting example is suggested in one of Gorgias' extant speeches, "A Defense on behalf of Palamedes." Plato in the *Phaedrus* attributes the origin of likelihood arguments not only to Tisias but also to Gorgias (267a). Of the works available to us, the Palamedes speech is the best candidate. Although Gorgias rarely uses the term *eikos* here, he clearly is weighing likelihoods in an effort to show that it was unlikely that Palamedes committed the treason of which he was accused.[15]

Gorgias has Palamedes offer a range of arguments to show that is was not likely that he did what Odysseus alleged. The treason must have begun in some way, and that would have involved a discussion. But a discussion requires a meeting, and how could there be a meeting unless either party sent a messenger to the other (DK 82 B1:1a.6)? But if a meeting takes place, how will they communicate? How can each understand the other's language? If they use an interpreter, that would require a third person for something that needs to be secret (11a.7). But suppose, contrary to the facts, that a meeting did take place. Why would Palamedes, a traitor, be trusted? His word would be worth little. Perhaps they could have exchanged hostages (a brother for a son) as a surety, but such an action would be evident to the jury (11a.8). But suppose it is alleged that the contract was made for money. It is not likely (*ouk eikos*) that a man would take a small sum for such a service. But how would a large sum be carried? If a number of people carried it, there would be witnesses; and if only one, then it could not be a large sum (11a.9). The argumentation continues in this vein, with one unlikelihood building upon another to give the weight to Palamedes' claim that he was not capable of that of which he was accused. Then he marshals a similar set of likelihoods to show that he also lacked the motive for the crime. Again, Gorgias' extensive use of likelihood argumentation does not exhaust the case he makes, he also employs ethotic argument, appealing to Palamedes' own credibility and against that of his accuser, and a pathotic appeal to the jury's emotions (even as he claims he will not [11a.33]). But it is the core of what is effectively *eikos* argumentation that should interest us. Matched to the details of the context, Gorgias' reasoning sifts through the claims and counter-claims to "prove" that the weight of likelihoods, as they would be measured against the standard of the jurors' experience, favours the innocence of the defendant.

The last example I will consider moves us into the wider range of texts that might be considered "Sophistic"[16] by nature or influence. It comes from one of Isocrates' speeches, "Against Euthynus, Without Witnesses." The title is significant; it is on account of the absence of witnesses that the speech relies so heavily

---

[14] Although the judgment on this is far from unanimous. Due (1980, p. 50) finds the evidence against the defendant to be strong and suggests further that since he is a Mytilinean accused of murdering an Athenian, his chances of acquittal would have been slim.

[15] This fictional speech, drawing on the myth of Odysseus and Palamedes, is another exercise in argumentation. At the siege of Troy, Palamedes has exposed Odysseus' feigned madness for what it was thus forcing him to join the expedition. In revenge, Odysseus frames Palamedes by forging a letter to him from Priam, in which he was to betray the Greeks, and hiding gold in his tent. On the basis of this evidence, Palamedes was found guilty and executed.

[16] As O'Sullivan (1996) argues, the break of the first Sophistic around the introduction of Socrates is arbitrary and unwarranted. It has the unfortunate consequence of excluding figures like Alcidamas and Isocrates, for whom we have much more in the way of representative works.

on likelihood arguments. And while Isocrates uses them freely elsewhere, the circumstances of this case brings them to the forefront.

The speech was written for a man named Nicias, who had attempted to hide his assets from the tyranny of the Thirty. He gave part of his money to a relative Euthynus, who allegedly failed to return a third of it when requested. The lawsuit was an attempt to recover the sum that was allegedly owed Nicias. Since no one, free or slave, witnessed the deposit or return of the money (21.4), other means than testimony must be used to argue the case.

Every one knows that those who are clever at speaking but lack money often try to bring malicious prosecutions against those who cannot speak for themselves and who have money. Nicias, however, is a poor speaker and has more money than Euthynus, so it is not likely that he is attacking Euthynus unjustly. Rather, it is more likely that Euthynus took the money and is denying it than that Nicias is false in his claim to have given it. It is not surprising that Euthynus would deny having stolen the money when even those who gave money in front of witnesses have trouble getting it back. Whereas it is not likely that Nicias would think he could get something through a false accusation when not even those who are rightly owed money can recover it. The greatest evidence (*tekmērion*), however, is that Nicias first told people that Euthynus owed him the money when the oligarchy had been established, and the jury should know this was a time when it was more dangerous to be wealthy than to do wrong. The city was run by those who did not punish wrongdoing but took money from the wealthy. In another case known to Euthynus, Nicias had had money extorted from him simply through a threat to take him to court. So it is not likely that he would bring a malicious prosecution when he was in danger. A series of questions then follow, each inviting the answer that Nicias would not have acted if the money was not really owed to him. While Euthynus might argue that if he wanted to do wrong he would not have returned two-thirds of the money, Isocrates uses this very argument against him, insisting that he only took the one-third because he knew this defense would be available to him. In summary, Isocrates shows that the weight of argument indicates there is no reason for Nicias to make a false accusation, but every reason for Euthynus to act unjustly (21.5-21).

17.5   Evaluating *eikos* arguments

The popularity of *eikos* arguments among the Sophists and their contemporaries is a testimony to their utility and power.[17] The examples considered above show them appearing in both artificial cases, like the Tetralogies and the 'Palamedes', apparently intended for the instruction of students or as exploratory demonstrations of argumentation, and real cases like those of the "Herodes" and "Nicias" speeches, as well as Antiphon's own trial speech. This suggests we are not dealing with a haphazard type of reasoning, but one that was reflected on, taught, and recognized as having better or worse instances. But this raises the question of evaluation and how such better cases were recognized. One obvious measure of quality might be the success of the case – good arguments are the ones that win the day. But this tells us little, and helps even less when the outcome of these cases is rarely known. Besides, where success does seem to have ensued, we cannot be sure that it was due to the *eikos* arguments used. As Gagarin hints, the apparent success of the "Herodes" case may have been due as much to

---

[17] The *Rhetoric to Alexander*, a pre-Aristotelian text attributed to Anaximenes of Lampsacus, puts *eikos* arguments at the top of a list of proofs drawn from words, actions or persons. Such arguments are deemed to be supported by examples present in the mind of the audience (7, 1428a25-34).

the illegality of the procedures employed by the prosecution as the likelihoods advanced by the defendant (although the illegalities are drawn into the likelihoods.) Adherence by an audience is too often taken as the only measure of rhetorical argumentation, and thus the grounds for criticism of it. Yet audiences are persuaded by some arguments and not by others, and that is the difference that interests us.

One thing clear from the cases examined is that *eikos* arguments are not arguments for every occasion. They stand in many cases as contributory arguments, working with others to strengthen a case in the way that we might today judge authority arguments to be contributory. And their power is judged relative to other types of direct evidence. It is not incidental to the detailed cases we have examined that first-hand witnesses are absent.[18] Both Tetralogy I and "Herodes" rely on the testimony of a deceased slave, and the "Palamedes" and "Nicias" cases are such that their very nature precludes the presence of witnesses. Athenian law strove to have the testimony of witnesses available. Hence, in the cases we have looked at, the testimony of witnesses has been rendered suspect in some way or removed altogether to allow the *eikos* arguments to reveal their power. Likewise, the Tetralogy I exercise suggests that facts, where they can be ascertained, will constitute stronger evidence than likelihoods. But, again, where these facts are being related to a jury, they are subject to the interpretation of the speakers, and jurors will always need to judge between competing *logoi* in deciding responsibility and punishment.[19]

One conclusion we might draw from this is that it would be a mistake to use *eikos* arguments when stronger types of evidence is available or expected. But this would be an error in the employment of the strategy itself, not a way to determine weak from strong *eikos* arguments. For this we need to look more closely at the arguments themselves and how they seem to have been understood.

From the cases examined, it is difficult to find corroboration of Plato's suggestion that what was *eikos* was what was accepted by the crowd (*Phaedrus* 273c). Granted, successful *eikos* arguments would have had this character, but this would have been no more than an incidental feature about them, and would also be the case with all arguments that gained adherence. Nor, unless we reduce all matters of dispute to a black-and-white opposition of truth and opinion, can we learn much by calling them arguments that reflected the opinion of the crowd. The careful working of details and weighing of likelihoods against each other indicates that there was much more involved than simply triggering preexistent opinions.

More appropriate seems Aristotle's judgment in which he favoured what was probable (or likely) in *general* over the Sophists' interest in what was probable given a particular set of circumstances involving the specificities of particular cases. On these terms, good *eikos* arguments would be those that made best use of the details of the case and the circumstances involved. The goodness or badness of an argument of this nature could not be determined in advance but depended entirely on how it was worked out and interacted with the opposing likelihoods. Thus, the authors of the exercise cases, Antiphon and Gorgias, took pains to demonstrate such interactions, along with the relative merits of *eikos* arguments against other available evidence. This evokes some of the ideas on Sophistic logic presented by (Poulakos 1997), particularly on the importance of the specific time

---

[18] The exception being the trial speech of Antiphon

[19] Tetralogy 2 makes this clear, since the facts of that case are not in question. One boy who was collecting javelins on the field fails to realize they are still being thrown and is killed by the javelin of another boy. We know who has died and who threw the instrument that killed. But the issue under dispute is the responsibility and subsequent punishment. The prosecution and defense present the facts in different ways that favour their interests in deciding responsibility.

and place of an argument and the circumstances in general. Crucial to this logic is the central idea of *kairos*. For Poulakos, this involves speakers responding to the fleeting situations marked by their unique features (p. 18). Kennedy (1963, p. 66), who connects the idea closely to Gorgias, refers to it specifically as the concept of the opportune.[20] Any given dispute involving opposing positions can only be resolved through consideration of time, place and circumstance. This accords with Aristotle's appreciation of what was at stake in the Sophists' practice.

Any *eikos* argument must be judged on how well it adjusts to the circumstances of the case and makes use of the materials presented. It cannot be transferred between cases, since each is unique.[21] So students can only learn the principles involved in identifying likelihoods and general rules, like not to oppose *eikos* to direct testimony. Their facility at employing the arguments will come only through practice as skill is developed. Aristotle, later in the Rhetoric (I.2.1355b35) will distinguish between two types of proof (*pisteis*), non-artistic (*atechnic*) and artistic (*entechnic*). The inartificial or non-artistic proofs are matters of direct evidence, like laws, and the testimony of free men or slaves. These are non-artistic because the arguer has little control over them. Although we have seen Antiphon using indirect evidence to attack direct evidence; he is not free to construct that evidence as he chooses. Artistic proofs, by contrast, rely on the arguer's skill and are freely chosen to suit the circumstances. Here Aristotle will divide the proofs among those related to *ethos*, *pathos*, and *logos*. But we can already see the germ of such categorizing alive in the thinking of arguers like Antiphon. And, of course, the artistic *logos* here is *eikos* argumentation.

17.6   Walton and the Plausibility argument

We no longer speak generally about *eikos* arguments,[22] but their evolved forms still have impact on the ways we reason and show up in modern argumentation theory. As noted in the Introduction, Douglas Walton ((Walton 1992c, p. 3); (Walton 1998b, p. 16); (Walton 2001c, p. 104); (Walton 2002a, chap. 4)) has on several occasions stressed how such reasoning fell out of favour in the history of logic, failing to have the rigor that was expected of good argumentation. On his terms, *eikos* arguments are now to be understood as plausible argumentation, traditionally "ignored, and even denounced, as belonging to the realm of sophists and slick persuaders" (Walton 1992c, p. 3). But elsewhere, Walton (2001c, 2002a) understands *eikos* arguments to appear to involve a kind of common knowledge, which on investigation is more a matter of plausibility than actual knowledge. "Eikotic arguments are arguments based on defeasible inferences or generalization" (Walton 2001c, p. 98), the latter understood as "everyday human experiences of the way things can be generally expected to go" (p. 100). This, as we have seen, is closer to the usage of the early Sophists. In using plausibility to explain some traditional problems with enthymemes, Walton notes again that this "concept of the way things can be normally expected to go in familiar situations was lost sight of in logic for two thousand years" (p. 104). For him, importantly, *eikos* arguments (or plausibility arguments) allow us to fill in the hidden premises

---

[20] Diogenes Laertius claims Protagoras was the first to stress the importance of *kairos* (DK 80 A1).

[21] I think Gagarin (Gagarin 1997, p. 123) overstates the case when he suggests that the *eikos* arguments in Tetralogy 1 could be transferred with only slight modification to any case where the issue was "who did it?" Perhaps he has in mind the parallel with the Corax/Tisias example. But most of the *eikos* arguments used by the prosecution and defense relate directly to the details of the case (the untouched clothing, the history between victim and defendant, and so on); any transference would involve far more than "slight" modification. Rather, "who did it?" disputes were naturally suited to argumentation based on likelihood.

in enthymematic arguments, because arguers implicitly expect audiences to contribute details of their common experience. Again, this does fit with the practice of the ancient Sophists and orators who employed this strategy to connect with the common experience of their audiences (jurors). A reference to the *Rhetorica Ad Alexandrum* (1428a25), suggests that something can be found plausible when hearers have examples in their own minds of what is being said (Walton 2002a, p. 135). But Walton also attributes the historical neglect of *eikos* arguments to their being based on a person's *subjective* understanding of how something can normally be expected to go in familiar situations. Hence, we have a clearer explanation for why such argumentation has lost favour: modern thinking finds the strategy too subjective and so not reliable for logical reasoning. But if this is so, it is a misunderstanding of the original intent behind *eikos* arguments. As we have seen, to stress a person's subjective understanding in this way–an interpretation that invokes some interpretations of Protagoras' measure maxim – is to overlook the very commonality that is at stake in such reasoning (and that Walton wishes to draw from it). Walton's final adoption of a model of plausible reasoning (Walton 2002a, p. 200) stresses that certain propositions are plausible in themselves (in a context), and hence not dependent on a hearer's subjective view. This understanding is consistent with the ways the early Sophists used *eikos* arguments. Propositions are plausible when they appear to be true because there is no overriding reason to think them false, and so they are acceptable to a rational person as plausibly true, although this acceptance is conditional and tentative.

As this fuller study will suggest, Walton's plausibility arguments have a rich history that extends back further than is generally acknowledged. Studying that history promises to provide details and examples that can enrich our current understandings and usage. What made, and makes, *eikos* arguments so persuasive in court settings is the appeal to what was customary in the experience of people generally. Innes (1991) suggests that the reason Antiphon (and Gorgias) consistently used anonymous, stock characters was to stress character types that could be generalized. To ask whether a particular action is the kind of thing that a man with a certain kind of character is likely to do in a certain situation is not to ask for the hearer's arbitrary opinion on the matter; it is to tap into their general fund of knowledge concerning the customs of their society, their community, and how they know people to generally behave. Indeed, the fact that people do sometimes behave out of character or in unexpected ways renders such reasoning always no more than plausible and vulnerable to counter argument. But in the kinds of situations we have seen the strategy used, it is a powerful tool of persuasive argument and, correctly employed, it will be the best available strategy in the circumstances. That it requires the arguer to know well the cognitive environment of his or her audience and use that in constructing the argumentation speaks more generally to the rhetorical nature of the activity involved.[23]

---

[22] A health care research and teaching web site with the name 'Evidence and Eikos', draws part of its name from the Greek: "Most arguments in health care employ evidence, usually statistical in nature, derived from empirical research. The results of these studies are usually provisional in nature, and reasoning from them implies provisional inference and plausible belief, hence *eikos*": http://individual.utoronto.ca/ecolak/EBM/evidence_and_eikos/index.htm (visited, July 29, 2007)

[23] I am grateful to an anonymous reviewer for comments that have led to several improvements in the paper.

CHAPTER 18

DEFEASIBLE REASONING

John Woods

"There is nothing to be learned from the second kick of a mule"

American saying

18.1 Ampliative Reasoning

The reasoning done by individuals is mainly ampliative. There is a flexible usage of the word in which all ampliative reasoning is defeasible. In the past forty years or so, there appear to have been no want of candidate logics for defeasibility — nonmonotonic logics, inheritance logics, default logics, autoepistemic logics, circumscription logics, logic programming and Prolog, preferential reasoning logics, abductive logics, theory revision logics, theory update logics, doxastic logics and whatever else. My general view of these logics is that, while they make some display of mathematical virtuosity, they are less impressive in matters of conceptual and empirical adequacy[1]. John Pollock remonstrates to the same effect:

> Unfortunately, their lack of philosophical training led AI researchers to produce accounts that were mathematically sophisticated, but epistemologically naïve. Their theories could not possibly be right as accounts of human cognition, because they could not accommodate the varieties of defeasible reasoning humans actually employ. (Pollock 2008, p. 452)

Pollock adds that although "there is still a burgeoning industry in AI studying nonmotonic logic, this shortcoming remains to the present day." (Pollock 2008, p. 452)

An exception to this criticism are Douglas Walton's contributions to defeasible reasoning in a number of works, including among others, *Plausible Argument in Everyday Conversation* (Walton 1992c), *Legal Argumentation and Evidence* (Walton 2002a), *Character Evidence* (Walton 2007a) and *Media Argumentation* (Walton 2007c). Walton is much struck by modes of reasoning that aren't well-captured by the dominant paradigms of deductive and inductive or Bayesian logic; and it is to his considerable credit to have advanced this cause by producing accounts that are factually informative and analytically engaging. The last thing that can be said of this work is that it sacrifices conceptual adequacy to mathematical versatility. There is no mathematics in Walton's defeasibility writings. This is not inadvertent. Walton's position — and mine, too, for that matter — is that in the

---

[1] Abduction is an exception among philosophically-minded logicians. See here Magnani (2010), Walton (2004) and Gabbay and Woods (2005).

absence of demonstrable impediments to doing so, the logic of reasoning, like the logic of argument, should aim at the elucidation of relevant concepts rather than the advancement of mathematically sophisticated formalisms. (Think here of the modal system S1, whose formal semantics is a brilliant solution of a murderously difficult technical problem, but which sheds no useful light on the meanings of 'necessary' and 'possible'.) What I would like to do here is to venture some opinions about defeasibility in this same spirit. While I would not eschew any technical methods that might be of use, I agree with Walton that, to the extent possible, conceptual clarity is king.[2]

My more particular purpose here is to draw attention to features of defeasible reasoning which have yet to receive the notice that is due them. This neglect — if that is not too strong a word for it — is widespread, and cannot be laid at the feet of particular persons, least of all Douglas Walton, to whom my paper is affectionately and respectfully dedicated. What I hope to be able to provide is a gentle corrective to omissions and oversights of the defeasibility literature *en large*.

18.2   Defeasible consequence

Let me say again that a position shared by Walton and me is the general undesirability of substituting technical finesse for analytical clarity. But this is far from saying that, when formal treatments of matters of interest are ready to hand, we should avoid effecting an acquaintance with them. So I will in this section make a (very) brief review of the state of play in the formal literature. In so doing, we will have developed some useful contact with the kind of thing that Pollock — and I daresay Walton, too — finds 'mathematically sophisticated, but epistemologically naive.'

Let D be the class of logics that focus primarily on a relation of defeasible consequence. The following is a representative list of the properties of such relations. Not all of them have all these properties, but many have most of them.

Nonmonotonicity. If α is a consequence of Γ, it need not be a consequence of the result of adding any sentence to Γ.

Cumulativity. If α is a consequence of Γ it is also a consequence of $\{\alpha\} \cup \Gamma$.

Cut. Let K(Γ) be the set of Γ defeasible consequences. Then if everything in Γ is in Δ and everything in Δ is in the set of defeasible consequences of Γ (i.e., K(Γ)), then all the defeasible consequences of Γ are defeasible consequences of Δ, i.e. K(Δ).[3]

Cautious Monotony. If, as before, everything in Γ is in Δ and everything in Δ is in K(Γ), then everything in K(Γ) is in K(Δ).[4]

---

[2]The tension between mathematical virtuosity and conceptual elucidation is explored in detail in (Woods 2011).

[3]Semantic inheritance networks lack Cut.

[4]Cautious Monotony doesn't hold in semantic inheritance networks. However, the system of Horty, Thomason and Touretsky (1990) mandates special versions of Cautious Monotony and Cut. Default logics in the manner of Reiter (1980) and Etherington and Reiter (1983) also fail Cautious Monotony. Preferential logics (Shoham 1987) fails Cut and Cautious Monotony unless they have a Limit Assumption which rules out the possibility that the logic not have a most preferred model. ($\Gamma \mid\sim \alpha$ just in case α is true in all the most preferred models of Γ.) Systems with the Limit Assumptions also have Full

Full Absorption. Let $Cl(\Gamma)$ be the classical closure of Γ. Then $Cl(K(\Gamma)) = K(\Gamma) = K(Cl\Gamma)$.[5]

Distribution. Everything common to $K(\Gamma)$ and $K(\Delta)$ is in what is common to the set of defeasible consequences of what's common to the classical closure of Γ and the classical closure of Γ and the classical closure of Δ; i.e., $K(\Gamma) \cap K(\Delta) \subseteq K(Cl(\Gamma) \cup Cl(\Delta))$.

Conditionalization. If α is a defeasible consequence of a set $\Gamma \cup \{\beta\}$, then the material conditional $\ulcorner \beta \supset \alpha \urcorner$ is a defeasible consequence of Γ itself.

Loop. If $|\sim$ denotes a defeasible consequence relation, then where $\alpha_1 |\sim \alpha_2 \ldots \alpha_{n-1}$ for α_n, then for any i and j, the α_i and α_j have just the same defeasible consequences.

A number of disputed properties have also been considered for defeasible consequence and are prefixed here with an asterisk

* *Disjunctive Rationality.* If $\Gamma \cup \{\alpha\} |\!\not\sim \beta$ and $\Gamma \cup \{\lambda\} |\!\not\sim$, then $\Gamma \cup \{(\alpha \vee \lambda)\} |\!\not\sim \beta$.[6]

* *Rational Monotony.* If $\Gamma |\sim \alpha$, then $\Gamma \cup \{\beta\} |\sim \alpha$ or $\Gamma |\sim \sim \beta$.[7]

* *Consistency Preservation.* If Γ is classically consistent so is $K(\Gamma)$[8].

Of course, it is clear on inspection that the orientation of these logics is mathematical rather than conceptual or empirical. The majority of the systems in D are elaborations of quite simple formal languages — more often than not propositional languages. The reason for this is that, with the except of OSCAR, automations of defeasible reasoners have not been successful for systems richer than propositional logic or some of its even less rich sublogics. In this regard OSCAR is a standout. OSCAR is an AI architecture for knowers — for cognitive agents — and can be thought of as a general-purpose defeasible reasoner (Pollock 1995). But, to date, OSCAR cannot handle defeasible reasonings that vary in degrees of goodness or strength. Indeed

> There are currently *no other* proposals in the AI or philosophical literature for how to perform defeasible reasoning with varying degrees of justification. (Pollock 2008, p. 459; emphasis added).

Part of the problem lies in the nature of the subject matter. D-reasoning is much more difficult than deductive reasoning to capture formally. A further part of the problem lies in inflexibility of the methods employed by D-logicians.

Absorption and Distribution. Such systems are also supraclassical: If β is a classical consequence of α, it is also included in α's defeasible consequences.

[5]What Full Absorption says is that defeasible consequence treats sets G of implying formulas and the classical closure of G in the same way.

[6]Some preferential logics lack Disjunctive Rationality, Rational Monotony and Consistency Preservation.

[7]Circumscription logics in the style of McCarthy (1980, 1986) and Lifschitz (1988) lack Rational Monotony but satisfy Consistency Preservation.

[8]Defeasible consequence in Lehmann and Magidor (1992) lacks Consistency Preservation, as does the 0-entailment relation in extreme probability systems in the manner of Pearl (1990).

Judging from the formalizing languages chosen and the character of the rules imposed, defeasibility inferences are widely seen in the D-community as variations of deductive inference. In particular, it is commonly assumed that, in the absence of information to the contrary, the connectives and quantifiers of defeasible languages are classical. But the greatest difficulty to date is the almost slovenly indifference shown by most going D-logics to the data of defeasible reasoning, to the very phenomenon that D-logics are supposed to elucidate and systematize. This tells us something important. A theory of reasoning is asking for trouble if it doesn't pay careful attention to its motivating data. We might say that the would-be theorist is subject to a look-before-you-leap principle. The-look-before- you-leap principle emphasizes the importance of this pre-theoretical reflectiveness. It offers some helpful counsel:

Proposition 1. *LOOK BEFORE YOU LEAP: In matters of human performances of kind K do not venture theoretical accounts of them without due regard for the K-data that the theory is to account for. In particular, be vigilant about data-bending. Be careful that the pre-theoretical senses of the terms used to describe the theory's motivating data aren't inappropriately overridden by contrary meanings embedded in the theorist's procedures and assumptions.*[9]

18.3 Misconceiving the interconnections

I want now to examine the logical connections, or want of them, between and among the three D-properties of nonmonotonicity, defeasibility (in the narrower sense, to be specified) and defaults, and the three classical properties of validity, soundness and inductive strength.

Nonmonotonicity and deductive validity. The universally accepted definition of a nonmonotive logic is one whose consequence relation is not nonmonotonic. A consequence relation is monotonic just in case whenever it obtains between a sentence α and sentence β or between a set of sentences Γ and a sentence β, it also holds between the result of adding any sentence to α or to Γ any number of times and that same consequent β. It is commonly held that monotonicity is a universal feature of deductive consequence. This is a mistake. Let L be a linear logic whose deductive consequence relation is $\leadsto$. Then, in a linear proof context, the derivation

1. α

2. $\alpha \leadsto \beta$

3. So, β

is L-valid. A distinguishing feature of L is that it is sensitive to premiss-redundancy. There are new premisses γ for which

1. α

2. $\alpha \leadsto \beta$

---

[9]Of course, there must be room for some give here. In well-defined scientific contexts, a theory is frequently unable to process its own raw data. Consider a case, which I borrow from Suppes (Suppes 1960, p. 297): "The maddeningly diverse and complex experience which constitutes an experiment is not the entity which is directly compared with a model of a theory. Drastic assumptions of all sorts are made in reducing the experimental experience, as I shall term it, to a simple entity ready for comparison with a model of the theory. ... [For example, on the one hand] we may have a rather set-theoretic model of the theory which contains continuous functions or infinite sequences, and, on the other hand, we have highly finitistic set-theoretical models of the data."

2′. γ

3. So, β

is invalid in L.[10] Hence,

Proposition 2. *MISCONCEIVING MONOTONICITY: Monotonicity is a typical but not intrinsic property of deductive consequence relations.*

Nonmonotonicity and inductive strength. The other part of the received wisdom about monotonicity is that the property of inductive consequence is intrinsically nonmonotonic. We say that β is an inductive consequence of α (or of Γ) just in case

1. α (or Γ)

2. So, β

is an inductively strong argument. The thesis that inductive consequence is nonmonotonic provides that there are inductively strong arguments such that for some proposition γ, adding γ to the premiss-set weakens (or even destroys) the argument's inductive strength. This happens whenever γ falsifies a premiss or directly conflicts with the argument's conclusion.

No doubt, nonmonotonicity is a typical feature of inductive strength, but not an invariable one. Consider the inductively strong argument

1. Most, but not more than 80%, of A are B

2. Object x is an A.

3. So, object x is a B.

Suppose now that we discover that more than 80% of A are B, and add this as a premiss. Then we have

1. Most, but not more than 80%, A are B

2. Object x is an A

2′. More than 80% A are B

3. So, object x is a B.

Of course, (1) and (2′) can't both be true, but at least one of them must be. How can the addition of a premiss that improves the odds of x's being a B be an inductive-strength spoiler? Purists might have an answer to this. They might object that, as we have it now, the argument has inconsistent premisses, and that this makes the conditional probability of its conclusion on that inconsistency undefinable. But this is precisely the kind of case that reveals the vulnerability of equating inductive strength with high conditional probability. *One* of those premisses is true; and, whichever it is, it together with premiss (2) lends considerable inductive strength to the conclusion. Consequently, the conclusion cannot be undermined by this new premiss. True, the new premiss (2′) falsifies (1) but it also compensates for its loss. So it is both technically and conceptually unsatisfying to insist that upon addition of (2′), the support for (3) collapses. Accordingly,

---

[10]Aristotle's syllogisms are also redundancy-intolerant. For example, ⟨"All A are B", "All B are C", "All A are C"⟩ is syllogistically valid. But ⟨"All D are A", "All A are B", "All B are C", "All A are C"⟩ is syllogistically invalid, made so by the redundancy of the new premiss. Linear logics in the modern sense originate in Girard (1987), which gives a semantics for a System F of the polymorphic lambda calculus. Connections to computer science were first discernible in the Curry-Howard isomorphism (Howard 1980).

Proposition 3. *MISCONCEIVING NONMONOTONICITY: Nonmonotonicity is a typical but not intrinsic feature of the property of inductive consequence*

Some intrinsically nonmonotonic properties. Anyone mindful of the importance of the difference between consequence-having and consequence-drawing[11] will already be predisposed not to draw every consequence he is able to recognize as such. Sometimes, to be sure, the addition of new information to a premiss-set (or data-base or knowledge-base) should change our minds about what now follows. In those cases, the failure to be a consequence is trivially a matter of whether it is a consequence anyone should (or could) *draw*. But a much commoner case against consequence-drawing has nothing to do with consequence-having. Perhaps the clearest example of this is one in which new information falsifies *without compensation* a premiss in an otherwise sound argument. By and large, we don't want (categorical) arguments from false premisses. A valid argument from true premiss is sound. But, even in those systems in which validity is monotonic — namely, most of them — soundness is not. Other such properties include: (at least in their intuitive senses): plausibility, likelihood and possibility.

When new information contradicts a premiss of an argument, the agent is faced with two tasks. One is to determine whether the new information undermines the argument's consequence relation. If it did, that would be a reason not to draw the argument's conclusion as a consequence of those altered premisses. If it didn't, the agent must turn his attention to the second task. He must determine whether the unsoundness of the argument is occasion to abandon it altogether. Excepting the far from trivial class of arguments from premiss-sets whose inconsistency it is beyond the powers of the agent to remove in a principled way, unsoundness is an argument-killer irrespective of whether it is a consequence-killer. It is simple economics that in the general case a reasoner's first task is to test for unsoundness, leaving the second task —— checking on the consequence relation —— to be performed only after positive finding with respect to the premisses (i.e., they are true, or plausible, or likely or presumable). Why would this be so? Because premiss inadequacy is decisive against consequence-drawing. Premiss adequacy underdetermines consequence-having. Non-consequence is decisive against consequence-drawing. Consequence-having underdetermines consequence-drawing.

Proposition 4. *THE PRIORITY OF UNSOUNDNESS: In the logic of reasoning, bad news about premisses trumps good news about consequence.*

Defeasibility and deductive validity. One of the most persistent claims made by defeasibility logicians is that deductive logic is one part of logic and defeasibility logic is all the rest of it. Pollock writes:

> What distinguishes defeasible arguments from deductive arguments is that the addition of information can mandate the retraction of the conclusion of a defeasible argument without mandating the retraction of any of the other conclusions from which the retracted conclusion was inferred. By contrast, you cannot retract the conclusion of a deductive argument without also retracting some of the premisses from which it was inferred. (p. 453)

Consider two cases, one a classically valid deductive argument

1. A

---

[11]This distinction is just about the whole basis for Harman's rejection of the implication schemes of deductive logic as rules of deductive inference (See (Harman 1986, chapter 1)). This same position was arrived at independently by Walton and me in our (Woods and Walton 1972).

2. ∴ B

and the other an inductively strong argument

1. E

2. ∴ C

The remarks below easily generalize to multi-premissed versions of such arguments.

Consider now the respective negations, ∼B and ∼E, of the conclusions of these arguments. Two points can be made straightaway.

   i. Even in the light of these negations, C remains an inductively strong consequence of E, and B remains a deductively valid consequence of A.

   ii. C is not an inductively strong consequence of "E ∧ ∼C", but B is a deductively valid consequence of "A ∧ ∼ B".

From (i) and (ii), we have it further that {A,∼B} is a deductively inconsistent set, whereas {,∼C} is not. Accordingly, if one's belief-revision goal (or dialectical goal) were the overall deductive consistency of one's retractions and commitments, then, in the first instance, the consistent simultaneous restriction of B and commitment to A is impossible and, in the second, the consistent simultaneous retraction of C and commitment to E is possible. Why should we let these facts fix the meaning of "defeasible"? Certainly they capture nothing like the original sense of the term, introduced into philosophy of H.L.A. Hart. Why would we invoke the notion of defeasibility to characterize the present set of facts, when they all flow from the monotonicity of classical consequence and the (typical) nonmonotonicity of non-deductive consequence? True, one could always stipulate a technical sense for 'defeasible', but that wouldn't change the fact that the ensuing usage would not enable us to state any facts about our two arguments not already stateable in terms of the monotonic-nonmotonic distinction. So, then,

Proposition 5. *DEFEASIBILITY AND NONMOTONICITY: With respect to soundness, they are co-extensive properties. Except for the comparatively rare cases of nonmonotonic deductive consequence already noted, in the matter of non-deductive consequence they are also co-extensive properties.*

Given Pollock's taxonomy, the intended extension of the term 'defeasible reasoning' is the class of deductively invalid arguments which in some manner or other are good arguments, that is, are arguments whose premises afford adequate reason to accept their conclusions. (So "defeasible" is a success-term). However, in the AI literature during the past forty years, the emphasis has fallen on patterns of reasoning involving a certain class of generalizations. Excluded from this focus are most types of ampliative reasoning: e.g., inference to the best explanation, abductive reasoning and analogical reasoning. In this narrower orientation, there is a direct link between the lack of guarantees that attends one's conclusion and the lack of universality that attends one's premises. In its most basic form defeasible reasoning in this narrower sense can be schematized as follows.

1. Such-and such is normally (typically, usually) the case.

2. So, thus-and-so is the case.

In the defeasibility literature, propositions of the form (1) are usually taken to be a kind of non-universal generalization. Widely called defaults or default-generalizations, they often have the character of generic propositions, typified by "Ocelots are four-legged". Generic defaults are the business of the next section.

The term "defeasible reasoning" has two senses, one passive and the other active. In its passive sense, reasoning of whatever kind is defeasible just when the elements required for its goodness are not subject to guarantees. Passively defeasible reasoning is simply reasoning attended by the possibility of error. In this sense, defeasible reasoning is not reasoning of a particular kind. It is reasoning under a certain kind of circumstance.

In its active sense, defeasible reasoning is reasoning that embeds certain features designed to offset or compensate for the possibility of error. It is what Rescher calls "a fall-back position in point of conclusion-drawing." (Rescher 2007, p. 163). Leading the list of these compensation devices are hedges, expressions that express a qualification, as with "perhaps", "possibly", "probably", and so on. Hedges operate in two dimensions. In one dimension they are operator-operators which modify the consequence relation. In another dimension that are sentence-operators which express propositional content. Consider the following argument schema:

1. α

2. β

3. γ

4. So probably δ

The schema is ambiguous with respect to the role of the hedge "probably". On one reading, "probably" qualifies the conclusion operator "so". It maps a conclusion operator of whatever strength to one that is weaker than it. It does this in a way that signifies that the underlying consequence relation has a corresponding lack of strictness. In standard deductive contexts, we have the equivalence of argument validity and premiss-to-conclusion entailment, a provision of the Deduction Theorem for classical logic. It might well appear that in the present situation we have a variation of the same thing. Putting "so^w" for the weakened conclusion operator and $\vDash_w$ (i.e., $|\sim$) for a correspondingly weakened consequence relation, the equivalence at hand would be that between

a. α

b. So^w, β

and

c. $\alpha \vDash_w \beta$ (i.e. $\alpha |\sim \beta$).

In fact, however, there is no full Deduction Theorem for logics of this sort. The imputed equivalence fails (Morgan 2000). Surprising as this might initially strike us, a little reflection shows that, given the nature of defeasible consequence, precisely this should follow. Consider our example. If the Deduction Theorem held for defeasible consequence then we would have the

Defeasibility Deduction Theorem

$\Gamma \cup \{\alpha\} | \sim \beta$ iff $\Gamma (\alpha \rightarrow \beta)$.

Let us suppose that in the absence of overriding considerations a implies the negation of β. Suppose that Γ also contains offsetting factors. Suppose it Γ contains γ, $\ulcorner(\gamma \wedge \alpha) \to \beta\urcorner$. Then Γ together with α gives β. The set $\Gamma \cup \{\alpha\}$ overrides the usual implication of $\ulcorner \sim \beta\urcorner$ by α. But Γ alone doesn't do this. So Γ is powerless to secure the desired implication $\ulcorner \alpha \to \beta\urcorner$.

For purists this is enough to give up on the idea that anything deserving the name of logic could be a logic of defeasibility. For others, it shows the depth of the divide between classical and defeasible consequence.

But a second and different reading of hedges is also possible. This is the reading in which "probably is a sentence-operator, signifying part of the content of the argument's terminal proposition. In the standard approaches to modal logic, α and $\ulcorner$Possibly $\alpha\urcorner$ are different propositions. On the present reading, the same is true for α and $\ulcorner$Probably $\alpha\urcorner$, α and $\ulcorner$Perhaps $\alpha\urcorner$, and so on. This allows for the underlying consequence relation to be both unqualified and —— in some cases at least —— as strong as it gets. Rescher holds to an extreme version of this second reading. He sees all defeasible arguments as enthymematic deductive validities of the following general kind:

1. In all ordinary (normal, standard, commonplace) cases, whenever α, then β.

2. α obtains in the case presently at hand.

3. ⟨The present case is an ordinary (normal, standard, etc.) one.⟩

4. So, β obtains in the present case. (Rescher 2007, p. 164)

in which (3) is the original argument's implicit premiss. We might note in passing a rather striking confusion on Rescher's part. Of the above argument schema he writes that arguments of that form "would be [deductively] valid if all their premises —— explicit and tacit alike —— were authentic truths, which they are not, since at least one of the critical premises of the argument is no more than a mere presumption." (Rescher 2007, p. 164). Rescher evidently thinks that the contents of presumptions aren't propositions or can't take truth values.

When a hedge is taken as a sentence-operator, it is in principle open to the consequence relation of defeasible reasoning to be both classically deductive and monotonic. However, the same is not true of the operator-operator sense of defeasibility. Accordingly

Proposition 6. *MONOTONIC AND NONMONOTONIC DEFEASIBILITY: There is a sense of the defeasible in which the consequence relation of defeasible reasoning can be monotonic, and a difference sense in which it cannot.*

Proposition 6 closes an important equivalency question for defeasibly hedged arguments. The question is whether pairs of identical arguments except for the different placement of identical hedge-symbols in the terminal line are equivalent, that is to say, necessarily good together or necessarily bad together. Consider the following pair for which the hedge-expression is "presumably".

Operator-operator	*Sentence-operator*
1. α	i. α
2. β	ii. β
3. So presumably, γ	iii. So, presumably γ

The answer is that they are not equivalent. The consequence relation of the argument on the left cannot be classically monotonic, whereas the consequence relation of the argument on the right can be. Suppose that we now augment the premiss-sets of the two arguments by adding, each time the sentence $\ulcorner \sim \gamma \urcorner$. Then in no sense is the γ of the left hand argument a consequence of these augmented premisses. But if the consequence relation of the original right hand argument is classical consequence, it remains classical consequence under this same premissory supplementation. Accordingly, $\ulcorner \sim \gamma \urcorner$ is a consequence-killer on the left hand side and a consequence-preserver on the right hand side. So the equivalency in question fails.

The distinction between the operator-operator and sentence-operator uses of hedges such as "possibly" and "presumably" is one that has legs. It offers the would-be theorist two models around which to construct a defeasibility logic. The one model is that of the mainstream modal logics, in which the symbols for possibility and necessity are sentence operators (and in quantificational extensions also open-sentence operators). This is a model in which statements of what is a consequent of what is not hedged. The other model is what you get in a straightforward adaptation of the classical probability calculus. Let Γ be a set of sentences and α a sentence. Let n be a real number. Let $\Pr (\alpha/\Gamma)$ denote the conditional probability of that sentence of those sentences. Suppose that number is n. Then we may speak in a derivative way of probabilistic consequence. When $\Pr(\alpha/\Gamma) = n$, then α is an nth degree probabilistic consequence of Γ. Let us denote this relation "$\models_{p/n}$", and its corresponding conclusion operator $\mathrm{So}^{p/n}$. Note that, although the probability calculus attaches probability values to sentences, $\Pr(\alpha)$ is not a sentence-operator. Accordingly,

1. Γ

2. $\mathrm{So}^{p/n}, \alpha$

is a probabilistically valid argument, in which the hedge expression serves as an operator-operator.

It is preferable, I think, to organize a theorist's defeasibility reflections around this second, or operator-operator, model, rather than around the sentence-operator model. The reason is largely pragmatic. It is more difficult to produce a semantics for content-hedges than inferential hedges. Not to say that either is easy.

18.4 Defaults

We come now to the *narrow* notion of defeasibility logics. A founding motivation was the interest taken by AI-theorists in the logic of planning. An early difficulty was the *frame problem*. The frame problem is implicit in the difficulty in distinguishing simple belief-revision from error-detection and correction. Belief-revision is required when new information contradicts an old belief that is no longer true. Error-correction is required when the new information shows that the old belief never was true. Relatedly, it is the problem that computers have in determining precisely where in the system's knowledge base to make updates or revisions when new information is made to flow through. The problem also takes an action-theoretic form. When an action is performed, some things change and some don't. The problem here is twofold. How do we distinguish changing from unchanging facts? And how do we represent the preservation of those unchanging facts in a natural way?[12]

---

[12]The frame problem is one of a trio. Another is the knowledge-organization problem. It is the problem of deciding on the various patterns in which knowledge should be stored, and has a clear bearing, for example, on how information gets organized around natural kinds. The third is the relevance problem. It is the problem of determining what information in a knowledge base would or might be of

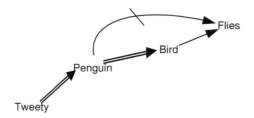

Figure 18.1: The Tweety Triangle

A second problem motivating the research programme in default reasoning is the *qualification problem*.[13] Imagine that we wanted to robotize a certain fairly simple task, such as the stacking and unstacking of some boxes in a warehouse. The problem besetting the program is that stacking and unstacking are open-ended processes. Any proposed procedures for performing these tasks will be subject to arbitrarily many possibilities of override. The question to which this commonplace fact gives rise is, "Given that one cannot embed all these possibilities of override as working qualifications on the system's algorithms, how does one determine the procedures which the stacking-unstacking enterprise should follow?"

A third motivator is Reiter's *closed-world assumption*.[14] In closed-world reasoning, it is assumed that information that would weaken or cancel the premises' connection to the conclusion is not true if not already contained in the premises. Of course, except for classically deductive contexts, the possibility is that the assumption is always false. So the problem is to determine whether the assumption is ever critically available to the reasoning agent and, if so, what are the conditions under which this is so?

A fourth inducement arises from *defeasible inheritance* difficulties associated with certain kinds of information-taxonomies.[15] The problem is exemplified by the famous inheritance network known as the Tweety Triangle. Another is the almost as famous Nixon Diamond.[16]

As figure 18.1 shows, although birds fly and Tweety is a bird, Tweety is a bird of the penguin kind; so Tweety doesn't fly.

Mr. Nixon was both a Quaker and a Republican. But Quakers are pacifists and Republicans not. In each case, the looming inconsistency is abetted by postulating that the links of birdness to flying, penguins to not-flying, Quakerhood to pacificism and Republicanism to non-pacifism is one of *default-consequence*. This triggers two problems. One is to characterize the default-consequence relation. The other is to show how inconsistency is averted by default conditionals with true antecedents and mutually inconsistent consequents. How can it be true that

---

assistance in handling a task that has been presented to a computer. We might note in passing that what philosophers mean by the frame problem is what computer scientists mean by the relevance problem. See Gabbay and Woods (2003).

[13]McCarthy (1977), pages 1038-1044.

[14]Reiter (1978). See also Etherington and Reiter (1983), McDermott and Doyle (1982). The closed world assumption is also a motivation for the logic of circumspection originated by McCarthy (1980). See also (McCarthy 1986).

[15]Touretsky (1986) and Horty, Thomason and Touretsky (1990).

[16]The geometrical descriptions are figures of speech, needless to say. The problem schematized by the Tweety Triangle could just as well be represented by a Tweety Diamond, and vice versa for the Nixon problem.

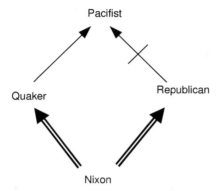

Figure 18.2: The Nixon Diamond

the correct default from Tweety's birdhood is that Tweety flies, that the correct default from Tweety's penguinhood is that Tweety does not fly, and that these facts do not constitute a contradiction?

Perhaps the best-known formal approach to defeasible reasoning in the narrow sense is the default logic pioneered by Reiter. In Reiter's approach, a default logic is an extension of first order logic, supplemented by default rules. Informally, a default rule is something like "From x is a bird, infer the default that x flies." In the AI literature, default logics — including Reiter's own — often embed a fixed point (or cognitive equilibrium) approach.[17] Also important are model-preference orientations developed by McCarthy's circumscription theory.[18]

18.5   Generality

A central interest of logicians is the consequence relation. A central interest of theorists of reasoning is the *drawing* of consequences. One of the standard ways in which human reasoners draw consequences is by making generalizations from samples. This is the issue I want to deal with here. There is a large consensus among mainstream logicians and psychologists alike that we aren't very good at drawing consequences of this sort, and that we are too prone for our own rational good to "jump to conclusions" or "yield to stereotypes". My own view is that this is an over-harsh judgement.

A focal theme of the default logics of inheritance reasoning is the generic proposition.[19] It is widely assumed that in the Tweety Triangle, "Birds fly" and "Penguins don't fly" are forwarded as generic claims. In the Nixon Rectangle, the same is true of "Quakers are pacificists" and "Republicans aren't pacifists". We will say in the section to follow why we think these are suspect assumptions, but there will be no harm in staying with them for now as an expository convenience. It will also prove instructive to examine this idea of genericity in

---

[17]McDermott and Doyle (1982) is a modal fixed point logic. Connections to default logics are worked out in Lifshitz (1994).

[18]Significant variations are Krautz (1986) and Shoham (1987).

[19]Generics also come in non-general forms, as with the kind-sentence, "Research-in-Motion invented the BlackBerry."

the context of a class of cases discussed by Gerd Gigerenzer and typified by the following example.

> Consider a 3-year-old who uses the phrase 'I gived' instead of 'I gave'. A child cannot know in advance which verbs are irregular; because irregular verbs are rare, the child's best bet is to assume the regular form until proved wrong. The error is 'good' —— that is, useful —— because if the 3-year-old did not try out new forms and occasionally make errors, but instead played it safe and used only those words it had already heard, she would learn a language at a very slow rate.[20]

Economic considerations are in play here. It is much more expensive to learn the irregular verbs "up front" and in one fell swoop (indeed for a three-year-old it is impossible) than to learn them one by one. A rule that purported to cover all the exceptions would be punishingly complex (and almost certainly wrong). Thanks to the efficiency of feedback, learning from one's mistakes is better economics than learning that is error-free and, in large ranges of cases, it is not an inferior breed of learning. A further consideration involves how individual agents handle generalizations. When the little girl said "I gived", she was instantiating a general rule —— the stem+–"ed" rule for the past tense.

As is well-known, logicians of even non-standard stripe usually default to the standard definitions of most of the logical particles. In the present instance, quantification is a case in point. Logicians have long since adapted their thinking to a particular view of generalizations. They see them as the universal quantifications of conditional propositions. Making true universal quantifications is risky business and, by and large, difficult to get right. Part of what makes this so is their brittleness. Universal quantifications are felled by a single true negative instance, that is, by the slightest deviation from invariability. Let us say, then, that a default instantiation is one in which the falsity of the instance is compatible with the truth of the generalization and the tenability of the inference. Consider a case. You are tramping in the wilds of Brazil and you espy your first ocelot, four legs and all. You say to your companion, "How interesting, I always imagined that ocelots were two-legged." As you now see, ocelots are four-legged, not two. Of course, some ocelots aren't four-legged, perhaps as a result of congenital mishap or injury. Suppose this other ocelot, Ozzie, turns out to be three-legged. Does this true negative instance falsify the claim that ocelots are four-legged? It does if your generalization is the universal quantification of the requisite conditional. It does not if your generalization is a generic proposition. Unlike universally quantified generalizations, which are brittle, generic generalizations are *elastic*.[21] There are true negative instances in the face of which they remain true. Elasticity facilitates economical generalization, both inferences from and inferences to. Accordingly,

Proposition 7. *FUNDAMENTAL LAW OF INSTANTIAL REASONING: Let α be a sentence in the form $\ulcorner \forall v(Gv \supset Hv) \urcorner$ and β an instantiation of it. Then beta's falsity entails α's falsity.*

---

[20](Gigerenzer 2005, p. 196).

[21]Compare here David Armstrong's characterization of laws of nature as 'oaken' rather than 'steel' (Armstrong 1997, pp. 230–231).

Proposition 8. *THE SEMANTICS OF GENERICS: Let α be a generic claim in the form $\ulcorner Gs\ H \urcorner$, and let β be an instantiation of it. Then β's falsity does not entail α's falsity.*[22]

Default logicians have a large stake in elastic generalizations. They play an essential role in defeating the Fundamental Law of instantial reasoning. The semantics of elasticity is an open problem in the genericity research programme.[23] Getting the *falsity* conditions right for "Ocelots are four-legged" lies at the heart of this enterprise (and is an open question in it.)

As a strictly practical matter, generic generalization offers a number of advantages. One is that it makes available to the reasoning agent an attractive way of reducing rates of error. When his instantiation from a universally quantified generalization is false, he has made *two* mistakes, one of which might be quite difficult to correct. He is mistaken about Ozzie's four-leggedness and he is wrong about "For all x, if x is an ocelot, then x is four-legged". So he is wrong in particular, and he is wrong in general. On the other hand, if his generalization is generic, he is wrong about Ozzie but not about ocelots. He is wrong in particular (which is a comparatively easy thing to correct) but he is right in general (leaving nothing that needs correcting). So there is a clear cognitive, as well as economic, advantage in not generalizing to universally quantified conditional propositions.

As we are telling it here, instantiations of generic generalizations are defaults. In the AI-literature defaults are often the major premisses of default inferences, that is to say, the generic propositions themselves. Here they are their conclusions. In some ways, the AI convention is unfortunate. It obscures the distinctive character of the relationship borne by the conclusions of default inferences to their premisses. It is further evidence of the sloppiness that runs through much of this literature. Everyone agrees that default reasoning incorporates — implicitly or otherwise — the qualification "unless there is information to the contrary." But the generic claim is not the default. For suppose it were; then it would be covered by a default rule, by something in the form $\ulcorner$Given α, conclude that ocelots are four-legged except where there is particular reason to the contrary.$\urcorner$ The trouble is that every generalization —— generic or universally quantified conditional —— is subject to it.

We could, if we liked, adopt the AI-habit of thinking of defaults as generic claims. But doing so wouldn't capture what is special about "Ozzie is four-legged." Ozzie's four-leggedness is inferred not in default of information that tells us that its major premiss is false, but rather in default of information that tells us that its conclusion is false *without also* telling us that that the premiss is false. What makes "Ozzie is four-legged" a default from "Ocelots are four-legged" is not that new information might falsify "Ocelots are four-legged", but rather that new information that doesn't falsify "Ocelots are four-legged" might falsify "Ozzie is four-legged". Here is idea that is not entirely new. The same characteristic also attaches to certain kinds of evidential reasoning. We might have it that evidence e supports a given conclusion; but future evidence could falsify that conclusion without falsifying any part of e. Even so, in the context of instantial reasoning, the present idea lays plausible claim to novelty.

There is little to be gained from terminological wrangles. The habit of calling generic premisses "defaults" is deeply dug in. So is the habit of calling them

---

[22]The idea of generic generalizations (if the pleonasm may be forgiven) is given early notice in Langford (1949) and Scriven (1959). Langford discusses the "institutional use of 'the' ", as in 'The horse is an equine'. Scriven speaks of "normic" propositions, that, is propositions about what is normally or typically the case.

[23]Concerning which, Carlson and Pelletier (Krifka et al. 1995) is required reading. Also important is Pelletier (2010).

rules, of seeing them as generic instantiations, and also of seeing the link between a generic and an instantiation as default-consequence. Fortunately, the uses are linked. A generic claim is a default because it mandates a default rule. (⌐Given α, conclude that β unless there is reason to the contrary⌐). An instantiation of a generic claim is default because it satisfies a rule the generic claim mandates. The tail of a default rule is a default consequence of its head. Better, then, that we follow the course of least resistance, and allow for the fourfold use, leaving it to context to sort out which is which in particular cases.

Defaults (in the instantial sense) can be false without damage to the generalizations which they instantiate. There is a lesson in this for hasty generalization, which is a form of what in the social scientific literature is called "thin-slicing".[24] It is that, when in ignorance of the particularities of the case, you default to "Ozzie is four-legged", then, since its falsity doesn't (just so) falsify "Ocelots are four-legged", neither does it (just so) spoil the reasoning that generated it. Proceeding from truths to falsehoods is precisely what defaulting is purpose-built for. This reinforces an interesting observation about error and reasonability.

Proposition 9. *APTNESS OF DEFAULTING: In reasoning to Ozzie's four-leggedness you made an error, but you made no error of reasoning. Although in reasoning that way an error was committed, reasoning in that way was not an error. Reasoning in that way was the apt thing to do.*

Accordingly

Proposition 10. *RIGHTLY REASONED FALSITY: Reasoning to false conclusions from generic premises is sometimes the right thing to do. It is the right thing to do when the proposition reasoned to is a default of the proposition reasoned from and the proposition reasoned from is true.*

18.6   Haste

The Ozzie example tells us something of importance about the instantiation of default generalizations or generics. But we should also attend to the converse relationship, in which generalizations are wrought from instances.

What would it make sense for beings having our interests, talents and wherewithal to generalize *to*? Universally quantified generalizations serve us well in domains in which literal invariability is required for generality. These are comparatively rare domains. They encompass mathematics and the various branches of science that harbour ambitions for mathematically strict lawlikeness. For most of the living world, however, non-trivial literal universality is unattainable. Its lawlike regularities are better-served by and more faithfully rendered as generic generalizations on the model of the four-footedness of ocelots.

A common setting in which individual agents generalize is one in which projections are made from samples. It is an established conviction among mainstream inductive logicians and philosophers of science that generalizations are worthless when they are projected from unrepresentative samples. They commit the dread fallacy of hasty generalization. It is also widely accepted that, beyond a certain point, the smaller the sample, the greater its unrepresentativeness. Even in polling contexts,[25] where samples can be very small in relation to test-populations, they must be quite large in comparison with the samples that very often trigger generalizations in everyday life. The requirement that a generalization be deferred until a sample of appropriate size and randomness is assembled

---

[24] Ambady and Rosenthal (1993), Carrre and Gottman (1999).
[25] Concerning which, see Walton (2007c).

is a costly one. In practical terms, it means that it is hardly ever the case that an individual agent is allowed to make a generalization when he needs to have it. Inductively robust generalizations are expensive for the likes of us, even when it is possible to pull them off. This is not significantly less true of generic generalizations than of universally quantified ones. Even so, holding the generic four-leggedness of ocelots to the strict standard of inductive strength is hardly less costly than doing the same for their universally quantified four-leggedness. When one reflects on the empirical record, it is clear that individual agents on the ground routinely omit to pay this cost in either case.

Generic claims are but one form of non-universal generalization. Similar advantages also accrue to other forms of non-universal generalizations, both those that carry quantifiers expressly ("For the most part, As are Bs") and those that are modified by non-quantificational sentential adverbs ("Normally As are Bs", "Usually As are Bs") Even so, there are significant differences which, apart from a brief mention in section 18.7, I won't have time for here.[26] For the present let us give formal recognition to a pair of central points:

Proposition 11. *ECONOMICAL INSTANTIATION: There are cognitive economies involved in instantiating from non-universal generalizations.*

Proposition 12. *ECONOMICAL GENERALIZATION: There are kindred economies involved in generalizing to non-universal generalizations.*

Thin-slicing turns essentially on our ability to recognize patterns. Gerald Edelman is instructive on this point:

> Selectionistic brains themselves show the effects of historical contingency, irreversibility, and the operation of non-linear processes. They consist of enormously complex and degenerate artworks that are uniquely embodied in each individual. Moreover, human brains operate fundamentally in terms of pattern recognition rather than logic[27] (Edelman 2006, p. 82).

In its hasty-generalization form, pattern recognition is tied to our ability to recognize natural kinds. Among philosophers such as Quine there is a certain scepticism about natural kinds.[28] But there are convincing approximations of them well-attested to in the empirical literature, whether *frames*[29], *prototypes*[30], or *exemplars*[31]. There is also some significant work in linguistics in which the semantics of natural kind noun phrases is intimately bound up with factors of genericity. It behoves a naturalized logic not to give the idea of natural kinds too short a shrift.[32] Even so, the idea of natural kinds is also linked to the temporal projectibility of propositions. Hasty generalization is thus implicated in the frame problem. The question whether it is all right to generalize on the proposition "This ocelot is young and frisky" resembles the question whether it is all right to temporally project the same proposition. The problem of temporal projection has no known solution in the logic of defeasibly. Analogously, the same applies to the thin-slicing/natural kinds problem.

---

[26]For plural quantification and like matters, see Sher (1991) and van Benthem (1999).
[27]What Edelman means here by logic are mainstream systems of mathematical logic.
[28]Quine (1969).
[29]Minsky (1975).
[30]Smith and Medin (1981).
[31]Rosch (1978).
[32]Krifka et al. (1995).

It is a matter of fact that individual agents are hell-bent-for-leather thin-slicers. On the received view, this convicts them of substantial error, if not of massive irrationality. But if the principle of the normativity of actual practice is given any standing here, we are wrong to confine our generalization to those that arise in an inductively strong way from the triggering samples. Certainly, it is difficult to believe that when the little girl of Gigerenzer's example generalized to the stem + "ed" rule she got the rule wrong or was in any way irrational. Such cases are legion. They are cases in which our hasty generalizations turn out to be right. Not every hasty generalization is right. But some are. In fact, lots are. There is a significant correlation here:

Proposition 13. *A BENIGN CORRELATION: The hasty generalizations that we are right to draw are by and large the ones that we do draw.*[33]

Seen this way, one of the individual agent's strongest contributions to his success in the cognitive economy is his capacity for determining when to thin-slice, when to leap to conclusions. This involves a capacity for seeing in samples that are sometimes as small as they get the requisite representativeness. One of the tasks of cognitive psychology and cognitive neurobiology is to uncover the mechanisms that provide for the individual reasoner so providentially. Even if it is conceded that for every mistake there is a perspective from which it can be seen as wrong, there are many cases in which the collateral benefits of actually committing them outweigh their wrongfulness. The little girl's error is a case in point. Her selection of "I gived" instead of "I gave" represents a default instanti-ation of the stem + "ed" rule. Considered generically, the rule is right. It is in this particular default that her mistake lies. But drawing it was not the wrong thing to do. It was the right thing to do. It is the only feasible way for a three-year-old to learn the irregular verbs.

Thin-slicing teaches a like lesson. The rule of which "I gived" is a default is something the little girl generalized to, from — at her tender age — a smattering, at most, of correct utterances in the form "I [stem]-ed." In generalizing to this rule, the child made no error. In generalizing to it hastily, she revealed her adept-ness in discerning generalization in small samples. Had she set out to find the universally quantified rules of lawlike linguistic correlations, she would never have learned her language. Accordingly,

Proposition 14. *THE ECONOMICS OF FEEDBACK: Learning from feedback is often cheaper than (and just as good as) learning by universal generalization.*

The vulnerability of haste is compensated for by the corrective wherewithal of feedback. In so saying, we are reminded of the widely-held view that a fal-lacy is a mistake which, among other things, displays a notable resistance to post-diagnostic avoidance. This now requires some qualification. Consider the large class of cases in the hasty generalizations of individual reasoners are be-nign but to which some exceptions have been found. With respect to those true negative instances — Ozzie, for example — the mechanisms of feedback operate smoothly and efficiently. We began by instantiating the four-legged generaliza-tion to Ozzie. When it turned out that Ozzie's four-leggedness is an error of fact,

---

[33]The cognitive science literature recognizes a related distinction between exploration and exploita-tion. When in an explorative state, an agent runs some experiments, tries to understand, asks (at times) stupid questions, and behaves suboptimally when judged against a short-term horizon. But this can be considered as an investment in a better understanding of the situation which will then — now adopting a long-term perspective — pay off in the future. The Secretary Problem (Gilbert and Mosteller 1966, Ferguson 1989) is famously one such example. In the first phase, people acquire information which will be used to set a threshold, and only in the second phase will they make a decision based on comparisons with objects against this threshold. It is suboptimal if phase one is too short or phase two is too long. The transition from exploration to exploitation is, however, quite often mathematically tractable.

we fixed it. Ozzie isn't four-legged; Ozzie is three-legged. In this there isn't the slightest indication of laggardness in our error-detection and correction capabilities. There is no sign of those measures in our application of those measures of our having struggled to perform their assigned roles. On the other hand, where our feedback mechanisms *would* be notably laggard — to the point of downing tools — is in the discouragement of the hasty generalization that ocelots are four-legged and of the default inference that any given ocelot is that way too. Clearly, it *would* take training, lots of it, to get people to stop doing these things. But this is as it should be. They shouldn't stop doing these things.

18.7   Errors of reasoning

What, then, are we to say that an untenable default-inference consists of? What would a counterexample be? What, if Tweety is a penguin, of the inference that Tweety flies from the premiss that Tweety is a bird? On some tellings, such an inference would be rendered untenable by a rival generic of greater specificity than the original.[34] So, although "Tweety flies" is correctly inferred from "Birds fly and Tweety is a bird", its negation, "Tweety does not fly" is correctly inferrable from "Penguins (are birds that) don't fly and Tweety is (a bird) that is a penguin". But there is something wrong with this. It does not show that the former inference is a mistake of reasoning. It does not show that its major premiss is false. It shows only that compatible generic claims can support inferences whose conclusions are incompatible with one another. The specificity criterion doesn't tell us which, if any, of these inferences is unreasonable. It merely stipulates a tie-breaker. It tells us to opt for the inference carrying the more specific information. This leaves the task of specifying untenability conditions for default-inferences still unperformed. It also leaves us with inferences in the form of the Triangle, supplemented by a standard "and"-introduction rule:

Birds fly

Penguins don't fly

Tweety is a bird

Tweety is a penguin

So^w, Tweety flies and Tweety doesn't fly.

Since the move from these premises to this conclusion is by default inference, we have it that although the conclusion is inconsistent it is not a reductio of its premises. The conclusion is a contradiction, but none of the premises is false.

This is not very satisfactory. It tells you what consequence to draw in the present case. But knowing that does not stop "Tweety flies and Tweety doesn't fly" from being a consequence of true premises. This helps us see that the specificity tie-breaker is hopelessly misguided if the more specific claim and the less specific claim are both true. The preference for specificity is right-headed, but not if the more general proposition that it contradicts is true. So it is unacceptably paradoxical for a theory of error to rule that the Tweety Triangle contains no error

---

[34]The specificity condition has had a chequered history in the AI literature. Pollock (1995) rejects it, except for statistical reasoning with projectable properties (pp. 64-66). Reiter (1980), McDermott and Doyle (1982), Moore (1985) and Konolige (1994) fail to adopt it. It also fails in McCarthy (1980, 1986) and Lifschitz (1988). The principle holds in Pearl (1988, 1990), Geffner and Pearl (1992) and Goldszmidt, Morris and Pearl (1993).

of reasoning. If we are to disentangle ourselves from this bad result, we must try to determine what it takes to falsify a generic claim.

Let us, then, ask for the relevant difference between "Ocelots are four-legged" and "Birds fly". So far as we know, negative instances of "Birds fly" are of two kinds. One is the one-off anomaly, as with a congenitally impaired robin.[35] The other is a countervailing natural kind: penguins, for example, or ostriches. As we see, one-offs aren't usually considered as generically overriding, not even when the one-offs statistically outrun the true positives.[36] There is no reason to cut the same slack to natural-kind exceptions. Although "Birds fly" would not be falsified by an anomalous exception — say a congenitally impaired robin — it is falsified by species of birds that don't fly.

Let us say of any two genericizations "Gs H" and "Js K" that the "head" of the first properly includes the "head" of the other if and only if all Gs are Js and some Js are not Gs. Similarly let us say that "Gs H" and "Js K" have incompatible "tails" if and only if K entails not-H. We can now suggest some thing official about falsifying the generic.

Proposition 15. *FALSIFYING THE GENERIC: Let α and α' be generic generalizations. Then α' falsifies α if α' is true, α''s head properly includes α's head and α and α' have incompatible tails.*

Our condition successfully captures the intuition that deference be shown to the more specific of any two incompatible generalizations. But it also leaves the question of what else would falsify a generic. Let us see.

There is something not quite right about the present condition. It seems to be open to a class of cases that would pull its teeth. We say that "Ocelots are four-legged" is a true generic. If we are right, our claim is not defeated by its anomalies. It is not defeated by the congenitally compromised Ozzie. It is not defeated by any three-legged specimen made so by surgery. But consider now the class of all those anomalies. We have it that leggedly anomalous ocelots aren't four-legged. If this is a generalization at least as strong as a generic indeed (it makes a fair claim to be a universally quantified conditional), then it certainly falsifies "Ocelots are four legged". Anomalous ocelots are properly included in the ocelots, and not-four-leggedness is incompatible with the four-leggedness.

I take it as given that no one seriously believes that anomalous ocelots overturn the generic four-leggedness of ocelots. If Proposition 15 is to hold water, we will have to exclude this kind of case. Can we do so in a plausible way? It would appear that we can. The falsification conditions require that more specific generics can overturn more expansive ones. What the present case suggests is that we cash an intuition of pages ago. So we venture to say that only a generic can falsify a generic. What is more, since anomalous ocelots are not a natural kind — certainly not a species — their non-four-leggedness is not a generic trait of them. Equally, since universally quantified conditionhood is not sufficient for genericity, the true sentence "$\forall x$ (x is an anomalous ocelot $\supset$ x is not four-legged"[37] does not overturn the generic default that ocelots are four-legged.

Proceeding in this way has some welcome advantages. One is that we have a shot at understanding when a false instantiation of a true generalization is a

---

[35] Or as with an albino raven in relation to "Ravens are black".

[36] Consider the following cases from Carlson and Pelletier (Krifka et al. 1995, p. 44): "A bird lays eggs", "An Anopheles mosquito carries malaria" and "A turtle lives a long life". Less than half the birds lay eggs, no more than five percent of Anopheles mosquitoes carry malaria, and hardly any turtles have long lives, owing to the efficiencies of anti-turtle predation.

[37] Cf. the circumscription default rule, with "ab_1", "$ab_2, \ldots$" as anomaly or abnormality predicates: For all x, (x is an ocelot $\wedge \sim ab_1(x) \wedge \sim ab_2(x) \wedge \cdots \wedge \sim ab_n(x)$) $\supset$ x is four-legged" (McCarthy 1980, 1986, Lifschitz 1988).

mistake of reasoning. The mistake in the reasoning that takes us from the true premisses "Birds fly" and "Tweety is a bird" to the false conclusion "Tweety flies" is now apparent. Contrary to what I (and the AI-literature) have been supposing, the general premiss is *not* true. It is falsified by a sentence one of whose default consequences is itself the negation of "Tweety flies". It is not in general sufficient for bad reasoning that it embody a false premiss. But when the way in which the premiss is false provides for the negation of its conclusion — call this 'clean-sweep' falsification — can we not fairly judge the reasoning as bad? I think not. There was a time when Europeans were unaware of Australia's black swans. Fred is a swan padding about in the pool in front of the Parliament in Canberra. Would a European have committed an error of reasoning had he predicted that Fred is white? Since when is provincialism an error of reasoning?

Of course, the world has been seized long since of the blackness of Australia's swans. Australia's black swans are a well-known counterexample to the white-ness of swans. Reasoning in disregard of that counterexample is subject to at least two different diagnoses. One is that the reasoning is defective at least to the extent that the agent's choice of premisses was careless or an offence to com-mon knowledge, or some such thing. No one doubts that "This swan is white" is a consequence of "Swans are white and this is a swan". But reasoning is consequence-drawing, and consequence-drawing is subject to constraints that don't inhibit consequence-having. No one thinks that drawing a consequence that a false proposition has is reasoning just as good as drawing a consequence that a true proposition has. This being so, let us say again that there is an ex-tent to which good reasoning — good consequence-drawing — is sensitive to the judiciousness of premiss-selection.

Another way of understanding the inference of the swan's whiteness is to re-interpret the premiss. Either the reasoner was not intending "Swans are white" as an unqualified generic, but rather as something in the form "Local swans are white" "; or he was intending it non-generically, as something in the form "For the most part, swans are white". The first possibility serves as a useful reminder of the appeal of Grice's Quantity Maxim: Don't over-commit; don't indulge in premissory over-kill. The value of the second interpretation is that it turns our attention to the importance of property-selection to generic utterance. It is well-known that colour is not invariant under speciation, but limbedness, for example, is. The safest context in which to thin-slice is one in which the triggering prop-erty is invariant under speciation. Of course, it is a large, and still largely open, question as to how we go about making these determinations.

It bears on those reflections that even in the face of non-flying penguins, "Birds fly" doesn't *feel* false. What would account for this feeling? It is that the truth of "Penguins don't fly" gets us to see that "Birds fly" does not, as we supposed, express a generic claim, but rather a quantificational claim in the form "For the most part, birds fly" or a claim in the form "Usually birds fly". These are not generic statements; they are non-universal plural quantifications, or for short, NUP-quantifications.

General propositions line up in a natural way depending on their degree of susceptibility to deviant instances. Another way of saying this is that general propositions are ordered by their brittleness, ranging from most to least, univer-sal quantifications, generic statements, the various forms of non-universal plural quantification themselves ordered by quantification range. The less brittle a gen-eralization is, the less likely that a deviant instance will overturn it. This gives rise to a sensible rule, something along the lines of the Quantity Maxim.

Proposition 16. *FAVOUR ELASTICITY: In matters of generality, and in the absence of indications to the contrary, there is a cognitive advantage in favouring elastic gener-*

alization over brittle generalization. In other words, do not embrace the brittle beyond necessity.

18.8 Ceteris paribus

Sometimes a generic claim is taken as a universal quantification conditional by a *ceteris paribus*. It is, I think, a suggestion of little merit. The *ceteris paribus* clause is an artificial universality-preserver. On a common reading, it is a place-holder for a principled and properly stand-alone specification of a generalization's exceptions. Its whole semantic content can be summed up as "except for exceptions". So it is a gesture. The *ceteris paribus* device has two disadvantages, both of them serious. One is that it is a vacuous qualification. The other, paradoxically, is that it is too strong a qualification. If told that all birds fly except for those that don't, one is given no indication of the statement's exceptions. (Vacuousness). If told that all frogs cause warts except for those that don't, one is told something that makes a mockery of the very idea of exceptions to the rule. And one is told it in a form that makes every false generalization true. (Triviality).

Devices similar in function to the *ceteris cerberis* clause have commended themselves to default logicians. More's the pity, since they are heavily implicated in a bad confusion about generic defaults. Consider again "Ocelots are four-legged" and "Birds fly". These tend to receive a common classification in the AI literature. They are both considered impervious to certain exceptions. The four-leggedness of ocelots is proof against falsification by anomalies (e.g., by a non-four-leggedness arising from congenital defect or injury) or irregularities (as in departures from the stem + "ed" rule). Whereupon we have it that, except for anomalies or irregularities, ocelots are four-legged and the past tense = infinitive stem + "ed". The exceptions to which the flyingness of birds is considered impervious is birds that don't fly, the nonflyingness of which is no anomaly, either with respect to the fliers or the non-fliers. The trouble with the anomaly-qualification is that, unlike the *ceteris paribus* qualification, it is both too strong and too weak. It is too strong in the sense that it obliterates the distinction between universally quantified and generic generalizations, leaving us nothing to choose between "Except for anomalies, for all x, if x is an ocelot then x is four-legged", and "Except for anomalies, ocelots are four-legged". It is too weak in the sense that it cannot save the flyingness of birds: "Except for anomalies, birds fly" is false. Penguins are birds. They are not anomalous birds. And they don't fly. This is not to say that "Birds fly" can't be saved exceptively (albeit incompletely). For "Except for penguins, ostriches, etc., etc., birds fly" is true. The trouble with this is the "etc." part. It is a gesture.

Proposition 17. *NUP-ECONOMIES: The economic advantage of NUP-quantifications elastic generalizations is that they are cheap to produce. Their cognitive advantage is that they are true and elastic. Like their generic cousins, only more so, they aren't necessarily felled by a false instance.*

There is a contrary view of the *ceteris paribus* qualification. On this view, the invocation of *ceteris paribus* is not in any sense a recognition of a claim's exceptions. To see why, consider a theory whose function is to provide causal explanations of its target phenomena P. Suppose a model of T in which it is possible to frame a causal law L for some event e in P. Not uncommonly L will involve an abstraction from facts f about e on the ground, not excluding causally salient facts. One says, then, that L holds of e *ceteris paribus*; that is, without regard to those f and notwithstanding their causal salience.[38] On this reading there is no notion of

---

[38] See here Michael Strevens (2008).

exceptions. The causally salient facts of that don't make the cut as parameters of the causal law L are not exceptions to it. That is, they are not not exceptions in the sense in which Ozzie's three-leggedness is an exception to the four-leggedness of ocelots.

18.9   Non-universal plural quantification

We can now see that all the advantages of generic generalizations pass to non-generic NUP-quantifications; and does so in a way that harmonizes with our definition of default instantiations. NUP-quantifications have cost-benefit advantages that generic claims don't have. One is that whereas the three-legged Ozzie is a true exception to the fact that ocelots are four-legged, it is not an exception to "Most ocelots are four-legged", or "Ocelots are for the most part ocelots" or "Usually ocelots are four-legged". Ozzie is not an exception to any sentence implicitly exempting a class of exceptions. "Ozzie is three-legged" is an exception to the "ocelots are four-legged"-clause of "Most ocelots are four-legged", but the apposition of the quantifier expressly provides that the ensuing general sentence is free of that exception. Since Ozzie is not an exception, "Ozzie is three-legged! is not a negative instance of "Most ocelots are four-legged." What, then, does our discovery of Ozzie's three-leggedness require us to do? It requires us to cease saying that Ozzie is four-legged. But it does not require that we do anything to amend, qualify or re-write the generalization from which the four-leggedness claim came. This is so irrespective of its form, irrespective of whether it was "Ocelots are four-legged" or "Most ocelots are four-legged."

People sometimes think that the primary fact about default reasoning is its dynamic character. A piece of reasoning is good or bad in a context, and contexts seriously involve what the reasoner then and there knows. Not knowing the particularities, you inferred that Ozzie the ocelot is four-legged. But now that you see that he isn't, you withdraw that inference. It takes little reflection to see that there is nothing in the claim of dynamical contextuality that is distinctive of non-monotonic reasoning. Doesn't it apply with equal force to monotonic reasoning? Not knowing the particularities, you inferred from "For all x, $Fx \supset Gx$" that a the F is G. But now that you see that a isn't in fact G, you retract your inference. Similarly, you infer from evidence e that H. Later evidence becomes available that counters or weakens your claim. So you withdraw it.

Dynamism is the hallmark of belief-change no matter what theory is proffered to account for it. Like all forms of belief-change, default reasoning generates conclusions that are revisable in the light of new information.

Proposition 18. *CHEAP CORRECTION: What is distinctive about default reasoning is that, when it arises, the need to revise carries comparatively slight costs. That which necessitates revision does almost no damage. Of course, once detected, every error has to be paid for. But what matters with default errors is how cheaply they are redeemed.*

18.10   Quantificational structure?

I am inclined to think that generic sentences in the manner of "Ocelots are four-legged" lack a quantificational structure; that is, that they embed in deep structure no quantifier of whatever strength. A contrary view is advanced by Krifka et al. (1995), in which it is proposed that the logical form of generic sentences is captured by the schema:

$$\text{GEN}[x_1 \ldots x_i] \, (\text{Restrictor} \, [x_1 \ldots x_i] \colon \exists [y_1 \ldots y_i] \, \text{Matrix} \, [\{x_1\} \ldots \{x_i\}; \, y_1 \ldots y_i].$$

GEN here is an unrestrictive quantifier, in the manner of Lewis (1975), which binds all the x_k variables simultaneously. Consider a pair of examples which I borrow from Pelletier (2010): "Bears with blue eyes are (normally) intelligent" and "Tabby the cat (usually) lands on her feet." Then on the present suggestion, the logical form of the first is:

GEN [x] (Bear (x) ∧ Blue-eyed [(x)] [Intelligent (x)]),

which is designed to say that, generically, blue-eyed bears are intelligent. The second example is more complex

GEN [x] (Event (x) ∧ Sub Event (y, x) ? Culmination (y, x) ∧ Agency (y, t) ∧ Land-on-feet (y, t)]),

which is designed to say that, generically, droppings of Tabby subsume a subevent whose culmination is Tabby's landing on its feet.

A good deal of effort has been devoted to finding a meaning for GEN. It is interesting to note what doesn't work. GEN is not "relevant quantification", GENs are not about arbitrary objects, GENs aren't about prototypes, GENs aren't about stereotypes and GENs don't express modal conditionals. "Perhaps", as Pelletier observes, "there are other available interpretations of GEN ? it is always possible that the whole approach under consideration ... is a flawed approach to meaning."[39]

For present purposes, we needn't press this question further, although I return to it in work under way. For now, I'll hang on to my non-quantificational intuition about "Ocelots are four-legged" until such time as the dust around GEN settles.

18.11   Final remarks

I began by saying that there are features of defeasible reasoning that haven't received the attention (or the understanding) that is due them. In this I hope to have made some modest headway. Even if that hope has been realized there can be no question of complete coverage. The present account is not free of its own omissions, one of which is particularly important and well-recognized in Douglas Walton's writings. This is the connection between defeasible reasoning and *presumption*. Its absence here is occasioned by the lack of space to deal with it, not for want of things to say about it. I hope to repair this omission in future writings.[40]

---

[39] Pelletier (2010, p. 15)
[40] Woods (2011)

ACKNOWLEDGEMENTS

This volume has involved a great deal of hard work from a dedicated team. The contributing authors have been patient when the process slowed, responsive when deadlines were pressing, and unfailingly good-humoured throughout, which has made the task of coordinating much easier than it might otherwise have been. All authors also acted as reviewers on other chapters, but in addition, we enlisted the help of a number of external reviewers, to whom we are greatly indebted. We would like to acknowledge the help of Floris Bex, Liz Black, Katarzyna Budzyńska, Kamila Dębowska, Geoff Goddu, Floriana Grasso, Dale Jacquette, Fred Kauffeld, Manfred Kienpointner, Sanjay Modgil, Nir Oren, Simon Wells and Frank Zenker in improving the quality of the contributions. Responsiveness and flexibility made coordinating their efforts that much easier. Finally, members of the Argumentation Research Group at Dundee put in a huge amount of effort into typesetting and proofing the final version. The editors owe an enormous debt of gratitude to Mark Snaith and John Lawrence for their hard work. Funding from the Engineering and Physical Sciences Research Council in the UK which supports the work of the Argumentation Research Group under grant EP/G060347/1 is also gratefully acknowledged.

BIBLIOGRAPHY

F. Abate and E. J. Jewell, editors. *New Oxford American Dictionary*. Oxford University Press, 2001.

J. Adler. Introduction. In J. Adler and L. Rips, editors, *Reasoning, Studies of Human Inference and Its Foundations*, pages 1–34. Cambridge University Press, Cambridge, 2008.

V. Aleven and K. D. Ashley. Evaluating a learning environment for case-based argumentation skills. In *ICAIL '97: Proceedings of the 6th international conference on Artificial intelligence and law*, pages 170–179. ACM Press, New York, 1997.

N. Ambady and R. Rosenthal. Half a minute: Prediction teacher evaluations from thin slices of behaviour and physical attractiveness. *Journal of Personality and Social Psychology*, 64:431–441, 1993.

R. Amossy and A. Herschberg-Perrot. *Stéréotypes et clichés. Langue, discours, société*. Nathan, 1997.

J. Andersen. The indulgence of reasonable presumptions: Federal court contractual civil jury trial waivers. *Michigan Law Review*, 102(1):104–124, 2003.

Aristotle. On sophisticated refutations. In *On Sophistical Refutations, On Coming-To-Be and Passing-Away*. Harvard University Press & William Heinemann, 1965. Translated by E.S. Forster.

D. Armstrong. *A World of States of Affairs*. Cambridge University Press, 1997.

K. J. Arrow. The economics of agency. In J. W. P. . R. J. Zeckhauser, editor, *Principals and Agents: The Structure of Business*, pages 37–51. Harvard Business School Press, Boston, MA, 1985.

K. D. Ashley. Arguing by analogy in law: A case-based model. In D. H. Helman, editor, *Analogical Reasoning*, pages 205–224. Kluwer, Dordrecht, 1988.

K. D. Ashley. *Modeling Legal Argument: Reasoning with Cases and Hypotheticals*. Artificial Intelligence and Legal Reasoning Series. MIT Press, Bradford Books, 1990.

K. Atkinson, T. Bench-Capon and P. McBurney. PARMENIDES: facilitating deliberation in democracies. *Artificial Intelligence and Law*, 14(4):261–275, 2006a.

K. Atkinson and T. J. M. Bench-Capon. Practical reasoning as presumptive argumentation using action based alternating transition systems. *Artificial Intelligence*, 171(10–15):855–874, 2007.

K. Atkinson, T. J. M. Bench-Capon and P. McBurney. Arguing about cases as practical reasoning. In *Proceedings of the Tenth International Conference on Artificial Intelligence and Law*, pages 35–44. ACM Press, 2005.

K. Atkinson, T. J. M. Bench-Capon and P. McBurney. Computational representation of practical argument. *Synthese*, 152(2):157–206, 2006b.

F. Baader, D. Calvanese, D. McGuinness, D. Nardi and P. Patel-Schneider, editors. *The Description Logic Handbook – Theory, Implementation and Applications.* Cambridge University Press, 2003.

E. M. Barth and E. C. W. Krabbe. *From axiom to dialogue. A philosophical study of logics and argumentation.* Walter de Gruyter, Berlin/New York, 1982.

D. T. Bastien. Change in organizational culture. the use of linguistic methods in a corporate acquisition. *Management Communication Quarterly,* 5(4):403–442, 1992.

M. C. Beardsley. *Practical Logic.* Prentice Hall, 1950.

U. Beck. *The Risk Society: Towards a New Modernity.* Sage Publications, London, 1992.

T. Bench-Capon, S. Doutre and P. Dunne. Asking the right question: Forcing commitment in examination dialogues. In P. Besnard, S. Doutre and A. Hunter, editors, *Computational Models of Argument: Proceedings of COMMA 2008,* pages 49–60. IOS Press, 2008.

J. van Benthem. Resetting the bounds of logic. *European Review of Philosophy,* 4:21–44, 1999.

A. Berger. *Encyclopedic Dictionary of Roman Law.* The American Philisophical Society, 1909.

L. Bermejo-Luque. Toulmin's model of argument and the question of relativism. In D. Hitchcock and B. Verheij, editors, *Arguing on the Toulmin Model: New Essays in Argument Analysis and Evaluation,* chapter 6. Springer, Dordrecht, 2006.

B. Bernake. Four questions about the financial crisis. Speech held at the Morehouse College (Atlanta), 2009.

T. Berners-Lee, J. Hendler and O. Lassila. The semantic web. *Scientific American,* 284(5):34–43, 2001.

P. Besnard and A. Hunter. *Elements of Argumentation.* MIT Press, 2008.

F. Bex. *Evidence for a Good Story.* Ph.D. thesis, University of Groningen, 2009.

F. Bex and H. Prakken. Investigating stories in a formal dialogue game. In P. Besnard, S. Doutre and A. Hunter, editors, *Computational Models of Argument (Proceedings of COMMA 2008),* pages 73–84. IOS Press, 2008.

F. Bex, H. Prakken, C. Reed and D. N. Walton. Towards a formal account of reasoning about evidence: Argument schemes and generalisations. *Artificial Intelligence and Law,* 11(2-3):125–165, 2003.

L. F. Bitzer. The rhetorical situation. *Philosophy and Rhetoric,* 1(1):1–14, 1968.

S. Blackburn. Morality and thick concepts: Through thick and thin. In *Proceedings of the Aristotelian Society.* 1992. Supplementary Volume.

J. A. Blair. Presumptive reasoning/argument: An overlooked class. *Protosociology,* 13:46–60, 1999.

J. A. Blair. A theory of normative reasoning schemes. In H. Hansen, C. Tindale and E. Sveda, editors, *Proceedings of the third OSSA conference: Argumentation at the century's turn.* Ontario Society for the Study of Argumentation, St. Catherines, ON, 2000.

J. A. Blair. Walton's argumentation schemes for presumptive reasoning: A critique and development. *Argumentation*, 15:365–379, 2001.

J. A. Blair. Towards a philosophy of argument. In *Informal Logic at 25: Proceedings of the Windsor Conference*. 2003.

J. A. Blair. The "logic" of informal logic. In H. Hansen, editor, *Dissensus and the Search for Common Ground (Proceedings of the OSSA Conference)*. 2007.

J. A. Blair and R. H. Johnson, editors. *Informal Logic: The First International Symposium*. Edgepress, Inverness, 1980.

J. A. Blair and R. H. Johnson. Misconceptions of informal logic: A reply to McPeck. *Teaching Philosophy*, 14:35–22, 1991.

J. A. Blair and R. H. Johnson. The logic of natural language. In J. McMurtry, editor, *Philosophy and World Problems*. Eolss Publishers, 2009.

S. Blum-Kulka. Rethinking genre: Discursive events as a social interactional phenomenon. In K. Fitch and R. Sanders, editors, *Handbook of language and social interaction*, pages 411–436. Lawrence Erlbaum, Mahwah, NJ, 2005.

A. Bondarenko, P. M. Dung, R. A. Kowlaski and F. Toni. An abstract argumentation theoretic framework for default reasoning. *Artificial Intelligence*, 93:63–101, 1997.

L. Boujour. *The structure of empirical knowledge*. Harvard University Press, Cambridge, MA, 1985.

M. Bratman. *Intentions, plans and practical reason*. Harvard University Press, 1987.

M. Bratman. Practical reasoning and acceptance in context. *Mind*, 101:1–15, 1992.

M. Bratman. *Faces of intention*. Cambridge University Press, 1999.

G. Brewka and T. F. Gordon. How to buy a Porsche: An approach to defeasible decision making. In *Working Notes of the AAAI-94 Workshop on Computational Dialectics*, pages 28–38. Seattle, Washington, 1994.

G. Brewka and S. Woltran. Abstract dialectical frameworks. In *Proceedings of the Twelfth International Conference on the Principles of Knowledge Representation and Reasoning*. in press, AAAI Press, 2010.

M. C. Bromby and M. J. J. Hall. The development and rapid evaluation of the knowledge model of advokate: An advisory system to assess the credibility of eyewitness testimony. In *Proceedings of Jurix 2002*, pages 143–52. 2002.

R. Bruner. *Deals from Hell. M&A Lessons that Rise Above the Ashes*. John WIley & Sons, 2005.

T. Buckles. *Laws of Evidence*. Delmar Cengage Learning, 2003.

R. Burchfield, editor. *Oxford English Dictionary*. Oxford University Press, Oxford, 1971.

E. Camp and J. Crowe. *The Encyclopdia of evidence*. L. D. Powell and Co, 1909.

J. D. Carney and R. K. Scheer. *Fundamentals of Logic*. The Macmillan Company, New York, 3rd edition, 1980.

S. Carrere and J. M. Gottman. Predicting divorce among newlyweds from the first three minutes of a marital conflict discussion. *Family Process*, 38:293–301, 1999.

S. Cartwright and C. L. Cooper. The role of culture compatibility in successful organizational marriage. *Academy of Management Executive*, 7(2):57–70, 1993.

C. Castelfranchi. Commitments: from individual intentions to groups and organizations. In V. Lesser, editor, *Proceedings of the 1st International Conferences on Multi-Agent Systems*, pages 528–535. AAAI Press, 1995.

C. Castelfranchi. Modelling social action for ai agents. *Artificial Intelligence*, 103:157–182, 1998.

C. Castelfranchi and F. Paglieri. The role of beliefs in global dynamics: prolegomena to a constructive theory of intentions. *Synthese*, 155:237–263, 2007.

C. Chesñevar, J. McGinnis, S. Modgil, I. Rahwan, C. Reed, G. Simari, M. South, G. Vreeswijk and S. Willmott. Towards an argument interchange format. *Knowledge Engineering Review*, 21(4):293–316, 2006.

S. Christopher-Guerra. Themen, thesen und argumente zur position des italienischen in der viersprachigen schweiz. *Studies in Communication Sciences*, 8(1):135–159, 2008.

Cicero. *On Invention*, volume 386 of *Loeb Classical Library*. Harvard University Press, Cambridge, 1949. Trans. H. M. Hubbell.

K. L. Clark. Negation as failure. In *Logic and Data Bases*, pages 293–322. 1977.

C. A. J. Coady. *Testimony: A Philosophical Study*. Clarendon Press, Oxford, 1992.

L. Cohen. *The probable and the provable*. Oxford University Press, Oxford, 1977.

L. J. Cohen. Belief and acceptance. *Mind*, 98:367–389, 1989.

P. Cohen and H. Levesque. Intention is choice with commitment. *Artificial Intelligence*, 42:213–261, 1990.

H. M. Collins and R. Evans. The third wave of science studies: Studies of expertise and experience. *Social Studies of Science*, 32(2):235–296, 2002.

R. Conte and C. Castelfranchi. *Cognitive and social action*. UCL Press, 1995.

I. Copi. *Symbolic Logic*. The Macmillan Company, New York, 1954.

I. Copi. *Introduction to Logic*. The Macmillan Company, New York, 6th edition, 1982.

B. Cottier and R. Palmieri. From trust to betrayal. *Studies in Communication Sciences*, 8(1):135–159, 2008.

J. Crosswhite, J. Fox, C. Reed, T. Scaltsas and T. Stumpf. Computational models of rhetorical argument. In Reed and Norman (2003), pages 175–209.

J. De Kleer. A general labeling algorithm for assumption-based truth maintenance. In *Proceedings of the 7th national conference on artificial intelligence*, pages 188–192. Morgan Kaufmanns Publishers, San Francisco, California, USA, 1988.

G. De la Dehesa. How to avoid further credit and liquidity confidence crises. In A. Felton and C. Reinhart, editors, *The First Global Financial Crisis of the 21st Century*, pages 151–154. CEPR, 2008.

F. Degeorge and E. Maug. Corporate finance in europe: A survey. In X. Freixas, P. Hartmann and C. Mayer, editors, *Financial Markets and Institutions: A European Perspective*. Oxford University Press, 2008.

H. Diels and W. Kranz. *Die Fragmente Der Vorsokratiker*. Weidmannsche Verlagsbuchhandlung, Berlin, 1952.

J. Doyle. A truth maintenance system. *Artificial Intelligence*, 12:231–272, 1979.

B. Due. *Antiphon: A Study in Argumentation*. Museum Tusculanum, Copenhagen, 1980.

P. M. Dung. On the acceptability of arguments and its fundamental role in nonmonotonic reasoning, logic programming and n-person games. *Artificial Intelligence*, 77(2):321–358, 1995.

P. E. Dunne, S. Doutre and T. J. M. Bench-Capon. Discovering inconsistency through examination dialogues. In *IJCAI-05 Proceedings*, pages 1680–1681. 2005.

R. K. Dybvig. *The Scheme Programming Language*. MIT Press, third edition edition, 2003.

G. Edelman. *Second Nature: Brain Science and Human Knowledge*. New Haven: Yale University Press, 2006.

F. H. van Eemeren. Democracy and argumentation. In D. Williams, editor, *Controversia Selected*. To be published, 2008.

F. H. van Eemeren. *Strategic maneuvering in argumentative discourse. Extending the pragma-dialectical theory of argumentation*. John Benjamins, Amsterdam-Philadelphia, 2010.

F. H. van Eemeren and B. Garssen. The fallacies of composition and division revisisted. *Cogency*, 1(1):23–42, 2009.

F. H. van Eemeren and R. Grootendorst. *Speech acts in argumentative discussions*. Foris, 1984a.

F. H. van Eemeren and R. Grootendorst. *Speech acts in argumentative discussions: A theoretical model for the analysis of discussions directed towards solving conflicts of opinion*. Foris/Walter de Gruyter, Dordrecht/Berlin, 1984b.

F. H. van Eemeren and R. Grootendorst. A transition stage in the theory of fallacies. *Journal of Pragmatics*, 13:99–109, 1989.

F. H. van Eemeren and R. Grootendorst. *Argumentation, communication and fallacies*. Lawrence Erlbaum Associates, 1992.

F. H. van Eemeren and R. Grootendorst. The fallacies of composition and division. In J. Gerbrandy, M. Marx, M. de Rijke and Y. Venema, editors, *Essays dedicated to Johan van Benthem on the occasion of his 50th birthday [CD ROM]*. Amsterdam University Press, 1999.

F. H. van Eemeren and R. Grootendorst. *A systematic theory of argumentation: The pragma-dialectical approach*. Cambridge University Press, Cambridge, 2004.

F. H. van Eemeren, R. Grootendorst, S. Jackson and S. Jacobs. *Reconstructing Argumentative Discourse*. The University of Alabama Press, 1993a.

F. H. van Eemeren, R. Grootendorst, S. Jackson and S. Jacobs. *Reconstructing argumentative discourse*. University of Alabama Press, Tuscaloosa, 1993b.

F. H. van Eemeren, R. Grootendorst and A. Snoek-Henkemans. *Argumentation: Analysis, evaluation, presentation*. Erlbaum, 2002.

F. H. van Eemeren, R. Grootendorst and F. Snoek Henkemans. *Fundamentals of Argumentation Theory*. LEA, 1996.

F. H. van Eemeren and P. Houtlosser. Strategic manoeuvring: Maintaining a delicate balance. In F. H. van Eemeren & P. Houtlosser, editor, *Dialectic and rhetoric: The warp and woof of argumentation analysis*, pages 131–159. Kluwer, Dordrecht, 2002.

F. H. van Eemeren and P. Houtlosser. Fallacies as derailments of strategic maneuvering: The argumentum ad verecundiam, a case in point. In C. A. W. . A. F. S. H. F. H. van Eemeren, J. A. Blair, editor, *Proceedings of the fifth conference of the International Society for the Study of Argumentation*, pages 289–292. SIC SAT, Amsterdam, 2003.

F. H. van Eemeren and P. Houtlosser. Theoretical construction and argumentative reality: An analytic model of critical discussion and conventionalised types of argumentative activity. In D. Hitchcock, editor, *The uses of argument: Proceedings of a conference at McMaster University*, pages 75–84. Hamilton, 2005.

F. H. van Eemeren and P. Houtlosser. The contextuality of fallacies. *Informal Logic*, 27(1):59–67, 2007a.

F. H. van Eemeren and P. Houtlosser. The study of argumentation as normative pragmatics. *Pragmatic & Cognition*, 15(1):161–177, 2007b.

K. M. Eisenhardt. Agency theory: An assessment and review. *The Academy of Management Review*, 14(1):57–74, 1989.

C. Z. Elgin. Non-foundationalist epistemology: Holism, coherence and tenability. In M. S. . E. Sosa, editor, *Contemporary debates in epistemology*, pages 156–167. Blackwell, Oxford, 2005.

J. Ellson, E. Gansner, L. Koutsofios, S. C. North and G. Woodhull. Graphviz — open source graph drawing tools. In *Proceedings of the 9th International Symposium on Graph Drawing (GD 2001)*, pages 483–484. Vienna, 2001.

P. Engel. Believing, holding true and accepting. *Philosophical Explorations*, 1:140–151, 1998.

R. L. Enos. Emerging notions of argument and advocacy in hellenic litigation: Antiphon's on the murder of herodes. *Journal of the American Forensic Association*, 16:182–191, 1980.

ESTRELLA Project. The legal knowledge interchange format (LKIF). Deliverable 4.1, European Commission, 2008.

D. W. Etherington and R. Reiter. On inheritance hierarchies and exceptions. In *Proceedings of the National Conference on Artificial Intelligence*. Los Altos, CA, 1983.

E. Fama. The behaviour of stock market prices. *Journal of business*, 38:34–105, 1967.

E. Fama. Efficient capital markets. a review of theory and empirical work. *The Journal of Finance*, 25(2):383–417, 1970.

P. Ferguson and F. Raitt. Reforming the scots law of rape: Redefining the offence. *Edinburgh Law Review*, 10(2):185–208, 2006.

T. S. Ferguson. Who solves the secretary problem? *Statistical Science*, 4:282–296, 1989.

R. Fikes. A commitment-based framework for describing informal cooperative work. *Cognitive Science*, 6:331–347, 1982.

M. Finocchiaro. Arguments, meta-arguments, and metadialogues: A reconstruction of Krabbe, Govier, and Woods. *Argumentation*, 21(3):253–268, 2007.

A. Fisher. *The Logic of Real Arguments*. Cambridge University Press, 1988.

R. J. Fogelin. *Understanding Arguments: An Introduction to Informal Logic*. Harcourt Brace Jovanovich, 4 edition, 1978.

B. van Fraassen. *The scientific image*. Oxford University Press, 1980.

D. M. Gabbay and J. Woods. Agenda relevance: A study in formal pragmatics. In *A Practical Logic of Cognitive Systems*, volume 1. Amsterdam: North Holland, 2003.

D. M. Gabbay and J. Woods. The reach of abduction: Insight and trial. In *A Practical logic of Cognitive Systems*, volume 2. Amsterdam: North Holland, 2005.

M. Gagarin, editor. *Antiphon The Speeches*. Cambridge University Press, Cambridge, 1997.

M. Gagarin. *Antiphon the Athenian: Oratory, Law, and Justice in the Age of the Sophists*. University of Texas Press, Austin, 2002.

B. Garssen. Argument schemes. In F. van Eemeren, editor, *Crucial Concepts in Argumentation Theory*, pages 81–99. Amsterdam University Press, Amsterdam, 2001.

J. Gasser. Social conceptions of knowledge and action: Dai foundations and open systems semantics. *Artificial Intelligence*, 47:107–138, 1991.

H. A. Geffner and J. Pearl. Conditional entailment: bridging two approaches to default reasoning. *Artificial Intelligence*, 53:209–244, 1992.

J. Gentzler. How to discriminate between experts and frauds: Some problems for socratic peirastic. *History of Philosophy Quarterly*, 12:227–246, 1995.

A. Gibbard. Morality and thick concepts. In *Proceedings of the Aristotelian Society*. 1992. Supplementary Volume.

G. Gigerenzer. I think before I err. *Social Research*, 1:195–217, 2005.

G. Gigerenzer and P. M. Todd. *Simple Heuristics That Make Us Smart*. Oxford University Press, 1999.

J. P. Gilbert and F. Mosteller. Recognising the maximum of a sequence. *American Statistical Association Journal*, 61:35–73, 1966.

M. Gilbert. *On social facts*. Routledge, 1989.

J. Y. Girard. Linear logic. *Theoretical Computer Science*, 50(1):1–101, 1987.

D. M. Godden. Editor's introduction to the special issue on Walton's work. *Informal Logic*, 27(1):1–4, 2007.

D. M. Godden and D. N. Walton. Advances in the theory of argumentation schemes and critical questions. *Informal Logic*, 27:267–292, 2007.

C. Goddu. What is a "real" argument? *Informal Logic*, 29:1:1–14, 2009a.

G. Goddu. Walton on argument structure. *Informal Logic*, 27:5–25, 2007.

G. Goddu. Against making the linked-convergent distinction. In F. van Eemeren & B. Garssen, editor, *Pondering on problems of argumentation*, pages 181–189. Springer, Dordrecht, 2009b.

P. Goldie. Seeing what is the kind thing to do: Perception and emotion in morality. *Dialectica*, 61:347–361, 2007.

M. Goldszmidt, P. Morris and J. Pearl. A maximum entropy approach to nonmonotonic reasoning. *IEEE Transaction on Pattern Analysis and Machine Intelligence*, 15:220–232, 1993.

J. Goodwin. Forms of authority and the real ad verecundiam. *Argumentation*, 12:267–280, 1998.

J. Goodwin. Cicero's authority. *Philosophy & Rhetoric*, 34:38–60, 2001.

J. Goodwin. The authority of wikipedia. In *Argument Cultures. Windsor*. OSSA, Hamilton, ONT, Forthcoming.

T. F. Gordon. *The Pleadings Game; An Artificial Intelligence Model of Procedural Justice*. Springer, New York, 1995. Book version of 1993 Ph.D. Thesis; University of Darmstadt.

T. F. Gordon. A computational model of argument for legal reasoning support systems. In P. E. Dunne and T. Bench-Capon, editors, *Argumentation in Artificial Intelligence and Law*, IAAIL Workshop Series, pages 53–64. Wolf Legal Publishers, Nijmegen, The Netherlands, 2005.

T. F. Gordon. Constructing arguments with a computational model of an argumentation scheme for legal rules. In *Proceedings of the Eleventh International Conference on Artificial Intelligence and Law*, pages 117–121. 2007a.

T. F. Gordon. Visualizing Carneades argument graphs. *Law, Probability and Risk*, 6(1-4):109–117, 2007b.

T. F. Gordon. Hybrid reasoning with argumentation schemes. In *Proceedings of the 8th Workshop on Computational Models of Natural Argument (CMNA 08)*, pages 16–25. The 18th European Conference on Artificial Intelligence (ECAI 2008), Patras, Greece, 2008.

T. F. Gordon, H. Prakken and D. N. Walton. The carneades model of argument and burden of proof. *Artificial Intelligence*, 171:875–896, 2007.

T. F. Gordon and D. N. Walton. The carneades argumentation framework. In P. Dunne and T. Bench-Capon, editors, *Computational Models of Argument: Proceedings of COMMA-2006*, pages 195–207. IOS Press, 2006.

T. F. Gordon and D. N. Walton. Legal reasoning with argumentation schemes. In C. D. Hafner, editor, *12th International Conference on Artificial Intelligence and Law (ICAIL 2009)*. ACM Press, New York, NY, USA, 2009a.

T. F. Gordon and D. N. Walton. Proof burdens and standards. In I. Rahwan and G. Simari, editors, *Argumentation in Artificial Intelligence*. Springer-Verlag, Berlin, Germany, 2009b.

T. Govier. *A Practical Study of Argument.* Wadsworth, 1985.

T. Govier. *Problems in Argument Analysis and Evaluation.* Foris, 1987.

T. Govier. *The Philosophy of Argument.* The Vale Press, 1999.

T. Govier. Emotion, relevance, and consolation arguments. In *Mistakes of Reason: Essays in Honour of John Woods.* University of Toronto Press, 2005a.

T. Govier. *A practical study of argument.* Thompson-Wadsworth, Toronto, 6th edition, 2005b.

W. S. Grennan. *Informal Logic.* McGill Queens University Press, 1997.

P. Grice. *Studies in the way of words.* Harvard University Press, 1989.

J. B. Grize. *De la logique á l'argumentation.* Librairie Droz, 1982.

B. N. Grosof, I. Horrocks, R. Volz and S. Decker. Description logic programs: Combining logic programs with description logics. In *Proceedings of the Twelfth International World Wide Web Conference (WWW 2003)*, pages 48–57. ACM, Budapest, Hungary, 2003.

J. J. Gumperz. Introduction. In J. J. G. . D. Hymes, editor, *Directions in sociolinguistics: The ethnography of communication*, pages 1–25. Holt, Rinehart and Winston, New York, 1972.

L. Gunnarson. The great apes and the severely disabled: Moral status and thick evaluative concepts. *Ethical Theory and Moral Practice*, 11:305–326, 2008.

J. C. Hage. *Reasoning with Rules – An Essay on Legal Reasoning and its Underlying Logic.* Kluwer Academic Publishers, Dordrecht, 1997.

U. Hahn and M. Oaksford. The rationality of informal argumentation: A bayesian approach to reasoning fallacies. *Psychological Review*, 114:704–732, 2007.

C. Hamblin. *Fallacies.* Methuen, 1970.

M. Hannibal and L. Mountford. *The Law of Criminal and Civil Evidence: Principles and Practice.* Longman, 2002.

R. Hardin. Conceptions and explanations of trust. In K. S. Cook, editor, *Trust in Society*, pages 3–39. Russell Sage Foundation, New York, NY, 2001.

G. Harman. *Change in View: Principles of Reasoning.* Cambridge, MA: MIT Press, 1986.

H. L. A. Hart. *The Concept of Law.* Clarendon Press, Oxford, 1961.

P. S. Hasper. Peirastic and pseudo-scientific arguments in aristotles *sophistical refutations*, 2009. 2009.

A. C. Hastings. *A Reformulation of the Modes of Reasoning in Argumentation.* Ph.D. thesis, Northwestern University', 1963.

G. A. Hauser. *Introduction to Rhetorical Theory.* Waveland Press, Prospect Heights, IL, 2nd edition, 2002.

A. F. S. Henkemans. *Analyzing Complex Argumentation. The reconstruction of multiple and coordinatively compound argumentation in a critical discussion.* Sic Sat, 1997.

A. F. S. Henkemans. State-of-the-art. the structure of argumentation. *Argumentation*, 14(4):447–473, 2000.

D. Hitchcock. Enthymematic arguments. *Informal Logic*, 7:83–97, 1985.

D. Hitchcock. Does the traditional treatment of enthymemes rest on a mistake? *Argumentation*, 12(1):15–37, 1998.

D. Hitchcock. Good reasoning on the Toulmin model. In D. Hitchcock and B. Verheij, editors, *Arguing on the Toulmin Model: New Essays in Argument Analysis and Evaluation*, chapter 12. Springer, Dordrecht, 2006.

D. Hitchcock. Informal logic and the concept of argument. In D. Jacquette, editor, *Philosophy of Logic*, volume 5 of *Handbook of the Philosophy of Science series, edited by Gabbay, D., Woods, J. and Thagard, P.* North-Holland Press, 2007a.

D. Hitchcock. On the generality of warrants, 2007b. Http://www.humanities.mcmaster.ca/ hitchckd/generality.pdf Last visited 19 February 2010.

M. Höpner and G. Jackson. An emerging market for corporate control? The Mannesmann takeover and German corporate governance. discussion paper 01/4. 2001.

M. Höpner and G. Jackson. Revisiting the Mannesmann takeover: how markets for corporate control emerge. *European Management Review*, 3:142–155, 2006.

J. F. Horty, R. H. Thomason and D. S. Touretsky. A sceptical theory of inheritance in nonmonotonic semantic networks. *Artificial Intelligence*, 42:311–348, 1990.

W. A. Howard. The formulas-as-types notion of construction. In J. Seldin and J. Hindley, editors, *To H. B. Curry: Essays on Combinatory Logic, Lambda Calculus, and Formalism*. Academic Press, 1980.

D. Hymes. The ethnography of speaking. In T. G. . W. C. Sturtevant, editor, *Anthropology and human behavior*, pages 13–53. Anthropolical Society of Washington, Washington DC, 1962.

D. Hymes. Models of the interaction of language and social life. In J. J. G. . D. Hymes, editor, *Directions in sociolinguistics: The ethnography of communication*, pages 35–71. Holt, Rinehart and Winston, New York, 1972.

D. C. Innes. Gorgias, Antiphon and Sophistopolis. *Argumentation*, 5:221–231, 1991.

S. Jackson. Building a case for claims about discourse structure. In D. G. E. . W. A. Donohue, editor, *Contemporary issues in language and discourse processes*, pages 129–147. Lawrence Erlbaum Associates, Hillsdale, NJ, 1986.

S. Jackson and S. Jacobs. Structure of conversational argument: Pragmatic bases for the enthymeme. *Quarterly Journal of Speech*, 66:251–265, 1980.

S. Jacobs. Messages, functional contexts, and categories of fallacy: Some dialectical and rhetorical considerations. In F. H. van Eemeren & P. Houtlosser, editor, *Dialectic and rhetoric: The warp and woof of argumentation analysis*, pages 119–130. Kluwer, Dordrecht, 2002.

N. Jennings. Commitments and conventions: the foundation of coordination in multi-agent systems. *The Knowledge Engineering Review*, 3:223–250, 1993.

R. H. Johnson. *Manifest Rationality*. Lawrence Erlbaum Associates, 2000.

R. H. Johnson. The alleged failure of informal logic. *Cogency*, 1.1 (Winter):59–88, 2009.

R. H. Johnson and J. A. Blair. The recent development of informal logic. In *Informal Logic: The First International Symposium*. Edgepress, Inverness, CA, 1980.

R. H. Johnson and J. A. Blair. The current state of informal logic. *Informal Logic*, 9:147–151, 1987.

R. H. Johnson and J. A. Blair. Informal logic: Past and present. In R. H. Johnson and J. A. Blair, editors, *New Essays in Informal Logic*. Informal Logic, 1994a.

R. H. Johnson and J. A. Blair. *Logical self-defense*. McGraw-Hill, Toronto, 3rd edition, 1994b.

R. H. Johnson and J. A. Blair. Informal logic and critical thinking. In van Eemeren et. al., editor, *Fundementals of Argumentation Theory*. Lawrence Erlbaum, Mahwah, NJ, 1996.

R. H. Johnson and J. A. Blair. Informal logic and the reconfiguration of logic. In D. Gabbay, R. H. Johnson, H.-J. Ohlbach and J. Woods, editors, *Handbook of the Logic of Argument and Inference: The Turn toward the Practical Reasoning*, pages 339–396. Elsevier, Amsterdam, 2002.

H. Kaptein. Rigid anarchic principles of evidence and proof. anomist panaceas against legal pathologies of proceduralism. In H. Kaptein, H. Prakken and B. Verheij, editors, *Legal Evidence and Proof: Statistics, Stories, Logic*, pages 195–221. Ashgate Publishing, 2009.

J. Katzav and C. Reed. On argumentation schemes and the natural classification of argument. *Argumentation*, 18(4):239–259, 2004.

F. J. Kauffeld and J. Fields. The commitments speakers undertake in giving testimony. In D. L. H. . D. Farr, editor, *The Uses of Argument*. OSSA, Hamilton, ONT, 2005.

A. Keane. The modern law of evidence. *Law, Probability & Risk*, 3:33–50, 2008.

G. Kennedy. *The Art of Persuasion in Greece*. Princeton University Press, Princeton, NJ, 1963.

M. Kienpointner. *Alltagslogik: Struktur und Funktion von Argumentationsmustern*. Frommann Holzboog, Stuttgart-Bad Canstatt, 1992a.

M. Kienpointner. How to classify arguments. In F. van Eemeren, R. Grootendorst, J. Blair and C. Willard, editors, *Argumentation Illuminated*, chapter 15, pages 178–188. SICSAT, 1992b.

I. Kirzner. *Competition and entrepreneurship*. University of Chicago Press, 1973.

D. B. Klein. *Assurance and Trust in a Great Society*. Foundation for Economic Education, Irving-on-Hudson, NY, 2000.

C. Kock. Is practical reasoning presumptive. *Informal Logic*, 27(1):91–108, 2007.

C. Kock. Arguing for different types of speech acts, 2009. Paper presented at the OSSA Conference, June 2009, Windsor, Ontario.

K. Konolige. Autoepistemic logic. In D. Gabbay, C. Hogger and J. Robinson, editors, *Handbook of Logic in Artificial Intelligence and Logic Programming*, volume III. Oxford: Clarendon Press, 1994.

C. Korsgaard. Realism and constructivism in twentieth century moral philosophy. *Journal of Philisophical Research*, pages 99–122, 2003.

J. Kovesi. *Moral Notions*. Routledge and Kegan Paul, 1967.

E. C. Krabbe and J. A. van Laar. About old and new dialectic: Dialogues, fallacies,
and strategies. *Informal Logic*, 27(1):27–58, 2007.

E. C. W. Krabbe. Metadialogues. In C. A. W. F. H. van Eemeren, J. A. Blair and
A. F. S. Henkemans, editors, *Anyone Who Has a View: Theoretical Contributions
to the Study of Argumentation*, pages 83–90. Kluwer, 2003.

M. Kraus. Early Greek probability arguments and common ground in dissensus.
In e. a. Hans V. Hansen, editor, *Dissensus and the Search for Common Ground*.
OSSA, Windsor, Ontario, 2007.

A. Krautz. The logic of persistence. In *Proceedings of the Fifth National Conference
on Artificial Intelligence*, pages 401–405. Philadelphia, PA: AAAI-86, 1986.

M. Krifka, F. J. Pelletier, G. N. Carlson, A. ter Meulen, G. Link and G. Chierchia.
Genericity: An introduction. In G. N. Carlson and F. J. Pelletier, editors, *The
Generic Book*. Chicago, IL: The University of Chicago Press, 1995.

C. H. Langford. The institutional use of 'the'. *Philosophy and Phenomenological
Research*, 10:115–120, 1949.

B. Latour and S. Woolgar. *Laboratory Life: The Construction of Scientific Facts*.
Princeton University Press, Princeton, NJ, 1979.

M. Lauritsen and T. F. Gordon. Toward a general theory of document modeling.
In C. D. Hafner, editor, *12th International Conference on Artificial Intelligence and
Law (ICAIL 2009)*. ACM Press, New York, NY, USA, 2009.

W. Lazonick. *Business Organization and the Myth of the Market Econimy*. Cambridge
University Press, 1991.

D. Lehmann and M. Magidor. What does a conditional knowledge base entail?
Artificial Intelligence, 55:1–60, 1992.

R. Lempert, editor. *A Modern Approach to Evidence*. West Publishing, 3 edition,
2000.

S. C. Levinson. Activity types and language. In P. D. . J. Heritage, editor, *Talk
at work. Interaction in institutional settings*, pages 66–100. Cambridge University
Press, Cambridge, 1992.

C. I. Lewis. *An analysis of knowledge and valuation*. Open Court, LaSalle, IL, 1946.

D. Lewis. Adverbs of quantification. In E. L. Keenan, editor, *Formal Semantics of
Natural Language*. Cambridge University Press, 1975.

V. Lifschitz. Circumscriptive theories: A logic-based framework for knowledge
representation. *Journal of Philosophical Logic*, 17:391–441, 1988.

V. Lifschitz. Minimal belief and negation as failure. *Artificial Intelligence*, 70(1-
2):53–72, 1994.

R. Loui. Process and policy: resource-bounded non-demonstrative reasoning.
Computational Intelligence, 14:1–38, 1998.

F. Macagno and D. N. Walton. Persuasive definitions, values, meaning, and implicit disagreements. *Informal Logic*, 28:203–228, 2008.

A. Macintosh, T. F. Gordon and A. Renton. Providing argument support for e-participation. *Journal of Information Technology & Politics*, 6(1):43–59, 2009.

L. Magnani. Abduction, affordances and cognitive niches, 2010. Forthcoming.

N. G. Mankiw, R. D. Kneebone, K. J. Mackenzie and N. Rowe. *Principles of Microeconomics*. T. Nelson, Toronto, 2006.

C. C. Marshall. Representing the structure of a legal argument. In *The Second International Conference on Artificial Intelligence and Law*, pages 121–27. ACM Press, 1989.

G. Martini, E. Rigotti and R. Palmieri. Teoria economica e razionalitá. *Atlantide*, 14:73–78, 2008.

P. McBurney and S. Parsons. Games that agents play: A formal framework for dialogues between autonomous agents. *Journal of Logic, Language and Information*, 11(3):315–334, 2002.

J. McCarthy. Circumspection a form of non-monotonic reasoning. *Artificial Intelligence*, 13:23–79, 1980.

J. McCarthy. Applications of circumscription to formalizing common sense knowledge. *Artificial Intelligence*, 26:89–116, 1986.

D. McDermott and J. Doyle. Non-monotonic logic i. *Artificial Intelligence*, 13:41–72, 1982.

D. L. McGuinness and F. van Harmelen. OWL Web Ontology Language overview. http://www.w3.org/TR/owl-features/, 2004.

M. Midgley. *Heart and Mind: The Varieties of Moral Experience*. Routledge, 1981.

M. Minsky. A framework for representing knowledge. In P. Winston, editor, *The Psychology of Computer Vision*. New York: McGraw-Hill, 1975.

B. M. Mitnick. Agency, theory of. In R. W. Kolb, editor, *Encyclopedia of Business Ethics and Society*, volume 1, pages 42–48. Sage Publications, Los Angeles, CA, 2008.

S. Modgil and J. McGinnis. Towards characterising argumentation based dialogue in the argument interchange format. In *Proceedings of the 4th International Workshop on Argumentation in Multi-Agent Systems (ArgMAS2007)*. Springer Verlag, 2008. To appear.

T. M. Moe. The new economics of organization. *American Journal of Political Science*, 28(4):739–777, 1984.

R. C. Moore. Semantic considerations on nonmonotonic logic. *Artificial Intelligence*, 25:75–94, 1985.

C. Morgan. The nature of nonmonotonic reasoning. *Minds and Machines*, 10:321–360, 2000.

R. Munson. *The Way of Words: An Informal Logic*. Houghton Mifflin, 1976.

P. Murphy. *Murphy on Evidence*. Oxford University Press, 2007.

S. Nielsen and S. Parsons. An application of formal argumentation: fusing Bayesian networks in multi-agent systems. *Artifical Intelligence*, 171:754–775, 2007.

E. Nowak. Recent developments in german capital markets and corporate governance. *Journal of Applied Corporate Finance*, 14(3):35–48, 2001.

D. Nute. Defeasible logic. In D. Gabbay, C. Hogger and J. Robinson, editors, *Handbook of Logic in Artificial Intelligence and Logic Programming*, pages 253–395. Clarendon Press, Oxford, 1994.

D. O'Keefe. Two concepts of argument. *Journal of the American Forensic Association*, 13:121–128, 1977.

D. J. O'Keefe. *Persuasion, Theory and Research*. Sage Publications, Thousand Oaks, CA / London / New Delhi, 2nd edition, 2002.

C. Oldenburg and M. Leff. Argument by anecdote. In H. Hansen, C. Tindale, R. Johnson and J. Blair, editors, *Argument Cultures. Proceedings of the OSSA 2009 Conference*. 2009.

N. O'Sullivan. Written and spoken in the first sophistic. In I. Worthington, editor, *Voice into Text: Orality and Literacy in Ancient Greece*, pages 115–127. E.J. Brill, Leiden, 1996.

F. Paglieri. No more charity, please! enthymematic parsimony and the pitfall of benevolence. In H. Hansen, C. Tindale, R. Johnson and J. Blair, editors, *Dissensus and the search for common ground*, pages 1–26. OSSA, CD-ROM, 2007.

F. Paglieri. Acceptance as conditional disposition. In A. Hieke and H. Leitgeb, editors, *Reduction: between the mind and the brain*. Ontos Verlag, 2009a.

F. Paglieri. Ruinous arguments: escalation of disagreement and the dangers of arguing. In H. Hansen, C. Tindale, R. Johnson and J. Blair, editors, *Argument cultures*. OSSA, CD-ROM, 2009b.

F. Paglieri and C. Castelfranchi. The Toulmin test: framing argumentation within belief revision theories. In D. Hitchcock and B. Verheij, editors, *Arguing on the Toulmin model. New essays in argument analysis and evaluation*, pages 359–377. Springer, 2006.

F. Paglieri and C. Castelfranchi. Why arguing? towards a cost-benefit analysis of argumentation. *Argument & Computation*, 1:71–91, 2010.

F. Paglieri and J. Woods. Enthymematic parsimony. *Synthese*, 2009. In press (doi 10.1007/s11229-009-9652-3).

C. Palmieri. *An integration of communication and financial models for the purpose of discovering manipulation: a qualitative application to Enron case*. Master's thesis, USI, 2007.

R. Palmieri. *The case of the 4you financial product: argumentative and legal aspects*. Master's thesis, USI, 2006.

R. Palmieri. Argumentative dialogues in mergers & acquisitions (m&as): evidence from investors and analysts conference calls. *Word meaning in argumentative dialogue*, 2:859–872, 2008a. Special issue of Lanalisi linguistica e letteraria XVI.

R. Palmieri. Reconstructing argumentative interactions in m&a offers. *Studies in Communication Sciences*, 8(2):279–302, 2008b.

R. Palmieri. Regaining trust through argumentation in the context of the current financial-economic crisis. *Studies in Communication Sciences*, 9(2):59–78, 2009.

R. Park, D. Leonard and S. Goldberg. *Evidence Law*. West Group, 1998.

P. F. Patel-Schneider and B. Swartout. Description logic knowledge representation system specification from the KRSS group of the ARPA knowledge sharing effort. Report, AT&T Bell Laboratories, Murray Hill, New Jersey, 1994.

J. Pearl. System z: A natural ordering of defaults with tractable applications to default reasoning. In R. Parikh, editor, *Proceedings of the Third Conference on Theoretical Aspects of Reasoning about Knowledge*. Morgan Kaufmann, 1988.

J. Pearl. System z: A natural ordering of defaults with tractable applications to default reasoning. In R. Parikh, editor, *Proceedings of the Third Conference of Theoretical Aspects of Reasoning about Knowledge*. San Mateo, CA: Morgan Kaufmann, 1990.

F. J. Pelletier, editor. *Kinds, Things and Stuff: Mass Terms and Generics*. New York: Oxford University Press, 2010.

G. J. Pendrick. *Antiphon the Sophist: The Fragments*. Cambridge University press, Cambridge, 2002.

C. Perelman. *The Idea of Justice and the Problem of Argument*. Routledge and Kegan Paul, 1963.

C. Perelman and L. Olbrechts-Tyteca. *Traité de l'argumentation : la nouvelle rhétorique*. Presses universitaires de France, Paris, 1958.

C. Perelman and L. Olbrechts-Tyteca. *The New Rhetoric: A Treatise on Argumentation*. University of Notre Dame Press, 1969.

K. Pike. *Language in relation to a unified theory of the structure of human behaviour*. Mouton, The Hague, 1967.

R. Pinto. Argument schemes and the evaluation of presumptive reasoning. In H. Hansen, editor, *Argument, inference and dialectic*, pages 98–104. Kluwer, Dordrecht, 2001.

R. C. Pinto. Evaluating inferences: The nature and role of warrants. In D. Hitchcock and B. Verheij, editors, *Arguing on the Toulmin Model: New Essays in Argument Analysis and Evaluation*, chapter 9. Springer, Dordrecht, 2006.

Plato. *Plato: Complete Works*. Hackett, Indianapolis, 1997.

J. Pollock. *Cognitive Carpentry*. MIT Press, 1995.

J. L. Pollock. *Contemporary theories of knowledge*. Rowan & Littlefied, Savage, MD, 1986.

J. L. Pollock. Defeasible reasoning. *Cognitive Science*, 11:481–518, 1987.

J. L. Pollock. Defeasible reasoning. In J. Adler and L. Rips, editors, *Reasoning: Studies in Human Influence and its Foundations*, pages 451–471. Cambridge and New York: Cambridge University Press, 2008.

J. Poulakos. The logic of greek sophistry. In D. Walton and A. Brinton, editors, *Historical Foundations of Informal Logic*. Ashgate, 1997.

H. Prakken. From logic to dialectic in legal argument. In *Proceedings of the Fifth International Conference on Artificial Intelligence and Law*, pages 165–174. Maryland, 1995.

H. Prakken. Analysing reasoning about evidence with formal models of argumentation. *Law, Probability and Risk*, 3:33–50, 2004.

H. Prakken. Coherence and flexibility in dialogue games for argumentation. *Journal of Logic and Computation*, 15:1009–1040, 2005.

H. Prakken. Combining modes of reasoning: An application of abstract argumentation. In *Proceedings of The 11th European Conference on Logics in Artificial Intelligence (JELIA 2008)*, volume 5293 of *Springer Lecture Notes in AI*, pages 349–361. Springer Verlag, Berlin, 2008a.

H. Prakken. A formal model of adjudication dialogues. *Artifical Intelligence and Law*, 16:305–328, 2008b.

H. Prakken. An abstract framework for argumentation with structured arguments. *Argument & Computation*, 1(2), 2010.

H. Prakken, C. Reed and D. N. Walton. Dialogues about the burden of proof. In *Proceedings of the 10th International Conference on AI & Law (ICAIL-05)*, pages 115–124. ACM Press, 2005.

H. Prakken and G. Sartor. A dialectical model of assessing conflicting argument in legal reasoning. *Artificial Intelligence and Law*, 4(3-4):331–368, 1996.

H. Prakken and G. Sartor. Presumptions and burdens of proof. In T. van Engers , editor, *Legal Knowledge and Information Systems. JURIX 2006: The Nineteenth Annual Conference*, pages 21–30. IOS Press, 2006.

H. Prakken and G. Sartor. A logical analysis of burdens of proof. In H. Kaptein, H. Prakken and B. Verheij, editors, *Legal Evidence and Proof: Statistics, Stories, Logic*, Applied Legal Philosophy Series, pages 223–253. Ashgate Publishing, 2009.

H. Prakken and G. Vreeswijk. Logics for defeasible argumentation. In D. Gabbay and F. Guenthner, editors, *Handbook of Philosophical Logic*, volume 4, pages 218–319. Kluwer, 2nd edition, 2002.

H. Putnam. *The Collapse of the Fact/Value Dichotomy and Other Essays*. Harvard University Press, 2002.

W. V. Quine. Natural kinds. In *Ontological Relativity and Other Essays*, pages 114–138. New York: Columbia University Press, 1969.

I. Rahwan and G. Simari, editors. *Argumentation in Artificial Intelligence*. Springer, 2009.

I. Rahwan, F. Zablith and C. Reed. Laying the foundations for a world wide argument web. *Artificial Intelligence*, 171:897–921, 2007.

V. Ramsey. *Construction Law Handbook*. Thomas Telford Limited, 2007.

A. Rao and M. Georgeff. Modeling rational agents within a bdi-architecture. In J. Allen, R. Fikes and E. Sandewall, editors, *Principles of knowledge representation and reasoning: proceedings of the second international conference (KR91)*, pages 463–484. Morgan Kaufmann, 1991.

A. Rao, M. Georgeff and E. Sonenberg. Social plans: a preliminary report. In E. Werner and Y. Demazeau, editors, *Decentralized Artificial Intelligence III*. Elsevier, 1992.

M. Redmayne. A corroboration approach to recovered memories of sexual abuse: A note of caution. *Law Quarterly Review*, 116:147–155, 2000.

C. Reed. Preliminary results from an argument corpus. In *Proceedings of the IX Symposium on Social Communication*, pages 576–580. 2005.

C. Reed. Representing dialogic argumentation. *Knowledge Based Systems*, 19(1):22–31, 2006.

C. Reed and T. J. Norman, editors. *Argumentation Machines: New Frontiers in Argument and Computation*. Kluwer, 2003.

C. Reed and G. W. A. Rowe. Araucaria: Software for argument analysis, diagramming and representation. *International Journal of AI Tools*, 14(3-4):961–980, 2004.

C. Reed and D. N. Walton. Towards a formal and implemented model of argumentation schemes in agent communication. *Autonomous Agents and Multi-Agent Systems*, 11(2):173–188, 2005.

C. Reed and D. N. Walton. Argumentation schemes in dialogue. In H. Hansen, C. Tindale, R. Johnson and J. Blair, editors, *Dissensus and the Search for Common Ground (Proceedings of OSSA 2007)*. 2007.

C. Reed, S. Wells, J. Devereux and G. Rowe. Aif+: Dialogue in the argument interchange format. In P. Besnard, S. Doutre and A. Hunter, editors, *Computational Models of Argument: Proceedings of COMMA-2008*, pages 311–323. IOS Press, 2008.

R. Reiter. A logic for default reasoning. *Artificial Intelligence*, 12:81–132, 1980.

N. Rescher. *Plausible Reasoning: An Introduction to the Theory and Practice of Plausibilistic Inference*. Van Gorcum, Amsterdam, 1976.

N. Rescher. *Error: Our Predicament When Things Go Wrong*. Pittsburgh: University of Pittsburgh Press, 2007.

L. Resnick. *Education and Learning to Think*. National Academy Press, 1987.

E. Rigotti. Zur rolle der pístis in der kommunikation. In S. C. et al., editor, *Dialoganalyse VI, Referate der 6*. Niemeyer, 1998.

E. Rigotti. Relevance of context-bound loci to topical potential in the argumentation stage. *Argumentation*, 20(4):519–540, 2006.

E. Rigotti. Locus a causa finali. *Word meaning in argumentative dialogue*, 2:559–576, 2008. Special issue of Lanalisi linguistica e letteraria XVI.

E. Rigotti. Whether and how classical topics can be revived in the contemporary theory of argumentation. In F. H. van Eemeren and B. J. Garssen, editors, *Pondering on problems of argumentation*, pages 157–178. Springer, 2009.

E. Rigotti and S. G. Morasso. Preliminaries of financial argumentation. argumentation for financial communication. argumentum elearning module. 2007.

E. Rigotti and S. G. Morasso. Argumentation as an object of interest and as a social and cultural resource. In N. Muller-Mirza and A. Perret-Clermont, editors, *Argumentation and Education*, pages 9–66. Springer, 2009.

E. Rigotti and S. G. Morasso. Comparing the argumentum-model of topics with other contemporary approaches to argument schemes; the procedural and the material components, (to appear). Forthcoming.

E. Rigotti and R. Palmieri. Argumentative interaction in corporate delibera-
tions: The case of carl icahn and motorola, 2008. Paper presented at the VIII
Amsterdam-Lugano Colloquium on Argumentation Theory, Lugano: Novem-
ber 28, 2008.

E. Rigotti, A. Rocci and S. G. Morasso. The semantics of reasonableness. In
Considering Pragma-Dialectics, pages 257–274. Lawrence Erlbaum Associates,
2006. Contribution to the Festschrift for Frans H. van Eemeren on the Occasion
of his 60th Birthday.

E. Rissland. Dimension-based analysis of hypotheticals from supreme court oral
argument. In *The Second International Conference on Artificial Intelligence and
Law*, pages 111–120. ACM Press, 1989.

D. Robertson. A lightweight coordination calculus for agent systems. In *Declar-
ative Agent Languages and Technologies II*, LNCS 3476, pages 183–197. Springer
Verlag, 2005.

A. Rocci. Manoeuvring with voices: the polyphonic framing of arguments in
an institutional advertisement. In *Examining argumentation in context. Fifteen
studies on strategic manoeuvring*, pages 257–283. John Benjamins, 2009.

A. Rocci and R. Palmieri. Financial news between narrative and predictions.
paper presented at the x ipra conference, 2007. Paper presented at the X Ipra
Conference. Göteborg: July 10, 2007.

E. Rosch. Principles of categorization. In E. Rosch and B. Lloyd, editors, *Cognition
and Categorization*, pages 27–48. Erlbaum, 1978.

P. E. Ross. The expert mind. *Scientific American*, pages 64–71, 2006.

S. A. Ross. Forensic finance. enron and others. fourth angelo costa lecture. *Rivista
Di Politica Economica*, 92(11-12):9–27, 2002.

G. W. A. Rowe, F. Macagno, C. Reed and D. N. Walton. Araucaria as a tool
for diagramming arguments in teaching and studying philosophy. *Teaching
Philosophy*, 29(2):111–124, 2006.

F. Schauer. *Profiles, Probabilities and Stereotypes*. Harvard University Press, 2003.

P. J. Schellens. *Redelijke Argumenten: Een Onderzoek naar Normen voor Kritische
Lezers*. Foris, Dordrecht, 1985.

D. Schum. *The Evidential Foundations of Probabalistic Reasoning*. Northwestern
University Press, Evanston, IL, 1994.

M. Scriven. Truisms as the grounds for historical explanations. In P. Gardiner,
editor, *Theories of History*, pages 443–475. Glencoe, IL: Free Press, 1959.

M. Scriven. *Reasoning*. McGraw Hill, 1976.

M. Scriven. The philosophical and pragmatic significance of informal logic. In
Informal Logic: The First International Symposium. 1980.

J. R. Searle. *Speech Acts: An Essay in the Philosophy of Language*. Cambridge University Press, 1969.

J. R. Searle. *The construction of social reality*. Free Press, 1995.

J. R. Searle and D. Vanderveken. *Foundations of Illocutionary Logic*. Cambridge
University Press, 1985.

M. Sergot. A query the user facitiy for logic programs. In M. Yazdani, editor, *New Horizons in Educational Computing*, pages 145–163. Ellis Horwood, Chichester, 1984.

S. P. Shapiro. The social control of impersonal trust. *American Journal of Sociology*, 93(3):623–658, 1987.

S. P. Shapiro. Agency theory. *Annual Review of Sociology*, 31:263–284, 2005.

A. Sharma. Professional as agent: Knowledge asymmetry in agency exchange. *Academy of Management Review*, 22(3):758–796, 1997.

D. Sharma. Epistemological negative dialectics of indian logic — abhava versus anupalabdhi. *Indo-Iranian Journal*, pages 291–300, 2004.

G. Sher. *The Bounds of Logic*. Cambridge, MA: MIT Press, 1991.

A. Shleifer. *Inefficient Markets: An Introduction to Behavioral Finance*. Oxford University Press, 2000.

Y. Shoham. *A Semantical Approach to Nonmonotonic Logics*. San Mateo, CA: Morgan Kaufmann, 1987.

G. Simari, C. Chesñevar and A. Garcia. The role of dialectics in defeasible argumentation. In *Proceedings of the XIV International Conference of the Chilean Computer Science Society*. 1994.

D. B. Skalak and E. L. Rissland. Arguments and cases: An inevitable intertwining. *Aritificial Intelligence and Law*, 1(1):3–45, 1992.

E. E. Smith and D. L. Medin. *Categories and Concepts*. Cambridge, MA: Harvard University Press, 1981.

R. G. Smith. The contract net protocol: High-level communication and control in a distributed problem solver. *IEEE Transactions on Computers*, C-29(12):1104–1113, 1980.

I. Snehota. Perspectives and theories of market. In D. H. H. and A. Waluszewski, editors, *Rethinking Marketing. Developing a new understanding of markets*. John Wiley & Sons, 2004.

F. Solmsen. The Aristotelian tradition in ancient rhetoric. In R. Stark, editor, *Rhetorika: Schriften zur aristotelischen und hellenistischen Rhetorik*, pages 312–349. Georg Olms Verlagbuchhandlung, Hildesheim, 1968.

D. Sperber and D. Wilson. *Relevance: communication and cognition*. Blackwell, 1986.

R. K. Sprague. *The Older Sophists*. University of South Carolina Press, Columbia, SC, 1972.

R. Stalnaker. *Inquiry*. MIT Press, 1984.

M. Strevens. *Depth*. Cambridge, MA: Harvard University Press, 2008.

P. Suppes. A comparison of the meaning and uses of models in mathematics and the empirical sciences. *Synthese*, 12:287–301, 1960.

C. Tapper. *Cross and Tapper on Evidence*. Oxford University Press, 2007.

C. Tappolet. Through thick and thin: Good and its determinates. *Dialectica*, 58:207–223, 2004.

J. Temkin. *Rape and the Legal Process*. Oxford University Press, 2002.

C. W. Tindale. Fallacies, blunders and dialogue shifts: Walton's contributions to the fallacy debate. *Argumentation*, 11:341–354, 1997.

P. Tolchinsky, S. Modgil, U. Cortes and M. Sanchez-Marre. Cbr and argument schemes for collaborative decision making. In P. Dunne and T. Bench-Capon, editors, *Computational Models of Argument (Proceedings of COMMA 2006)*, pages 171–182. IOS Press, 2006.

D. Tollefsen. Collective intentionality and the social sciences. *Philosophy of the Social Sciences*, 32:25–50, 2002.

S. E. Toulmin. *The Uses of Argument*. Cambridge University Press, 1958.

D. S. Touretsky. *The Mathematics of Inheritence Systems*. Los Altos: Morgan Kaufmann, 1986.

L. Tummolini and C. Castelfranchi. The cognitive and behavioral mediation of institutions: Towards an account of institutional actions. *Cognitive Systems Research*, 7:307–323, 2006.

R. Tuomela. *The importance of us*. Stanford University Press, 1995.

R. Tuomela. Belief versus acceptance. *Philisophical Explorations*, 2:122–137, 2000.

R. Tuomela. *The philosophy of social practices*. Cambridge University Press, 2002.

S. P. Turner. What is the problem with experts? *Social Studies of Science*, 31(1):123–149, 2001.

S. Usher. *Greek Oratory: Tradition and Originality*. Oxford University Press, Oxford, 1999.

D. Velleman. How to share an intention. *Philosophy and Phenomenological Research*, 57:29–50, 1997.

B. Verheij. *Rules, Reasons, Arguments. Formal Studies of Argumentation and Defeat*. Ph.d., Universiteit Maastricht, 1996.

B. Verheij. Dialectical argumentation with argumentation schemes: An approach to legal logic. *Artificial Intelligence and Law*, 11(2):167–195, 2003a.

B. Verheij. Dialectical argumentation with argumentation schemes: Towards a methodology for the investigation of argumentation schemes. In *Proceedings of the Fifth International Conference on Argumentation (ISSA 2002)*, pages 1033–1037. SicSat, 2003b.

B. Verheij. *Virtual Arguments: On the Design of Argument Assistants for Lawyers and Other Arguers*. T.M.C. Asser Press, 2005.

B. Verheij. Evaluating arguments based on Toulmin's scheme. In D. Hitchcock and B. Verheij, editors, *Arguing on the Toulmin Model: New Essays in Argument Analysis and Evaluation*, chapter 12. Springer, Dordrecht, 2006.

G. Vreeswijk. Abstract argumentation systems. *Artificial Intelligence*, 90(1-2):225–279, 1997.

W3C. Web service choreography interface (WSCI) 1.0. 2002.

D. N. Walton. *Ethics of Withdrawal of Life Support Systems*. Greenwood Press, 1983.

D. N. Walton. *Arguer's Position: A Pragmatic Study of Ad Hominem Attack, Criticism, Refutation and Fallacy.* Greenwood Presses universitaires de France, 1985a.

D. N. Walton. *Physician-Patient Decision-Making.* Greenwood Press, 1985b.

D. N. Walton. *Informal Fallacies: Towards a Theory of Argument Criticisms.* John Benjamins, 1987.

D. N. Walton. Burden of proof. *Argumentation,* 2:233–254, 1988.

D. N. Walton. *Informal Logic: A Handbook for Critical Argumentation.* Cambridge University Press, Cambridge, 1989.

D. N. Walton. *Practical reasoning: goal-driven, knowledge-based, action-guiding argumentation.* Rowman and Littlefield, 1990a.

D. N. Walton. What is reasoning? what is an argument? *Journal of Philosophy,* 87:399–419, 1990b.

D. N. Walton. *Begging the Question: Circular Reasoning as a Tactic of Argumentation.* Greenwood, 1991a.

D. N. Walton. Hamblin and the standard treament of fallacies. *Philosophy and Rhetoric,* 24:353–361, 1991b.

D. N. Walton. *Arguments from Ignorance.* The Pennsylvania University Press, 1992a.

D. N. Walton. *The Place of Emotions in Argument.* Pennsylvania State University Press, 1992b.

D. N. Walton. *Plausible Argument in Everyday Conversation.* Albany: State of New York Press, 1992c.

D. N. Walton. Types of dialogue, dialectical shifts and fallacies. In F. H. van Eemeren et al., editor, *Argumentation illuminated,* pages 133–147. SicSat, Amsterdam, 1992d.

D. N. Walton. The speech act of presumption. *Pragmatics & Cognition,* 1:125–148, 1993.

D. N. Walton. *A Pragmatic Theory of Fallacy.* University of Alabama Press, 1995.

D. N. Walton. *Argument Structure: A Pragmatic Theory.* University of Toronto Press, 1996a.

D. N. Walton. *Argumentation Schemes for Presumptive Reasoning.* Lawrence Erlbaum Associates, 1996b.

D. N. Walton. *Arguments from ignorance.* Pennsylvania State University Press, 1996c.

D. N. Walton. *Appeal to Expert Opinion: Arguments from Authority.* The Pennsylvania State University Press, University Park, PA, 1997a.

D. N. Walton. *Appeal to pity.* State University of New York Press, 1997b.

D. N. Walton. How can logic best be applied to arguments? *Logic Journal of the International Conference in Logic Programming,* 5(4):603–614, 1997c.

D. N. Walton. *Ad hominem arguments.* University of Alabama Press, 1998a.

D. N. Walton. *The New Dialectic*. University of Toronto Press, 1998b.

D. N. Walton. *Appeal to popular opinion*. Pennsylvania State University Press, 1999a.

D. N. Walton. Applying labelled deductive systems and multi-agent systems to source-based argumentation. *Journal of Logic and Computation*, 9:63–80, 1999b.

D. N. Walton. Case study of the use of a circumstantial ad hominem in political argumentation. *Philosophy and Rhetoric*, 33:101–115, 2000a.

D. N. Walton. The place of dialogue theory in logic, computer science and communication studies. *Synthese: An International Journal for Epistemology, Logic and Philosophy of Science*, 123:327–346, 2000b.

D. N. Walton. *Scare tactics*. Dordrecht Kluwer, 2000c.

D. N. Walton. Abductive, presumptive and plausible arguments. *Informal Logic*, 21:141–169, 2001a.

D. N. Walton. Definitions and public policy arguments. *Argumentation and Advocacy*, 37:117–132, 2001b.

D. N. Walton. Enthymemes, common knowledge, and plausible inference. *Philosophy and Rhetoric*, 34:93–112, 2001c.

D. N. Walton. *Legal Argumentation and Evidence*. Penn State Press, 2002a.

D. N. Walton. The sunk costs fallacy or argument from waste. *Argumentation*, 16:473–503, 2002b.

D. N. Walton. The interrogation as a type of dialogue. *Journal of Pragmatics*, 35:1771–1802, 2003.

D. N. Walton. *Abductive Reasoning*. Tuscaloosa: University of Alabama Press, 2004.

D. N. Walton. *Argumentation methods for artificial intelligence in law*. Springer, 2005a.

D. N. Walton. An automated system for argument invention in law using argumentation and heuristic search procedures. *Ratio Juris*, 18:434–463, 2005b.

D. N. Walton. Deceptive arguments containing persuasive language and persuasive definitions. *Argumentation*, 19:159–186, 2005c.

D. N. Walton. How to evaluate argumentation using schemes, diagrams, critical questions and dialogues. *Studies in Communication Sciences, Argumentation in Dialogic Interaction*, Special Issue:51–74, 2005d.

D. N. Walton. How to evaluate argumentation using schemes, diagrams, critical questions and dialogues. *Argumentation in dialogic interaction*, pages 51–74, 2005e. Special issue of Studies in Communication Sciences.

D. N. Walton. Justification of argumentation schemes. *The Australasian Journal of Logic*, 3:1–13, 2005f.

D. N. Walton. Practical reasoning and proposing: Tools for e-democracy. In M. F. Morens and S. P, editors, *Legal Knowledge and Information Systems*, pages 113–114. IOS Press, 2005g.

D. N. Walton. Argument from appearance: A new argumentation scheme. *Logique et Analyse*, 195:319–340, 2006a.

D. N. Walton. Examination dialogue: An argumentation framework for critically questioning an expert opinion. *Journal of Pragmatics*, 38:745–777, 2006b.

D. N. Walton. *Fundamentals of Critical Argumentation*. Cambridge University Press, 2006c.

D. N. Walton. How to make and defend a proposal in deliberation dialogue. *Artifical Intelligence and Law*, 14:77–239, 2006d.

D. N. Walton. The impact of argumentation on artificial intelligence. In P. Houtlosser and A. van Rees, editors, *Considering Pragma-Dialects*. Erlbaum, Mahwah, New Jersey, 2006e.

D. N. Walton. *Character Evidence: An Abductive Theory*. Berlin: Springer, 2007a.

D. N. Walton. Evaluating practical reasoning. *Synthése*, 157:197–240, 2007b.

D. N. Walton. *Media Argumentation: Dialectic: Persuasion and Rhetoric*. Cambridge: Cambridge University Press, 2007c.

D. N. Walton. Metadialogues for resolving bruden of proof disputes. *Argumentation*, 21:291–316, 2007d.

D. N. Walton. *Informal Logic*. Cambridge University Press, 2 edition, 2008a.

D. N. Walton. Presumption, burden of proof and lack of evidence. *L'Analisi Linguistica e Letteraria*, 16:291–316, 2008b.

D. N. Walton. *Witness Testimony Evidence: Argumentation, Artificial Intelligence and Law*. Cambridge University Press, 2008c.

D. N. Walton. Argument visualization tools for corroborative evidence. In *Proceedings of the 2nd International Conference on Evidence Law and Forensic Science*, pages 32–49. Institute of Evidence Law and Forensic Science, Beijing, 2009a.

D. N. Walton. A dialogical theory of presumption. *Intelligence and Law*, 16:209–243, 2009b.

D. N. Walton. Enthymemes and argumentation schemes in health product ads. In *Proceedings of the Workshop W5: Computational Models of Natural Argument, Twenty-First International Joint Conference on Artifical Intelligence, Pasadena*, pages 49–56. 2009c.

D. N. Walton. Explanations and arguments based on practical reasoning. In *Proceedings of Workshop W10: Explanation-Aware Computing, Twenty-First International Joint Conference on Artificial Intelligence, Pasadena*, pages 72–83. 2009d.

D. N. Walton and D. M. Godden. Persuasion dialogue in online dispute resolution. *Artificial Intelligence and Law*, 12(2):273–295, 2005.

D. N. Walton and D. M. Godden. Using conversation policies to solve problems of ambiguity in argumentation and artificial intelligence. *Pragmatics and Cognition*, 14:3–36, 2006.

D. N. Walton and E. C. W. Krabbe. *Commitment in Dialogue*. SUNY Press, 1995.

D. N. Walton and F. Macagno. Common knowledge in argumentation. *Studies in Communication Sciences*, 6:3–26, 2006.

D. N. Walton and F. Macagno. Types of dialogue, dialectical relevance and textual congruity. *Anthropology & Philosophy*, 8(1-2):101–119, 2007.

D. N. Walton and F. Macagno. Argument from analogy in law, the classical tradition, and recent theories. *Philosophy and Rhetoric*, 42(2):154–182, 2009.

D. N. Walton and C. Reed. Diagramming argumentation schemes and critical questions. In F. van Eemeren, J. Blair, C. Willard and A. Snoeck Henkemans, editors, *Anyone Who Has a View: Theoretical Contributions to the Study of Argumentation*, pages 195–211. Kluwer, 2003.

D. N. Walton and C. Reed. Argumentation schemes and enthymemes. *Synthese*, 145(3):339–370, 2005.

D. N. Walton and C. Reed. Evaluating corroborative evidence. *Argumentation*, 22:531–553, 2008.

D. N. Walton, C. Reed and F. Macagno. *Argumentation Schemes*. Cambridge University Press, 2008.

M. Weber. *Philosophy of Experimental Biology*. Cambridge University Press, Cambridge, 2005.

S. Wells. *Formal Dialectical Games in Multi-Agent Argumentation*. Ph.D. thesis, University of Dundee, 2007.

J. H. Wigmore. *A Treatise on the System of Evidence in Trials at Common Law: Including the Statutes and Judicial Decisions of all Jurisdictions of the United States*. Little, Brown and Company, Boston, Massachusetts, USA, 1908.

J. H. Wigmore. *The principles of judicial proof as given by logic, psychology and general experience and illustrated in judicial trials*. Little, Brown & Company, Boston, 1913.

T. Winograd. A language/action perspective on the design of cooperative work. *Human-Computer Interaction*, 3:3–30, 1987.

J. Woods. How philosophical is informal logic? *Informal Logic*, 20:139–167, 2000.

J. Woods. Seductions and shortcuts: Error in the cognitive economy, 2011. Forthcoming in 2011.

J. Woods and D. Walton. On fallacies. *Journal of Critical Analysis*, 4:103–112, 1972. Reprinted in Woods and Walton, *Fallacies: Selected Papers 1972-1982*, 2nd edition, pages 1-10. London: College Publications, 2007.

J. Woods and D. N. Walton. Informal logic and critical thinking. *Education*, 95:84–86, 1974.

J. Woods and D. N. Walton. Arresting circles in formal dialogues. *Journal of Philosophical Logic*, 7:73–90, 1978.

J. Woods and D. N. Walton. *Argument: The Logic of Fallacies*. McGraw Hill Ryerson, 1982.

J. Woods and D. N. Walton. *Fallacies: Selected Papers 1972-1982*. Foris, Dordrecht, Holland, and Providence, RI, 1989.

M. Wooldridge. *Reasoning about rational agents*. MIT Press, 2000.

M. Wooldridge and W. van der Hoek. On obligations and normative ability: Towards a logical analysis of the social contract. *Journal of Applied Logic*, 3:396–420, 2005.

M. Wooldridge, P. McBurney and S. Parsons. On the metalogic of arguments. In *Proceedings of AAMAS-2005*, pages 560–567. ACM, 2005.

K. Wray. Collective belief and acceptance. *Synthese*, 129:319–333, 2001.

A. Wyner and T. J. M. Bench-Capon. Argument schemes for legal case-based reasoning. In *JURIX 2007: The Twentieth Annual Conference on Legal Knowledge and Information Systems*. 2007.

www.ingramcontent.com/pod-product-compliance
Lightning Source LLC
LaVergne TN
LVHW012328060326
832902LV00011B/1770